QUANTUM CHEMISTRY SIMULATION OF BIOLOGICAL MOLECULES

Nanobiotechnology crosses the boundaries between physics, biochemistry, and bioengineering, and has profound implications for the biomedical engineering industry. This book describes the quantum chemical simulation of a wide variety of molecular systems, with detailed analyses of their quantum chemical properties, individual molecular configurations, and cutting-edge biomedical applications.

Topics covered include the basic properties of quantum chemistry and its conceptual foundations, the nanoelectronics and thermodynamics of DNA, the optoelectronic properties of the five DNA/RNA nucleobase anhydrous crystals, and key examples of molecular diode prototypes.

A wide range of important applications are described, including protein binding of drugs – such as cholesterol-lowering, anti-Parkinson, and anti-migraine drugs. Recent developments in cancer biology are also discussed.

This modern and comprehensive text is essential reading for graduate students and researchers in multidisciplinary areas of biological physics, chemical physics, chemical engineering, biochemistry, and bioengineering.

EUDENILSON L. ALBUQUERQUE is Professor of Physics and Biophysics at Federal University of Rio Grande do Norte, Brazil. His PhD in Physics is from Essex University, UK. He was Visiting Scientist at ICTP-Trieste, Italy; Visiting Scholar at Harvard University; Visiting Professor at University of Western Ontario, Canada and ETH-Zurich; and Fulbright Visiting Professor at Boston University.

UMBERTO L. FULCO is Professor of Biophysics at Federal University of Rio Grande do Norte (UFRN), Brazil. He obtained his BSc, MSC and PhD in Physics at UFRN. He has more than 100 research papers published in diverse fields such as nonequilibrium phase transitions, nanobiostructures devices, and DFT calculations applied to biological molecules and proteins.

EWERTON W. S. CAETANO gained his BSc, MSc and PhD in Physics at the Federal University of Ceará, becoming Professor of Physics in 2006. He has more than 100 research papers published in diverse fields such as semiconductor physics, quantum confinement in nanostructures, and DFT calculations applied to molecules of biological interest and proteins.

VALDER N. FREIRE is Professor of Physics at the Federal University of Ceará. He obtained his PhD in Physics at UNICAMP, Brazil. He was a visiting scholar at CNRM, Toulouse, France. He has more than 300 research papers in diverse fields, mainly in semiconductor physics and nanobiotechnology.

QUANTUM CHEMISTRY SIMULATION OF BIOLOGICAL MOLECULES

EUDENILSON L. ALBUQUERQUE

Federal University of Rio Grande do Norte

UMBERTO L. FULCO

Federal University of Rio Grande do Norte

EWERTON W. S. CAETANO

Federal Institute of Education, Science and Technology of Ceará

VALDER N. FREIRE

Federal University of Ceará

CAMBRIDGE
UNIVERSITY PRESS

CAMBRIDGE
UNIVERSITY PRESS

University Printing House, Cambridge CB2 8BS, United Kingdom

One Liberty Plaza, 20th Floor, New York, NY 10006, USA

477 Williamstown Road, Port Melbourne, VIC 3207, Australia

314–321, 3rd Floor, Plot 3, Splendor Forum, Jasola District Centre, New Delhi – 110025, India

79 Anson Road, #06–04/06, Singapore 079906

Cambridge University Press is part of the University of Cambridge.

It furthers the University's mission by disseminating knowledge in the pursuit of
education, learning, and research at the highest international levels of excellence.

www.cambridge.org
Information on this title: www.cambridge.org/9781108477796
DOI: 10.1017/9781108774956

First published 2021

Printed in the United Kingdom by TJ Books Ltd. Padstow Cornwall

A catalogue record for this publication is available from the British Library.

Library of Congress Cataloging-in-Publication Data
Names: Albuquerque, Eudenilson L., author.
Title: Quantum chemistry simulation of biological molecules / Eudenilson Albuquerque,
Universidade Federal do Rio Grande de Norte, Umberto Fulco,
Universidade Federal do Rio Grande de Norte, Ewerton Caetano, Instituto Federal
de Educação, Ciência e Tecnologia do Ceará, Valder Freire, Universidade Federal do Ceará.
Description: New York, NY : Cambridge University Press, 2021. |
Includes bibliographical references and index.
Identifiers: LCCN 2020016734 (print) | LCCN 2020016735 (ebook) |
ISBN 9781108477796 (hardback) | ISBN 9781108774956 (epub)
Subjects: LCSH: Quantum biochemistry–Computer simulation. | Biomolecules–Computer simulation.
Classification: LCC QP517.Q34 A43 2020 (print) | LCC QP517.Q34 (ebook) | DDC 572.801/51–dc23
LC record available at https://lccn.loc.gov/2020016734
LC ebook record available at https://lccn.loc.gov/2020016735

ISBN 978-1-108-47779-6 Hardback

Contents

Contents

Preface

The twentieth century began with physics dominating the science landscape. The publication of the groundbreaking theories of relativity and quantum physics in the 1910s and 1920s defined the frontiers of knowledge at that time. The world was divided into two domains – the macroscopic and the microscopic – putting an end to the conviction that it would be governed by a single set of laws valid for all scales and magnitudes, as preached by the classical physics of the previous century. Apparently, everything was settled, except for a fundamental question that did not want to be kept silent: what is the secret of life?

It was in this environment that the book *What Is Life?* by E. Schrödinger (Cambridge University Press, 1944) suggested that life could be conceived in terms of the storage and transmission of biological information confined in a molecular structure, as opposed to depending on something spiritual. The book had great repercussion and stimulated the search for a "code of life" so perfect as to convey the exuberance of the living world. At that time, it was believed that proteins, with their structure composed from the polymerization of 20 amino acids, were the true carriers of genetic instructions. When, in 1944, the American bacteriologist O. Avery and co-workers discovered that it was the DNA molecule (not the proteins) that was responsible for storing the genetic information, a great rush was then made to decipher it. Its highest achievement occurred when J. Watson and F. Crick, in 1953 at the Cavendish Laboratory of the University of Cambridge, unveiled the double-helix structure of DNA, a discovery that gave new directions to science.

As a result, the interconnection and complementarity between different areas were more and more necessary to describe living organisms; becoming increasingly less important the approach of specific themes, as it was done before. The maturing of knowledge together with the impressive speed of the scientific production of research groups in the most diverse areas of scientific knowledge became a pattern to be copied. The current stage of the genomic science is, in itself, a great example of the powerful role that physics and chemistry, associated with molecular biology, will have in the future. Medical research progressively uses electronic equipments and biosensors searching for a more precise clinical insight of the state of a patient.

In this sense, the multidisciplinary field of nanobiotechnology emerged as one of the most important research fields in recent years. Although scientists still face great difficulties in controlling and manipulating structures at the nanoscale, nature has been performing these tasks with great precision and high efficiency using biological systems based on the nucleic acids (DNA/RNA), amino acids, and proteins. As a consequence, nowadays there is a great deal of interest in developing theoretical concepts and experimental techniques for self-organized biological systems in search of their technological applications. Biology, physics, and chemistry, among other areas of knowledge, provide models and mechanisms to promote this approach.

Thus, the sophisticated molecular recognition machinery of various natural biological systems has been used in the formation of a complex network of structures potentially useful for a variety of optical, electronic, biosensory, and pharmaceutical applications. In fact, they offer solutions to many of the obstacles that need to be overcome, due to their capacity for self-assembly and self-replication, making possible the production of nanostructures with accuracy not feasible with the technologies usually available in materials science.

Within this context, the purpose of this book is to present a comprehensive and up-to-date account of the main electronic, optical, structural, vibrational, and pharmaceutical properties of some biological systems and their interactions, using a theoretical/computational framework based on quantum chemistry. Computer simulations allow us to study processes that cannot be analyzed directly through experiments, speeding the development of new products and devices since costly industrialized tools can be avoided. Additionally, the quality of products can be improved by the investigation of phenomena not previously accessible; this plays a decisive role especially in the development of new devices in material sciences as well as in biotechnology and pharmacology.

Because of these potential applications, our intention here is to provide a graduate-level text for students, researchers, and academic staff working in this field who have interest in understanding the unique interrelationship of the physical, chemical, and biological properties of living organisms. Additionally, the book addresses questions in biological systems, keeping in mind that – because experimental reality is approaching theoretical models and assumptions – detailed analysis and precise predictions are being made possible.

The book is organized so that we start with the basic properties of quantum chemistry, highlighting its main conceptual foundations as well as some recent developments and advances (Chapter 1). Next, we present an investigation of the nanoelectronics of the biological DNA molecule, stressing its charge transport (Chapter 2) and electrical conductivity (Chapter 3), taking into account several different topologies to depict it. Our quantum simulations are based on an effective tight-binding Hamiltonian model, describing an electron moving in a chain with a single orbital per site and nearest-neighbor interactions, together with a transfer-matrix approach to simplify the algebraic manipulations, which can be otherwise quite complex. Its thermodynamic properties then follow, considering not only the conventional classical (Maxwell-Boltzmann) and quantum (Fermi-Dirac) statistics but also the so-called non-extensive statistics, a scenario referred to as weak chaos, involving long-range interactions characterized by a power-law instead of the usual exponential behavior. A description of the DNA denaturation thermodynamics properties is also presented (Chapter 4). To improve our understanding of devices based on the nucleic acids DNA and RNA, Chapter 5 deals with the dispersion-corrected density functional theory (DFT) and time-dependent DFT (TDDFT) calculations to obtain the optimized geometries, Kohn–Sham band structures and orbitals, charge distribution, optical absorption, Frenkel exciton binding energies, and complex dielectric functions of the five DNA/RNA nucleobase anhydrous crystals – namely cytosine, guanine, adenine, thymine, and uracil. As an attempt to integrate electronic circuits with biological systems, with the consequent impact on a wide variety of medical and biological applications, we present in Chapter 6 some examples of molecular diode prototypes in which nanoscale electronic and optoelectronic structures must be designed to function in biological environments. The structural, electronic, optical, and vibrational (infrared and Raman spectra) properties of some amino acids anhydrous crystals is the theme discussed in Chapter 7. Chapter 8 is devoted to the presentation of the protein-protein systems. In this chapter, we introduce the so-called protein data bank (PDB), an international database for structural biology obtained mainly by X-ray crystallography and NMR spectroscopy, emphasizing the importance of its interaction and the role played by its dielectric function.

From that point in the book, a wide-ranging and interesting multidisciplinary (physical, chemical, biological, and pharmaceutical) set of applications of quantum biochemistry investigating the protein binding of several well-known drugs is discussed. The starting point is the widely used drugs, ascorbic acid and ibuprofen (Chapter 9). More sophisticated drugs follow, such as the cholesterol-lowering ones (Chapter 10), collagen-based biomaterials (Chapter 11), anti-migraine medications (Chapter 12), and anti-Parkinson drugs (Chapter 13). Here, important questions regarding their efficiency and delivery mechanisms are addressed. The role played by the different types of willardiines as a partial agonist in AMPA receptor to mitigate disorders in the central nervous system, which affect approximately one billion people around the world and lead to more hospitalizations than any other group of diseases, is depicted in Chapter 14. The subject of Chapter 15 is the biology of cancer – including immunotherapy, a revolutionary treatment based on the activation or suppression of our immune system, allowing it to recognize and more efficiently attack only the cancer cells, and avoiding the destruction of healthy cells as in conventional chemotherapy and radiotherapy treatments. Finally, in Chapter 16, we present some concluding remarks, and discuss future directions for this research field.

We have been heavily engaged for many years in research programs focused on nanobiotechnology. We believe that our book, devoted to this burgeoning area, will be valuable in covering the new developments that have occurred in the last decade or so. Furthermore, as this field is rapidly changing, a good comprehension of the fundamental concepts developed so far should be worthwhile for readers interested in this subject. These latest developments have provided the motivation and focus for the book.

We are indebted to a large number of friends and collaborators who directly or indirectly influenced this book and provided ideas. Last but not least, we would like to express our thanks to our families for the invaluable support and unfailing encouragement they give to us.

Figure Credits

Figures 2.2; 2.3; 2.4a; 2.4b; 2.6a; 2.6b; 2.6c; 2.8; 2.9; 2.10; 3.1; 3.2; 3.3; 3.9a; 3.9b; 3.10a; 3.10b; 3.11; 4.1a; 4.1b; 4.1c; 4.2a ; 4.2b; 4.2c; 4.3a; 4.3b; 4.4a; 4.4b; 4.4c; 4.5a; 4.5b; 4.6a; 4.6b; 4.7a; 4.7b; 4.8a; 4.8b; 4.9a; 4.9b; 4.9c; 4.10; 4.11a; 4.11b.
Reprinted from *Physics Reports* **535**, 139–209 (2014).
Authors: E. L. Albuquerque, U. L. Fulco, V. N. Freire, E. W. S. Caetano, M. L. Lyra, and F. A. B. F. Moura.
Title: "DNA-based nanobiostructured devices: the role of the quasiperiodicity and correlation effects"
With permission from Elsevier.

Figures 3.4; 3.5a; 3.5b; 3.5c; 3.6a; 3.6b; 3.6c.
Reprinted from *Chemical Physics* **478**, 48–54 (2016).
Authors: M. L. de Almeida, G. S. Ouriques, U. L. Fulco, E. L. Albuquerque, F. A. B. F. de Moura, and M. L. Lyra.
Title: "Charge transport properties of a twisted DNA molecule: a renormalization approach"
With permission from Elsevier.

Figure 3.8. Table 3.1.
Reprinted from *Applied Physics Letters* **107**, 203701 (2015).
Authors: M. L. de Almeida, J. I. N. Oliveira, J. X. Lima Neto, C. E. M. Gomes, U. L. Fulco, E. L. Albuquerque, V. N. Freire, E. W. S. Caetano, F. A. B. F. de Moura, and M. L. Lyra.
Title: "Electronic transport in methylated sites of DNA"
With permission of AIP Publishing.

Figures 4.12; 4.13a; 4.13b; 4.13c; 4.13d.
Reprinted from *Physica A* **404**, 234–241 (2014).
Authors: D. X. Macedo, I. Guedes, and E. L. Albuquerque.
Title: "Thermodynamics of a DNA denaturation with solvent interaction"
With permission from Elsevier.

Figures 5.1; 5.2; 5.3; 5.4; 5.5; 5.6; 5.7; 5.8; 5.9; 5.10; 5.11. Table 5.1.
Reprinted from *Physical Review B* **96**, 085206 (2017), Copyright 2017.
Authors: M. B. da Silva, T. S. Francisco, F. F. Maia, Jr., E. W. S. Caetano, U. L. Fulco, E. L. Albuquerque, and V. N. Freire
Title: "Improved description of the structural and optoelectronic properties of DNA/RNA nucleobase anhydrous crystals: experiment and dispersion corrected density functional theory calculations"
With permission of the American Physical Society.

Figures 6.1; 6.2; 6.3a; 6.3b; 6.4.
Reprinted from *Chemical Physics Letters* **612**, 14–19 (2014).
Authors: J. I. N. Oliveira, E. L. Albuquerque, U. L. Fulco, P. W. Mauriz, and R. G. Sarmento
Title: "Electronic transport through oligopeptide chains: an artificial prototype of a molecular diode"
With permission from Elsevier.

Figures 6.5; 6.8; 6.9a; 6.9b; 6.9c; 6.10a; 6.10b; 6.10c.
Reprinted from *Chemical Physics Letters* **542**, 123–127 (2012).
Authors: G. A. Mendes, E. L. Albuquerque, U. L. Fulco, L. M. Bezerril, E. W. S. Caetano, and V. N. Freire.
Title: "Electronic specific heat of an alpha3-helical polypeptide and its biochemical variants"
With permission from Elsevier.

Figure 6.7. Table 6.1.
Reprinted from *Applied Physics Letters* **98**, 053702 (2011).
Authors: L. M. Bezerril, U. L. Fulco, J. I. N. Oliveira, G. Corso, E. L. Albuquerque, V. N. Freire, and E. W. S. Caetano.
Title: "Charge transport in fibrous/not fibrous α_3-helical and $(5Q,7Q)\alpha_3$ variant peptides"
With permission of AIP Publishing.

Figures 6.11; 6.12; 6.13. Tables 6.2; 6.3.
Reprinted from *Europhysics Letters* **107**, 68006 (2014).
Authors: J. I. N. Oliveira, E. L. Albuquerque, U. L. Fulco, P. W. Mauriz, R. G. Sarmento, E. W. S. Caetano, and V. N. Freire.
Title: "Conductance of single microRNAs chains related to the autism spectrum disorder"
With permission of the European Physical Society.

Figures 7.1a; 7.1b; 7.2.
Reprinted from *Chemical Physics Letters* **512**, 208–210 (2011).
Authors: J. R. Candido-Junior, F. A. M. Sales, S. N. Costa, P. Lima Neto, E. W. S. Caetano, D. L. Azevedo, E. L. Albuquerque, V. N. Freire.
Title: "Monoclinic and orthorhombic cysteine crystals are small gap insulators"
With permission from Elsevier.

Figures 7.3; 7.4a; 7.4b; 7.5; 7.6; 7.7.
Reprinted with permission from *Crystal Growth and Design* **13**, 2793–2802 (2013).
Authors: S. N. Costa, F. A. M. Sales, V. N. Freire, F. F. Maia Jr., E. W. S. Caetano, L. O. Ladeira, E. L. Albuquerque, and U. L. Fulco.
Title: "L-serine anhydrous crystals: structural, electronic and optical properties by first-principles calculations, and optical absorption measurement"
American Chemical Society. Copyright 2013.

Figures 7.8; 7.9; 7.10a; 7.10b; 7.11; 7.12; 7.13; 7.14.
Reprinted from *Journal of Physics and Chemistry of Solids* **121**, 36–48 (2018).
Authors: E. W. S. Caetano, U. L. Fulco, E. L. Albuquerque, A. H. de Lima Costa, S. N. Costa, A. M. Silva, F. A. M. Sales, V. N. Freire.
Title: "Anhydrous proline crystals: Structural optimization, optoelectronic properties, effective masses and Frenkel exciton energy"
With permission from Elsevier.

Figure 7.15. Table 7.1.
Reprinted from *Physical Review B* **86**, 195201 (2012), Copyright 2012.
Authors: A. M. Silva, B. P. Silva, F. A. M. Sales, V. N. Freire, E. Moreira, U. L. Fulco, E. L. Albuquerque, F. F. Maia, Jr., and E. W. S. Caetano.
Title: "Optical absorption and DFT calculations in L-aspartic acid anhydrous crystals: Charge carrier effective masses point to semiconducting behavior"
With permission of the American Physical Society.

Figures 7.16; 7.17; 7.18; 7.19; 7.20a; 7.20b; 7.21; 7.22.
Reprinted with permission from *The Journal of Physical Chemistry A* **119**, 11791-11803 (2015).
Authors: A. M. Silva, S. N. Costa, F. A. M. Sales, V. N. Freire, E. M. Bezerra, R. P. Santos, U. L. Fulco, E. L. Albuquerque, and E. W. S. Caetano.
Title: "Vibrational spectroscopy and phonon-related properties of the L-aspartic acid anhydrous monoclinic crystal"
American Chemical Society. Copyright 2015.

Figures 7.23; 7.24.
Reprinted with permission from *Crystal Growth and Design* **13**, 4844–4851 (2013).
Authors: A. M. Silva, S. N. Costa, B. P. Silva, V. N. Freire, U. L. Fulco, E. L. Albuquerque, E. W. S. Caetano, and F. F. Maia, Jr.
Title: "Assessing the role of water on the electronic structure and vibrational spectra of monohydrated L-aspartic acid crystals"
American Chemical Society. Copyright 2013.

Figures 9.1a; 9.1b; 9.2; 9.3a; 9.3b; 9.3c; 9.4a; 9.4b; 9.4c; 9.5. Table 9.1.
Reprinted with permission from *The Journal of Physical Chemistry B* **112**, 14267–14272 (2008).
Authors: S. G. Santos, J. V. Santana, F. F. Maia, Jr., V. Lemos, V. N. Freire, E. W. S. Caetano, B. S. Cavada, and E. L. Albuquerque.
Title: "Adsorption of ascorbic acid on the C60 fullerene"
American Chemical Society. Copyright 2008.

Figures 9.6a; 9.6b; 9.7a; 9.7b; 9.8; 9.9; 9.10; 9.11.
Reprinted with permission from *The Journal of Physical Chemistry C* **115**, 24501–24511 (2011).
Authors: A. Hadad, D. L. Azevedo, E. W. S. Caetano, V. N. Freire, G. L. F. Mendonça, P. Lima Neto, E. L. Albuquerque, R. Margis, and C. Gottfried.
Title: "Two-level adsorption of ibuprofen on C60 fullerene for transdermal delivery: Classical molecular dynamics and density functional theory computations"
American Chemical Society. Copyright 2011.

Figures 9.12; 9.13; 9.14; 9.15; 9.16; 9.17a; 9.17b.
Reprinted from *RSC Advances* **5**, 49439–49450 (2015).
Authors: D. S. Dantas, J. I. N. Oliveira, J. X. Lima Neto, R. F. da Costa, E. M. Bezerra, V. N. Freire, E. W. S. Caetano, U. L. Fulco and E. L. Albuquerque.
Title: "Quantum molecular modelling of ibuprofen bound to human serum albumin"
Reproduced by permission of The Royal Society of Chemistry.

Figures 10.1; 10.2; 10.3; 10.4; 10.5; 10.6; 10.7.
Reprinted from *Physical Chemistry Chemical Physics* **14**, 1389–1398 (2012).
Authors: R. F. da Costa, V. N. Freire, E. M. Bezerra, B. S. Cavada, E. W. S. Caetano, J. L. de Lima Filho, and E. L. Albuquerque
Title: "Explaining statin inhibition effectiveness of HMG-CoA reductase by quantum biochemistry computations"
Reproduced by permission of the PCCP Owner Societies.

Figures 11.1; 11.2; 11.3; 11.4; 11.5; 11.6; 11.7; 11.8; 11.9; 11.15. Tables 11.1; 11.2; 11.3.
Reprinted from *RSC Advances* **7**, 2817–2828 (2017).
Authors: J. I. N. Oliveira, K. S. Bezerra, J. X. Lima Neto, E. L. Albuquerque, U. L. Fulco, E. W. S. Caetano, and V. N. Freire
Title: "Quantum binding energy features of the T3-785 collagen-like triple-helical peptide"
Reproduced by permission of The Royal Society of Chemistry.

Figures 11.10; 11.11; 11.12; 11.13; 11.14. Table 11.4.
Reprinted from *New Journal of Chemistry* **42**, 17115–17125 (2018).
Authors: K. S. Bezerra, J. X. Lima Neto, J. I. N. Oliveira, E. L. Albuquerque, E. W. S. Caetano, V. N. Freire, and U. L. Fulco
Title: "Computational investigation of the $\alpha 2\beta 1$ integrin-collagen triple-helix complex interaction"
Reproduced by permission of The Royal Society of Chemistry (RSC) on behalf of the Centre National de la Recherche Scientifique (CNRS) and the RSC.

Figures 12.1; 12.2; 12.3; 12.4; 12.5; 12.6; 12.7; 12.8; 12.9. Table 12.1.
Reprinted from *New Journal of Chemistry* **42**, 2401–2412 (2018).
Authors: J. X. Lima Neto, V. P. Soares-Rachetti, E. L. Albuquerque, V. M. Vieira and U. L. Fulco.
Title: "Outlining migrainous through dihydroergotamine-serotonin receptor interactions using quantum biochemistry"
Reproduced by permission of The Royal Society of Chemistry (RSC) on behalf of the Centre National de la Recherche Scientifique (CNRS) and the RSC.

Figures 13.1; 13.2; 13.3a; 13.3b; 13.3c; 13.4; 13.5; 13.6; 13.7; 13.8a; 13.8b; 13.8c; 13.9a; 13.9b; 13.9c; 13.10; 13.11; 13.12; 13.13; 13.14; 13.15. Tables 13.1; 13.2; 13.3; 13.4.
Reprinted from *RSC Advances* **2**, 8306–8322 (2012).

Authors: N. L. Frazão, E. L. Albuquerque, U. L. Fulco, D. L. Azevedo, G. L. F. Mendonça, P. Lima Neto, E. W. S. Caetano, J. V. Santana, and V. N. Freire.
Title: "Explaining statin inhibition effectiveness of HMG-CoA reductase by quantum biochemistry computations"
Reproduced by permission of The Royal Society of Chemistry.

Figures 14.2; 14.3; 14.4; 14.5; 14.6; 14.7; 14.8; 14.9.
Reprinted from *Physical Chemistry Chemical Physics* **17**, 13092 (2015).
Authors: J. X. Lima Neto, U. L. Fulco, E. L. Albuquerque, G. Corso, E. M. Bezerra, E. W. S. Caetano, R. F. da Costa, and V. N. Freire.
Title: "A quantum biochemistry investigation for willardiine partial agonism in AMPA receptors"
Reproduced by permission of the PCCP Owner Societies.

Figures 15.1; 15.2; 15.3; 15.4; 15.5.
Reprinted from *Computational and Theoretical Chemistry* **1089**, 21–27 (2016).
Authors: K. B. Mota, J. X. Lima Neto, A. H. Lima Costa, J. I. N. Oliveira, K. S. Bezerra, E. L. Albuquerque, E. W. S. Caetano, V. N. Freire, and U. L. Fulco.
Title: "A quantum biochemistry model of the interaction between the estrogen receptor and the two antagonists used in breast cancer treatment"
with permission from Elsevier.

Figures 15.6; 15.7; 15.8; 15.9; 15.10; 15.11.
Reprinted from *New Journal of Chemistry* **41**, 11405–11412 (2017).
Authors: J. X. Lima Neto, K. S. Bezerra, D. N. Manso, K. B. Mota, J. I. N. Oliveira, E. L. Albuquerque, E. W. S. Caetano, V. N. Freire, U. L. Fulco.
Title: 'Energetic description of a cilengitide bound to integrin"
Reproduced by permission of The Royal Society of Chemistry (RSC) on behalf of the Centre National de la Recherche Scientifique (CNRS) and the RSC.

Figures 15.12a; 15.12b; 15.12c; 15.21; 15.22a; 15.22b; 15.22c.
Reprinted from *New Journal of Chemistry* **43**, 7185–7189 (2019).
Authors: Ana Beatriz M. L. A. Tavares, J. X. Lima-Neto, U. L. Fulco, and E. L. Albuquerque.
Title: "A quantum biochemistry approach to investigate immunotherapeutic drugs for cancer"
Reproduced by permission of The Royal Society of Chemistry (RSC) on behalf of the Centre National de la Recherche Scientifique (CNRS) and the RSC.

1

Basic Properties of Quantum Chemistry

1.1 Introduction

Physics can be divided into two realms: classical and quantum. In classical physics, the properties of a system in a given moment can be represented geometrically as a point in a phase space with a number of dimensions related to the number of degrees of freedom of the system [1]. For example, if one considers a spring–mass system – i.e., an object of mass m coupled to a fixed string and constrained to move along one direction, the so-called one-dimensional (1-D) harmonic oscillator – its state can be represented as a point in a 2-D plane. If a coordinate system is constructed to make a correspondence between each point to a pair of numbers, (q, p), the first number can be the distance of the object to the position at which the string is relaxed (x), while the second number can be the product of the mass of the object by dx/dt (the object velocity). In other words, the second number is the linear momentum of the object along x, $p = m\, dx/dt$.

The dynamics of the system is represented, in the phase space, as a curve whose points are states successively occupied by the system as the time goes on. In terms of coordinates, we have a trajectory given by the time-dependent state vector $\mathbf{S} = [x(t), p(t)]$. The temporal evolution of $\mathbf{S}$ can be obtained by considering the Hamiltonian function $H(\mathbf{S})$:

$$H(\mathbf{S}) = \frac{p^2}{2m} + V(x). \qquad (1.1)$$

Here, $p^2/2m$ is the kinetic energy of the object and $V(x)$ its potential energy, which in the case of the spring–mass system is given by $V(x) = kx^2/2$, where k is the elastic constant of the spring. The time evolution of the state is found by solving Hamilton equations:

$$\frac{dx}{dt} = \frac{\partial H}{\partial p}, \qquad (1.2)$$

1

$$-\frac{dp}{dt} = \frac{\partial H}{\partial x},$$

(1.3)

or, considering the state vector **S**,

$$\frac{d\mathbf{S}}{dt} = \{\mathbf{S}, H\},$$

(1.4)

where $\{\mathbf{A}, B\}$ is the Poisson bracket,

$$\{\mathbf{A}, B\} = \frac{\partial \mathbf{A}}{\partial x}\frac{\partial B}{\partial p} - \frac{\partial B}{\partial x}\frac{\partial \mathbf{A}}{\partial p}.$$

(1.5)

Thus, the Hamiltonian rules the evolution of the classical state **S** in time.

As a matter of fact, this method can be extended to systems much more complex than the spring–mass system. If the system has N particles, with independent spatial coordinates $(q_1, q_2, \ldots, q_N)$ and respective conjugated momenta $(p_1, p_2, \ldots, p_N)$, the Hamiltonian is a function of the state:

$$\mathbf{S} = (q_1, q_2, \ldots, q_N, p_1, p_2, \ldots, p_N),$$

(1.6)

which evolves in time according to the Eq. (1.4) provided we redefine the Poisson bracket as

$$\{\mathbf{A}, B\} = \sum_{i=1}^{N}\left(\frac{\partial \mathbf{A}}{\partial q_i}\frac{\partial B}{\partial p_i} - \frac{\partial B}{\partial q_i}\frac{\partial \mathbf{A}}{\partial p_i}\right).$$

(1.7)

At every moment, the classical system is completely described by the classical state **S**, and one can always in principle solve Hamilton equations to evaluate its past and future states. Classical physics is completely deterministic: knowing **S** at some time t_0 and the form of its Hamiltonian function, it is possible to disclose the whereabouts of the system at any given time t. This picture, however, is radically changed in quantum mechanics.

In a quantum system, the physical state is described not by a position-momentum vector **S** but by a unit complex vector $|\Psi\rangle$ in an abstract geometric space called Hilbert space. The x and p coordinates of our spring-mass oscillator must be replaced by Hermitian operators $\hat{x}$ and $\hat{p}$, whose eigenstates correspond, respectively, to states with well-defined values of the position x and the momentum p.

Unfortunately, these eigenstates do not belong to Hilbert space, and, therefore, one cannot find the spring–mass system with well-defined position or momentum! Besides, the eigenstates of $\hat{x}$ and $\hat{p}$ always "point" along different directions in such a way that a well-defined value of the position coordinate is physically described as a somewhat homogeneous superposition of infinitely many momentum eigenstates and vice versa. This leads directly to the Heisenberg uncertainty principle, which affirms the impossibility of measuring accurately the position and the momentum of a quantum particle at the same time.

1.2 The Schrödinger Equation

The time evolution of the complex vector $|\Psi\rangle$, however, is governed by the Hamiltonian operator $\hat{H}$ (obtained from the classical Hamiltonian by replacing x and p with the operators $\hat{x}$ and $\hat{p}$, respectively), according to

$$\hat{H}\,|\Psi\rangle = i\hbar \frac{\partial}{\partial t}\,|\Psi\rangle, \tag{1.8}$$

which, projected along an arbitrary eigenstate of the position operator $\hat{x}$, leads to the differential equation

$$-\frac{\hbar^2}{2m}\frac{\partial^2 \Psi(x,t)}{\partial x^2} + V(x)\Psi(x,t) = i\hbar\frac{\partial \Psi(x,t)}{\partial t} \tag{1.9}$$

if one uses the classical Newtonian form given by Eq. (1.1).

Equation (1.9) is the time-dependent one-dimensional Schrödinger equation. The function $\Psi(x,t)$ is the wave function of the spring–mass oscillator, and it contains all the information about its physical state. Unfortunately, however, one cannot evaluate from it the exact position and momentum of the 1-D harmonic oscillator at a given moment, as already explained. The wave function $\Psi(x,t)$ allows one only to find the probability of observing the oscillator in a given region of space at a given moment t. Such probability is given by

$$P(a \leq x \leq b) = \int_a^b \Psi^*(x,t)\Psi(x,t)dx = \int_a^b |\Psi(x,t)|^2 dx. \tag{1.10}$$

As the complex vector $|\Psi\rangle$ is unitary, the wave function is normalized to the unity,

$$\int_{-\infty}^{+\infty} |\Psi(x,t)|^2 dx = 1, \tag{1.11}$$

which means that the probability of finding the oscillator at some place is one.

For a given classic dynamical variable $A(x,p,t)$, one assigns the Hermitian operator $\hat{A}\left(\hat{x},\hat{p},t\right)$. This Hermitian operator, together with the wave function $\Psi(x,t)$, allows the evaluation of the average value of A at instant t according to

$$\langle A \rangle = \int_{-\infty}^{+\infty} \Psi^*(x,t)\hat{A}\left(\hat{x},\hat{p},t\right)\Psi(x,t)dx. \tag{1.12}$$

In the representation of the position operator $\hat{x}$ eigenstates, the two operators $\hat{x}$ and $\hat{p}$ are $\hat{x}=x$ and $\hat{p}=-i\hbar\,\partial/\partial x$. If the complex vector $|\Psi\rangle$ is expanded in the eigenstates of the Hermitian operator $\hat{A}$, the absolute square of the inner product $\langle a|\Psi\rangle$ gives the probability (probability density) of finding the system in the eigenstate $|a\rangle$ if the eigenvalue a is discrete (continous). Therefore, in quantum mechanics, one introduces a statistical element that is intrinsic to the theory. Before performing a measurement on the physical system, its quantum state evolves deterministically obeying Schrödinger's equation, but when a measurement is carried out, all one can do is to evaluate the probability of obtaining a particular outcome. The quantum state collapses to a single eigenstate of the operator representing the measurement being performed, and its time evolution proceeds deterministically from this eigenstate until another measurement is done.

For a quantum system consisting of a single particle of mass m moving in 3-D space, the wave function becomes a function of three spatial coordinates and time, $\Psi(\mathbf{r}, t)$, where $\mathbf{r} = (q_1, q_2, q_3)$ is a three-coordinate vector. In Cartesian coordinates, $\mathbf{r} = (x, y, z)$. The quantity $|\Psi(\mathbf{r}, t)|^2 d\mathbf{r}$ is the probability of finding the system within the infinitesimal volume $d\mathbf{r}$. Schrödinger's equation then becomes

$$-\frac{\hbar^2}{2m}\nabla^2\Psi(\mathbf{r}, t) + V(\mathbf{r})\Psi(\mathbf{r}, t) = i\hbar\frac{\partial\Psi(\mathbf{r}, t)}{\partial t}, \tag{1.13}$$

where $V(\mathbf{r})$ is the 3-D potential in which it moves, and ∇^2 the Laplacian operator. Equation (1.13) can be solved by separating the space and time coordinates according to

$$\Psi(\mathbf{r}, t) = \psi(\mathbf{r})\varphi(t), \tag{1.14}$$

which leads to two differential equations – one for $\varphi(t)$ and another for $\psi(\mathbf{r})$, namely

$$i\hbar\frac{d\varphi(t)}{dt} = E\varphi(t), \tag{1.15}$$

$$-\frac{\hbar^2}{2m}\nabla^2\psi(\mathbf{r}) + V(\mathbf{r})\psi(\mathbf{r}) = E\psi(\mathbf{r}). \tag{1.16}$$

Equation (1.15) has a simple solution,

$$\varphi(t) = \exp\left(-\frac{iEt}{\hbar}\right), \tag{1.17}$$

while Eq. (1.16) is the time-independent Schrödinger equation. The integration constant E is the energy of the particle. It is an eigenenergy equation for the energy eigenfunctions $\psi(\mathbf{r})$ – i.e., $\hat{H}\psi(\mathbf{r}) = E\psi(\mathbf{r})$. If the energy spectrum of the system is discrete, one can label the eigenenergies E using a vector with integer components $\mathbf{n}$, E becoming $E_\mathbf{n}$. The time-independent solutions can be written as $\psi_\mathbf{n}(\mathbf{r})$, while the time-dependent solutions are

$$\Psi_\mathbf{n}(\mathbf{r}, t) = \psi_\mathbf{n}(\mathbf{r})\exp\left(-\frac{iE_\mathbf{n}t}{\hbar}\right). \tag{1.18}$$

The probability density assigned to $\Psi_n(\mathbf{r}, t)$ does not depend on time, so the $\Psi_n(\mathbf{r}, t)$ are labeled stationary states. The general solution of Eq. (1.13) is a superposition of stationary states:

$$\Psi(\mathbf{r}, t) = \sum_n c_n \Psi_n(\mathbf{r}, t), \qquad (1.19)$$

where $|c_n|^2$ gives the probability of finding the system with energy E_n at instant t.

If the eigenenergies of Eq. (1.16) are continuous, one can label the eigenfunctions using a vector with continuous components $\mathbf{k}$. In this case, one finds the general solution:

$$\Psi(\mathbf{r}, t) = \int_{V_\mathbf{k}} c(\mathbf{k}) \Psi_\mathbf{k}(\mathbf{r}, t) d\mathbf{k}. \qquad (1.20)$$

Here, $|c(\mathbf{k})|^2 d\mathbf{k}$ is the probability of finding the system with energy $E(\mathbf{k})$ within the volume $V_\mathbf{k}$.

Lastly, if the eigenenergies have a mixed spectrum, the general solution of the time-dependent Schrödinger equation is a combination of Eqs. (1.19) and (1.20).

1.2.1 The Born-Oppenheimer Approximation

In a molecule or crystal, we have mutual interactions between electrons and nuclei. In this case, one can write the following interaction potentials to describe the system:

$$V_{nn}(\mathbf{R}_i, \mathbf{R}_j) = \frac{k Z_i Z_j e^2}{|\mathbf{R}_i - \mathbf{R}_j|}, \qquad (1.21)$$

$$V_{en}(\mathbf{r}_i, \mathbf{R}_j) = -\frac{k Z_j e^2}{|\mathbf{r}_i - \mathbf{R}_j|}, \qquad (1.22)$$

$$V_{ee}(\mathbf{r}_i, \mathbf{r}_j) = \frac{k e^2}{|\mathbf{r}_i - \mathbf{r}_j|}. \qquad (1.23)$$

Here, $V_{nn}(\mathbf{R}_i, \mathbf{R}_j)$ is the Coulomb potential between two nuclei i and j at the positions $\mathbf{R}_i$ and $\mathbf{R}_j$ with atomic numbers Z_i and Z_j, respectively; $V_{en}(\mathbf{r}_i, \mathbf{R}_j)$ is the Coulomb potential between the ith electron at the position $\mathbf{r}_i$ and the atomic nuclei at $\mathbf{R}_j$ with atomic number Z_j; and $V_{ee}(\mathbf{r}_i, \mathbf{r}_j)$ is the Coulomb potential between the ith and jth electron at positions $\mathbf{r}_i$ and $\mathbf{r}_j$.

The wave function for this system is $\psi(\mathbf{r}_1, \mathbf{r}_2, \ldots, \mathbf{r}_N, \mathbf{R}_1, \mathbf{R}_2, \ldots, \mathbf{R}_N, t)$ or, put more simply, $\Psi(\mathbf{r}, \mathbf{R}, t)$, where $\mathbf{r}$ and $\mathbf{R}$ stand for the electron and nuclei coordinates, respectively. The time-dependent Schrödinger equation for a system with N_e electrons and N_n nuclei is

$$\left[T_e + T_n + V_{nn}\left(\mathbf{R}\right) + V_{en}\left(\mathbf{r},\mathbf{R}\right) + V_{ee}\left(\mathbf{r}\right) \right] \Psi\left(\mathbf{r},\mathbf{R},t\right) = i\hbar \frac{\partial \Psi\left(\mathbf{r},\mathbf{R},t\right)}{\partial t}. \quad (1.24)$$

Here, we identify the electron and nuclei kinetic energy terms, T_e and T_n, the total interaction energy between the nuclei $V_{nn}\left(\mathbf{R}\right)$, the total interaction energy between the electrons and the nuclei $V_{en}\left(\mathbf{r},\mathbf{R}\right)$, and the total interaction energy between the electrons $V_{ee}\left(\mathbf{r}\right)$.

If we separate the spatial and time-dependent parts,

$$\Psi\left(\mathbf{r},\mathbf{R},t\right) = \Phi\left(\mathbf{r},\mathbf{R}\right)\varphi(t), \quad (1.25)$$

we obtain the time-independent Schrödinger equation:

$$\left[T_e + T_n + V_{nn}\left(\mathbf{R}\right) + V_{en}\left(\mathbf{r},\mathbf{R}\right) + V_{ee}\left(\mathbf{r}\right) \right] \Phi\left(\mathbf{r},\mathbf{R}\right) = E_T \Phi\left(\mathbf{r},\mathbf{R}\right), \quad (1.26)$$

where E_T is the total energy of the system.

We now decouple the kinetic nuclear energy according to

$$\Phi\left(\mathbf{r},\mathbf{R}\right) = \Phi_e\left(\mathbf{r};\mathbf{R}\right)\Phi_n\left(\mathbf{R}\right). \quad (1.27)$$

Putting Eq. (1.27) into Eq. (1.26), we have

$$\frac{\left[T_e + V_{nn}\left(\mathbf{R}\right) + V_{en}\left(\mathbf{r},\mathbf{R}\right) + V_{ee}\left(\mathbf{r}\right) \right] \Phi_e\left(\mathbf{r};\mathbf{R}\right)}{\Phi_e\left(\mathbf{r};\mathbf{R}\right)} + \frac{T_n \Phi_n\left(\mathbf{R}\right)}{\Phi_n\left(\mathbf{R}\right)} = E_T. \quad (1.28)$$

Making the assumption that the nuclear coordinates change much more slowly than the electron coordinates due to the large difference between the masses of electrons and nuclei, $\{\mathbf{R}\}$ becomes a set of constant parameters for the motion of the electrons. This is known as the Born-Oppenheimer approximation. It allows us to convert Eq. (1.28) into two independent equations, namely

$$\left[T_e + V_{nn}\left(\mathbf{R}\right) + V_{en}\left(\mathbf{r},\mathbf{R}\right) + V_{ee}\left(\mathbf{r}\right) \right] \Phi_e\left(\mathbf{r};\mathbf{R}\right) = E_e\left(\mathbf{R}\right)\Phi_e\left(\mathbf{r};\mathbf{R}\right), \quad (1.29)$$

$$T_n \Phi_n\left(\mathbf{R}\right) = \left[E_T - E_e\left(\mathbf{R}\right) \right]\Phi_n\left(\mathbf{R}\right). \quad (1.30)$$

From Eq. (1.29), for a given configuration of nuclear coordinates $\{\mathbf{R}\}$, one obtains a set of electron eigenstates $\{\Phi_{e,v}\left(\mathbf{r};\mathbf{R}\right)\}$ with eigenenergies $E_{e,v}\left(\mathbf{R}\right)$. Of particular interest is the lowest-energy eigenstate, $\Phi_{e,0}\left(\mathbf{r};\mathbf{R}\right)$, with eigenenergy $E_{e,0}\left(\mathbf{R}\right)$ (assuming a non-degenerate case). This ground state defines, as we change $\mathbf{R}$ slowly, a potential energy hypersurface (or adiabatic surface) for the nuclear motion, which allows one to minimize the total energy of the electron–nuclei system (geometry optimization procedure).

1.2.2 Slater Determinant

Let us consider now what happens when one has a system formed by N identical particles interacting with each other in a fixed external potential $V(\mathbf{r})$. The interaction potential between the particles i and j is $v(\mathbf{r}_i, \mathbf{r}_j)$, while the kinetic energy operator of the ith particle is $-(\hbar^2/2m)\nabla_i^2$. The wave function of the system depends on the coordinates $\{\mathbf{r}_i\}$ and can be written as $\Psi = \Psi(\mathbf{r}_1, \mathbf{r}_2, \ldots, \mathbf{r}_N, t)$. Hence, the time-dependent Schrödinger equation for this system is

$$\left[\sum_{i=1}^{N} \left(-\frac{\hbar^2}{2m}\nabla_i^2 + V(\mathbf{r}_i) + \sum_{j=1}^{N} v(\mathbf{r}_i, \mathbf{r}_j) \right) \right] \Psi = i\hbar \frac{\partial \Psi}{\partial t}. \tag{1.31}$$

One can perform a separation between space and time coordinates, $\Psi = \Psi(\mathbf{r}_1, \mathbf{r}_2, \ldots, \mathbf{r}_N, t) = \psi(\mathbf{r}_1, \mathbf{r}_2, \ldots, \mathbf{r}_N)\varphi(t)$, to obtain stationary states with well-defined energy E, yielding

$$\left[\sum_{i=1}^{N} \left(-\frac{\hbar^2}{2m}\nabla_i^2 + V(\mathbf{r}_i) + \sum_{j=1}^{N} v(\mathbf{r}_i, \mathbf{r}_j) \right) \right] \psi = E\psi, \tag{1.32}$$

which is the time-independent Schrödinger equation for the system with N identical particles.

As each vector $\mathbf{r}_i$ has three components, one has a partial differential equation with $3N$ independent variables without analytical solution in most cases. However, one can move forward from Eq. (1.32) in the case of non-interacting particles, in which $v(\mathbf{r}_i, \mathbf{r}_j) = 0$. For such situation, one obtains

$$\left[\sum_{i=1}^{N} \left(-\frac{\hbar^2}{2m}\nabla_i^2 + V(\mathbf{r}_i) \right) \right] \psi(\mathbf{r}_1, \mathbf{r}_2, \ldots, \mathbf{r}_N) = E\psi(\mathbf{r}_1, \mathbf{r}_2, \ldots, \mathbf{r}_N). \tag{1.33}$$

Here, a separation of variables is feasible:

$$\psi(\mathbf{r}_1, \mathbf{r}_2, \ldots, \mathbf{r}_N) = u_1(\mathbf{r}_1)u_2(\mathbf{r}_2)\ldots u_N(\mathbf{r}_N), \tag{1.34}$$

which leads to a set of N equations of the form:

$$\left[-\frac{\hbar^2}{2m}\nabla_i^2 + V(\mathbf{r}_i) \right] u_i(\mathbf{r}_i) = E_i u_i(\mathbf{r}_i). \tag{1.35}$$

The normalized eigenstates u_i can have a discrete, continuous, or mixed eigenenergy spectrum. We will assume the first case and write the eigenstates as $u_{\mathbf{n}(i)}(\mathbf{r}_i)$ and the eigenenergies as $E_{\mathbf{n}(i)}$. Then one finds

$$\psi(\mathbf{r}_1, \mathbf{r}_2, \ldots, \mathbf{r}_N) = u_{\mathbf{n}(1)}(\mathbf{r}_1)u_{\mathbf{n}(2)}(\mathbf{r}_2)\ldots u_{\mathbf{n}(N)}(\mathbf{r}_N), \tag{1.36}$$

with energy

$$E = E_{\mathbf{n}(1)} + E_{\mathbf{n}(2)} + \cdots + E_{\mathbf{n}(N)}. \tag{1.37}$$

Note that an eigenfunction with exactly the same energy can be obtained if the coordinates of the ith and jth particles are permuted in Eq. (1.36). As a matter of fact, identical quantum particles are indistinguishable, so the permutation of spatial coordinates does not change the probability density obtained from $\psi(\mathbf{r}_1, \mathbf{r}_2, \ldots, \mathbf{r}_N)$. This leads to

$$\hat{P}_{ij} \psi(\mathbf{r}_1, \mathbf{r}_2, \ldots, \mathbf{r}_N) = \pm \psi(\mathbf{r}_1, \mathbf{r}_2, \ldots, \mathbf{r}_N), \tag{1.38}$$

where $\hat{P}_{ij}$ is the operator that permutes the coordinates $\mathbf{r}_i$ and $\mathbf{r}_j$ in the wave function. Also, $\psi(\mathbf{r}_1, \mathbf{r}_2, \ldots, \mathbf{r}_N)$ is an eigenfunction of $\hat{P}_{ij}$ with only two eigenvalues, $+1$ and -1. In the former, the identical particles are called bosons, and in the latter, they are called fermions. Some examples of two-particle bosonic solutions are

$$\psi(\mathbf{r}_1, \mathbf{r}_2) = u_{\mathbf{n}(1)}(\mathbf{r}_1) u_{\mathbf{n}(2)}(\mathbf{r}_2), \tag{1.39}$$

$$\psi(\mathbf{r}_1, \mathbf{r}_2) = (1/\sqrt{2}) \left[u_{\mathbf{n}(1)}(\mathbf{r}_1) u_{\mathbf{n}(2)}(\mathbf{r}_2) + u_{\mathbf{n}(1)}(\mathbf{r}_2) u_{\mathbf{n}(2)}(\mathbf{r}_1) \right]. \tag{1.40}$$

From Eqs. (1.39) and (1.40), one can see that bosons can occupy the same quantum states. Indeed, statistically they prefer this kind of configuration. For fermions, however, the opposite is true. In the case of a two-particle fermionic state, we have

$$\psi(\mathbf{r}_1, \mathbf{r}_2) = (1/\sqrt{2}) \left[u_{\mathbf{n}(1)}(\mathbf{r}_1) u_{\mathbf{n}(2)}(\mathbf{r}_2) - u_{\mathbf{n}(1)}(\mathbf{r}_2) u_{\mathbf{n}(2)}(\mathbf{r}_1) \right]. \tag{1.41}$$

If the states are the same – i.e., $\mathbf{n}(1) = \mathbf{n}(2)$ – then the wave function vanishes! So it is impossible for two fermions to occupy the same quantum state at the same time (Pauli's exclusion principle). Because of this, the quantum state occupation statistics for fermions is different from the statistics for bosons.

In general, the normalized wave function for N indistinguishable non-interacting fermions is given by the Slater determinant:

$$\psi(\mathbf{r}_1, \mathbf{r}_2, \ldots, \mathbf{r}_N) = \frac{1}{\sqrt{N!}} \begin{vmatrix} u_{\mathbf{n}(1)}(\mathbf{r}_1) & u_{\mathbf{n}(1)}(\mathbf{r}_2) & \cdots & u_{\mathbf{n}(1)}(\mathbf{r}_N) \\ u_{\mathbf{n}(2)}(\mathbf{r}_1) & u_{\mathbf{n}(2)}(\mathbf{r}_2) & \cdots & u_{\mathbf{n}(2)}(\mathbf{r}_N) \\ \vdots & \vdots & \ddots & \vdots \\ u_{\mathbf{n}(N)}(\mathbf{r}_1) & u_{\mathbf{n}(N)}(\mathbf{r}_2) & \cdots & u_{\mathbf{n}(N)}(\mathbf{r}_N) \end{vmatrix}. \tag{1.42}$$

If $\mathbf{n}(i) = \mathbf{n}(j)$ for $i \neq j$, two lines of the determinant are equal, and it becomes zero, ensuring that two identical one-particle quantum states cannot be occupied.

1.3 Chemical Bonds

When atoms interact, they change their electronic structures in such a way that an effective attraction takes place between them, forming a chemical bond [2]. This attraction can be either very strong, creating a new stable chemical compound (a molecule), or less strong, forming unstable structures that can be dismantled with relatively small amounts of energy. The nature of the interaction depends on the electronic structure of the atoms involved. The stable atomic arrangements have a constant characteristic nuclear geometry, making it possible to obtain different configurations with the same set of atoms (isomers).

This fact was already known at the beginning of the nineteenth century, as the atomic theory of J. Dalton was being embraced by chemists. After the discovery of the electron and the beginnings of quantum mechanics, the American physical chemist G. N. Lewis proposed, in 1916, an electronic theory to explain the chemical bonds between atoms in molecules. According to him, two atoms can form a bond by transferring one electron to another (the electron-pair model); by sharing two electrons, one from each atom (his covalent bond model); or by one atom sharing two electrons with another.

With the progress of quantum mechanics, however, it became clear that Lewis's original idea could not reflect accurately the reality of the interatomic forces in molecules, as the electrons are truly indistinguishable and delocalized, meaning that one cannot point out them and say which electrons in particular effectively are involved in a chemical bond.

However, there are ways to establish a correspondence between the original concepts of Lewis and the quantum mechanics postulates. It can be done by analyzing the nuclear geometry, which allows the definition of a concept of nuclear vicinities, and the definition of space domains, where there is a large probability of finding two electrons with opposite spins. One can also consider an evaluation of the differential electron density, meaning the difference between the electron density of the compound and the superposition of electron densities from the isolated atoms, to see which regions are electronically populated or depopulated when a molecule is formed.

1.3.1 The Ionic Bond

The electron affinity of an atom is essential to determine the character of a chemical bond. If two elements, one with high electronegativity and the other with low electronegativity, approach each other, one or more electrons of the latter can be effectively transferred to the first. This electron migration ionizes each original atom, and the resulting ionic compound forms a strong electric dipole (see Figure 1.1).

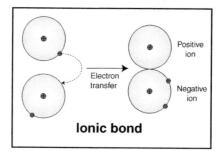

Figure 1.1 Pictorial view of the ionic bond.

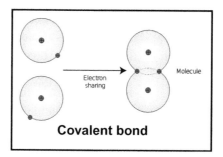

Figure 1.2 Pictorial view of the covalent bond.

The potential energy between the ions can be approximated by a superposition of the attractive Coulomb potential with a repulsive term due to the exchange quantum interaction:

$$V(r) = -\frac{ke^2}{r} + \frac{\alpha}{r^s}, \tag{1.43}$$

where α and s are constant dependent on the elements involved.

The ionic bond is the strongest of all chemical bonds, with binding energies per bond between 150 and 1,000 kcal/mol.

1.3.2 The Covalent Bond

The covalent bond occurs between two atoms with similar electronegativities, which prevents the transfer of electrons from one atom to the other. Instead, both atoms reach a compromise and share pairs of electrons (see Figure 1.2), leading to binding energies between 50 (the weakest single covalent bond) and 250 kcal/mol (the strongest triple covalent bond).

An approximate description of these bonds can be given using quantum mechanics: molecular orbitals are formed from the linear combination of original atomic orbitals, employed as a basis set to represent them, and these molecular orbitals are filled with all electrons available (valence bond theory).

The bonding state of a hydrogen molecule, for instance, with two electrons can then be given by

$$\Psi_0 = (1/2)\Big[\psi_{1s,A}(\mathbf{x}_1)\psi_{1s,B}(\mathbf{x}_2) + \psi_{1s,B}(\mathbf{x}_1)\psi_{1s,A}(\mathbf{x}_2)\Big]\Big[\alpha1\beta2 - \alpha2\beta1\Big], \quad (1.44)$$

where $\psi_{1s,A}$ and $\psi_{1s,B}$ are 1s spatial atomic orbitals centered at the hydrogen atoms A and B, respectively. The α and β are the up and down spin orbitals, respectively. As the spin states in Eq. (1.44) are in opposition (minus sign), the resulting spin of this molecular orbital is $S = 0$ (singlet state). If the plus and minus signs are interchanged, the molecular orbital becomes a triplet, with $S = 1$ (antibonding state).

Another possibility is to form ionic states where the electrons are completely transferred from the hydrogen atoms A to B or B to A, whose wave functions yield

$$\Psi_A = (1/\sqrt{2})\Big[\psi_{1s,A}(\mathbf{x}_1)\psi_{1s,A}(\mathbf{x}_2)\Big]\Big[\alpha1\beta2 - \alpha2\beta1\Big], \quad (1.45)$$

$$\Psi_B = (1/\sqrt{2})\Big[\psi_{1s,B}(\mathbf{x}_1)\psi_{1s,B}(\mathbf{x}_2)\Big]\Big[\alpha1\beta2 - \alpha2\beta1\Big]. \quad (1.46)$$

Therefore, a more adequate description of the molecular orbital of the bonded electrons in the H_2 molecule is something like

$$\Psi = \Psi_0 + c_A\Psi_A + c_B\Psi_B. \quad (1.47)$$

Using a variational procedure to minimize the total energy for the wave function Ψ, one can show that the ionic contribution coefficients c_A and c_B are much smaller than 1, indicating the dominant covalent character in the bond between the two hydrogen atoms.

Another approach (molecular orbital theory) is to build a single electron molecular orbital,

$$\Psi_0(\mathbf{x}) = (1/\sqrt{2})\Big[\psi_{1s,A}(\mathbf{x}) + \psi_{1s,B}(\mathbf{x})\Big], \quad (1.48)$$

and, from it, to define the wave function for the two electrons as

$$\Psi = (1/\sqrt{2})\Psi_0(\mathbf{x}_1)\Psi_0(\mathbf{x}_2)\Big[\alpha1\beta2 - \alpha2\beta1\Big]. \quad (1.49)$$

The main difference between the valence bond and the molecular orbital theories is the fact that in the former, the electrons can be localized at different atoms while, for the latter, they are necessarily extended through the hydrogen molecule. It can be shown that Eq. (1.49) leads to an overestimation of the ionic character of the interatomic interaction – see Eqs. (1.45) and (1.46) – while Eq. (1.44) only neglects any ionic contribution. So the molecular orbital theory of the covalent bond is unable to accurately predict dissociation energies. However, both theories converge to the same results as one increases the sizes of their basis sets including more atomic orbitals, as described by the configuration interaction (CI) approach.

The electron density associated to the wave function Ψ in the covalent bond in the hydrogen molecule indicates a concentration of electrons along the line connecting the two protons. This electron cloud attracts both nuclei, overcoming their Coulomb repulsion and stabilizing the molecule.

Extending this reasoning for more complicated molecules is relatively straight-forward if one permits the hybridization of different atomic orbitals, allowing for up to three covalent bonds between neighbor atoms. The occupied states with highest energy (valence states) are mostly responsible for the bonds, while the lowest-energy occupied states (core states) barely contribute to it. In particular, for highly symmetrical structures, such as benzene, the electron density associated to the chemical bond must be obtained from the superposition of distinct extended wave functions with same energy (resonances), creating an aromatic ring.

1.3.3 Hydrogen Bond

Hydrogen bonds are fundamental for biochemistry. They are responsible for the unusual properties of water and the most important structural features of proteins, as they are also essential for the transcription of the genetic information contained in DNA (see Figure 1.3 for a pictorial view of a hydrogen bond in water). Their binding energies are small, ranging from 0.2 to 45 kcal/mol.

They are formed between an electronegative atom Y and a hydrogen atom H covalently bound to another electronegative atom X: $X - H \cdots Y$. They have a strong ionic character and can occur between different molecules or within the same molecule. When two molecules interact, the angular geometry of the hydrogen bonds thus formed can often be reasonably predicted from electrostatic considerations. The distance between the heavy atoms X and Y is shorter than the sum of their van der Waals radii and the length of the $X - H$ covalent bond increases, leading to alterations in its stretching vibrational frequencies.

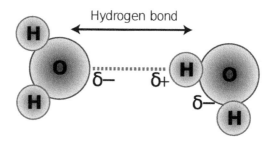

Figure 1.3 Pictorial view of a single hydrogen bond between two water molecules.

Changes in the electron density and polarizability of the groups forming hydrogen bonds lead to an interesting reinforcement when a chain of hydrogen bonds occur. In other words, hydrogen bonds are cooperative and for this reason, long strings of hydrogen bonds are frequent in most systems. A correlation between the hydrogen bond strength and the electronegativities of X and Y indicates a greater dependence on the nature of the first (donor) in comparison with the second (acceptor).

Recent experimental and theoretical results, however, have challenged the traditional picture of the hydrogen bond as being simply the interaction between electric dipoles (it can occur between apolar groups). It seems to involve the occurrence of a partial intermolecular resonance $X-H\cdots Y \leftrightarrow X\cdots H-Y$, leading to the formation of a fractional covalent chemical bond.

1.3.4 van der Waals Interaction

In chemical physics, a van der Waals force, named after the Dutch scientist J. D. van der Waals, is the sum of all attractive or repulsive forces derived neither from covalent bonds between molecules (or between parts of the same molecule) nor from Coulomb's electrostatic interaction. They are ubiquitous and anisotropic, being much weaker than hydrogen bonds, with binding energies from 0.1 to 1 kcal/mol.

These interactions occur when there are electrical charges in the molecules. These charges, in turn, arise when there is polarization in one or a few bonds in the molecule, i.e., when there is a concentration of electrons in a region of the molecule caused by the difference in the electronegativity between the bonded atoms. When the electrons are concentrated in a region of the molecule, a slightly negative region is created there. The region from which the electrons have moved becomes slightly positive, so the overall charge of the molecule is zero. For example, the water molecule has a large polarization in the H–O (hydrogen–oxygen) bonds, due to the great difference in the electronegativity between the H and O atoms. Since oxygen is a much more electronegative element than hydrogen, an accumulation of electrons occurs in the O atom, which becomes slightly negative, and a consequent accumulation of positive charges occurs in the H atoms. This difference generates what we call permanent dipole because, at all times, there will be a difference of electronegativity between the atoms of hydrogen and oxygen. The van der Waals forces originating from a permanent dipole are called dipole–dipole forces.

However, it is possible to have a dipole even in molecules formed by a single element, as is the case of the iodine molecules (I_2). In this case, we have two iodine atoms and, therefore, no difference in electronegativity. Since the iodine atoms are voluminous, there may be a temporary polarization for a brief moment, with the

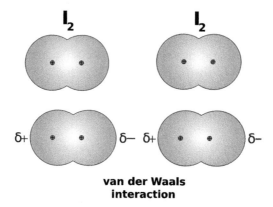

van der Waals interaction

Figure 1.4 van der Waals interaction between two iodine (I_2) molecules due to transient induced dipoles, creating a London force.

formation of an induced dipole much shorter in duration and intensity than the permanent dipole presenting in the water molecule (see Figure 1.4). The van der Waals forces originating from such induced dipole are called London forces, in honor of the German American physicist F. W. London.

The potential energy defining the terms of attraction and repulsion of the van der Waals interaction is given by the Lennard-Jones potential

$$V(r_{ij}) = 4\epsilon \left[\left(\frac{\sigma}{r_{ij}} \right)^{12} - \left(\frac{\sigma}{r_{ij}} \right)^{6} \right].$$
(1.50)

Here, ϵ is the depth of the potential well between the attractive and repulsive barrier, σ is the finite distance at which the interparticle potential is zero, and $r_{ij} = |\mathbf{r}_i - \mathbf{r}_j|$ is the distance between the particles at positions i and j. The parameters ϵ and σ are adjusted experimentally and by theoretical calculations. The well is responsible for both the cohesion of the condensed phases and the conformational stabilization of the macromolecules.

1.4 Classical Calculations

Classical models allow the description of systems with millions of atoms, including the search for energetically optimized geometries, molecular dynamics at a given temperature, and Monte Carlo simulations. If a molecule is composed of a large number of atoms, making it impossible to use a quantum model, it is still possible to model its behavior by obtaining properties of the system using a classical treatment. The simplest of these model is so-called molecular mechanics (MM), which makes use of a simple classical algebraic expression for the total energy of a compound.

Molecular mechanics uses molecular force fields to predict the geometry of a new molecule using data obtained from known molecules. Potential energy terms are obtained to describe bond stretchings, bond bendings, dihedral motions, out-of-plane motions, nonbonded interactions, and Coulomb interactions with or without solvation effects. Adding up all these contributions, one obtains the full mathematical expression of the force field, which can be parameterized through experimental data or by using quantum mechanical calculations for specific systems, such as amino acids in solution (to investigate proteins).

The advantage of molecular mechanics is that it allows the modeling of large molecules, such as proteins and DNA segments, making it the main tool of computational biochemists. Its disadvantage is that there are many chemical properties that are not well defined within the method. Molecular mechanics software packages also include powerful graphical interfaces to help their users visualize extremely large and complicated systems.

Among their approaches, we can cite so-called molecular dynamics (MD) as one of the main tools for the study of atomic-molecular systems through computational simulation, for which the effects of temperature can not be neglected. Physical, chemical, and biological systems consisting of a large number of atoms or molecules whose properties are governed by processes involving thermal energy of the order of $k_B T$ (k_B = Boltzmann's constant; T = temperature) can be studied by MD. Such systems are intrinsically thermodynamic, and their understanding is based on statistical mechanics.

The use of MD can be done in three important situations: First, the simulated properties are compared with the experimental results, and when they agree on results, it is reasonable to say that the experimental data can be explained by a computational simulation model. In the second situation, the MD simulations are used to interpret experimental results, the inverse process of the first case. Finally, in the third situation, the simulations are implemented as a tool to help the initial understanding of a given problem and provide possible ways of investigation, both theoretical and experimentally.

Computer simulation by using MD is now an increasingly common approach to describe many bodies within the physics of condensed matter. Since the first tests, the simulations of MD have proved to be efficient in relation to the experimental results and to provide consistent results for the problems analyzed theoretically.

Another classical tool is the so-called molecular docking, a key tool in structural molecular biology for the design of computer assisted drugs. The strategy used in this approach predicts the interaction between molecules with biological activity (called ligands – small organic molecules) and their macromolecular targets (called receptors – especially the proteins, nucleic acids, carbohydrates, and lipids) specific

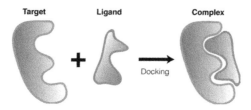

Figure 1.5 Schematic illustration of the molecular docking of a ligand (small molecule) with a protein target (large molecule) producing a stable complex.

to an organism when the structure of both – ligand and receptor – are already known experimentally (see Figure 1.5). The site where the ligand binds to the receptor is called the binding site and has unique chemical characteristics, which are determined – in the case of proteins – by the specific properties of the amino acids that comprise it. Furthermore, the reason why a particular molecule with biological activity binds to a specific receptor, and not to all the molecules that they find in the organism, is directly related to the structure and the chemical properties of the two molecules.

Molecular docking can be used in situations where the structure of the receptor is known without knowledge of how the ligand interacts with the receptor. In this way, it is possible to obtain the best possible conformations of the ligand–receptor interaction that serve as a starting point for subsequent studies of the binding site.

1.5 Quantum Calculations

Ab initio methods aim to determine the physical and chemical properties of a system of many electrons interacting with the atomic nuclei. They do this from approximations of the exact solution of the Schrödinger equation without any empirical input, except fundamental physical constants, together with the initial nuclear geometry and the number of electrons. They are invaluable tools for scientists of diverse fields such as chemistry, physics, nanoscience, and nanobioscience.

A common feature of all ab initio calculations is the use of one-electron wave functions expanded into a basis set of atomic orbitals, Gaussian packets, or plane waves, to mimic the ground state many-electron wave function of systems with one to thousands of electrons from a variational principle (total energy minimization). From the ground state wave function, one obtains optimal nuclear geometries, energies, electronic and vibrational quantum states, electron densities, charge and bond populations, mechanical parameters, optical characteristics and nuclear magnetic resonant (NMR) spectra.

1.5.1 Hartree Method

If one ignores the exclusion principle, the ground state wave function $\Phi_{e,0}(\mathbf{r})$ (the nuclear coordinates $\mathbf{R}$ are omitted here for the sake of simplicity) of a system with N_e electrons in an external nuclear potential can be written as

$$\Phi_{e,0}(\mathbf{r}) = \chi_1(\mathbf{r}_1)\chi_2(\mathbf{r}_2)\ldots\chi_{N_e}(\mathbf{r}_{N_e}). \tag{1.51}$$

Replacing Eq. (1.51) in the expression corresponding to the total energy, and applying the variational principle with respect to the one-electron states $\chi_i(\mathbf{r}_i)$,

$$E_{e,0}\{\chi_i(\mathbf{r}_j)\} = \min\langle\Phi_{e,0}(\mathbf{r})|T_e + V_{nn} + V_{en} + V_{ee}|\Phi_{e,0}(\mathbf{r})\rangle, \tag{1.52}$$

with the orthonormalization constraint,

$$\langle\chi_i(\mathbf{r}_j)|\chi_k(\mathbf{r}_j)\rangle = \delta_{ik}. \tag{1.53}$$

The Hartree differential equation can then be obtained as

$$-\frac{\hbar^2}{2m_e}\nabla_\xi^2\chi_i(\xi) + \sum_{i=1}^{N_n} V_{en}(\xi,\mathbf{R}_j)\chi_i(\xi) + V_H(\xi)\chi_i(\xi) = \varepsilon_i\chi_i(\xi), \tag{1.54}$$

where the Hartree potential V_H is given by

$$V_H(\xi) = ke^2\sum_{k=1}^{N_e}\int d\xi'\frac{|\chi_k(\xi')|^2}{|\xi-\xi'|}. \tag{1.55}$$

The Hartree potential gives the average Coulomb potential produced on the electron in the χ_i state by the other electrons occupying the states χ_k with $k \neq i$ (excluding self-interaction). Equation (1.54) must be solved self-consistently, as the Hartree potential depends on the same states one wishes to determine. It has the obvious disadvantage of not taking into account the exclusion principle that electrons must obey.

1.5.2 Hartree-Fock Method

In order to consider the fact that electrons are fermions, one can write the ground state electronic wave function $\Phi_{e,0}(\mathbf{r})$ for a system with N_e interacting electrons in an external nuclear potential as a Slater determinant:

$$\Phi_{e,0}(\mathbf{r}) = \frac{1}{\sqrt{N_e!}}\begin{vmatrix} \chi_1(\mathbf{r}_1) & \chi_1(\mathbf{r}_2) & \cdots & \chi_1(\mathbf{r}_{N_e}) \\ \chi_2(\mathbf{r}_1) & \chi_2(\mathbf{r}_2) & \cdots & \chi_2(\mathbf{r}_{N_e}) \\ \vdots & \vdots & \ddots & \vdots \\ \chi_{N_e}(\mathbf{r}_1) & \chi_{N_e}(\mathbf{r}_2) & \cdots & \chi_{N_e}(\mathbf{r}_{N_e}) \end{vmatrix}. \tag{1.56}$$

The one-electron orbitals are again obtained by minimizing the expected value of the total electron energy functional with the orthonormalization constraint, given by Eqs. (1.52) and (1.53), which leads to the following set of differential equations for the $\chi_i(\mathbf{r}_j)$ one-electron orbitals:

$$\left[\hat{H} + \sum_{k=1}^{N_e}\left(\hat{J}_k - \hat{K}_k\right)\right]\chi_i = \hat{F}\chi_i = \varepsilon_i\chi_i. \tag{1.57}$$

Here, $\hat{H}(\xi)$ is the Hamiltonian for a single electron in the external potential of the nuclei:

$$\hat{H}(\xi) = -\frac{\hbar^2}{2m_e}\nabla_\xi^2 + \sum_{i=1}^{N_n} V_{en}(\xi, \mathbf{R}_j). \tag{1.58}$$

The other two terms, on the left-hand side of Eq. (1.57), are the Coulomb operator, defined as

$$\hat{J}_k(\xi) = ke^2\int d\xi'\frac{\left|\chi_k(\xi')\right|^2}{|\xi - \xi'|}, \tag{1.59}$$

which describes the interaction of one electron with the charge density produced by the remaining electrons and the nonlocal exchange operator

$$\hat{K}_k(\xi)\chi_i(\xi) = ke^2\left[\int d\xi'\frac{\chi_k^*(\xi')\chi_i(\xi')}{|\xi - \xi'|}\right]\chi_k(\xi), \tag{1.60}$$

which is a quantum effect due to the antisymmetry of the Slater wave function. Also, $\hat{F}$ is the Fock operator and $k \neq i$.

The energy levels ε_i of the one-electron orbitals give the energy of the Hartree-Fock (HF) ground state:

$$E_{e,0}^{HF} = \varepsilon_1 + \ldots + \varepsilon_{N_e}. \tag{1.61}$$

The energy index i range goes from the one-electron ground state $i = 1$ to the $(N_e - 1)$th excited state (assuming non-degeneracy). One must note that the solutions of Eq. (1.57) depend on themselves, so they must be found iteratively to produce a self-consistent field for the electronic ground state. Even so, the solution found after this procedure is only an approximation to the exact ground state, which in fact demands an expansion of the wave function $\Phi_{e,0}(\mathbf{r})$ in an infinite series of Slater determinants depending on the infinite set of excited states for the one-electron orbitals χ_i. The difference between the exact true ground state energy $(E_{e,0}^{exact})$ and the ground state energy estimated through the Hartree-Fock variational approach $(E_{e,0}^{HF})$ is called the correlation energy, namely

$$E_{\text{correlation}} = E_{e,0}^{exact} - E_{e,0}^{HF}. \tag{1.62}$$

As the Hartree-Fock energy is a superior limit for the exact energy, the correlation energy is negative. In order to obtain it, several improvements on the Hartree-Fock approach were implemented. The first – and, conceptually, the most simple – is the configuration interaction method (CI)[3], which consists in the diagonalization of the N_e electron Hamiltonian in a basis of Slater determinants using the variational method. Unfortunately, even for small molecules, the number of Slater determinants that can be built is huge. Thus, it is necessary to truncate the wave function at some point. Other approximation schemes can be used, the most successfully one being DFT, which we will now explore.

1.6 Density Functional Theory

Density functional theory (DFT) is one of the most used quantum mechanical computational methods nowadays to solve the electronic structure of the most diverse systems [4]. It is a very versatile method, relying on the electron density instead of the many-electron wave function to evaluate or predict many physical and chemical properties. It can often be considered as an ab initio method while, in other cases, it incorporates some empirical fitting or combination with wave-function-based techniques. Some drawbacks, such as the lack of accuracy to take into account dispersive forces and the incapacity to describe excited states and systems where strong correlation effects are present, are limitations to the use of DFT calculations. Notwithstanding, it is expected that improvements on DFT methods, in the next 25 years or so, could be lead to the development of strategies able to describe with chemical accuracy all types of chemical reactions, physical states, and spectroscopies with or without solvation, as well as an accurate description and simulation of very large systems, such as proteins and other macromolecules of biological interest. DFT owes its versatility to the general aspects of its fundamental concepts as well as the flexibility one has in implementing them. It is based on quite a rigid conceptual framework.

In this section, we intend to present the main theoretical tools behind DFT. We will start by introducing the Thomas-Fermi approximation, the very beginning of this method. Then the two core elements of this theory will be discussed, namely the Hohenberg-Kohn theorems and the Kohn-Sham approach.

1.6.1 Thomas-Fermi Approximation

If one determines the electron ground state wave function $\Phi_{e,0}(\mathbf{r})$, all physical properties related to it can be obtained in principle. However, it demands a high computational cost, as the wave function for a system with N_e electrons has $3N_e$ spatial coordinates and N_e spin coordinates. Systems of interest in physics, chemistry, biochemistry, and materials science have many atoms and large numbers of electrons. So any treatment dealing directly with the electron wave function in these systems is very difficult, if not unfeasible, while also preventing an intuitive interpretation of the physical processes underlying their behavior.

On the other hand, their Hamiltonians contain only operators that act upon one (T_e, V_{en}) or two (V_{ee}) particles simultaneously, no matter the system size, which suggests the possibility of adopting a less computationally expensive approach that avoids the explicit determination of the electron wave function and replaces it with the electron density $\rho_e(\mathbf{r}, s)$. This density measures the number of electrons with a given spin component s ($\pm 1/2$) within an infinitesimal volume dV at $\mathbf{r} = (x, y, z)$ in Cartesian coordinates. As a matter of fact, the ground state electron density has every ingredient required to represent the Hamiltonian: its integral over all space gives the number of electrons N_e, and their maxima reveal the nuclear coordinates and the magnitude of each nuclear charge present.

Preliminary attempts to use the electron density instead of the wave function are almost as ancient as quantum mechanics. In the oldest model, due to Thomas and Fermi, there is a statistical modeling of the electron distribution that only takes into account the electron kinetic energy.

The Thomas-Fermi (TF) kinetic energy for a uniform electron gas with spatial density $\rho_e(\mathbf{r})$ is given by

$$T_{\text{TF}}[\rho_e(\mathbf{r})] = \frac{3}{10}(3\pi^2)^{2/3} \int [\rho_e(\mathbf{r})]^{5/3} d\mathbf{x}. \tag{1.63}$$

Combining this expression with the energy terms due to the electron–nuclei and electron–electron interactions, the Thomas-Fermi energy for an atom with atomic number Z is

$$E_{\text{TF}}[\rho_e(\mathbf{r})] = \frac{3}{10}(3\pi^2)^{2/3} \int [\rho_e(\mathbf{r})]^{5/3} d\mathbf{r} -$$
$$- Zke^2 \int \frac{\rho_e(\mathbf{r})}{|\mathbf{r}|} d\mathbf{r} + \frac{1}{2}ke^2 \int \frac{\rho_e(\mathbf{r})\rho_e(\mathbf{r}')}{|\mathbf{r} - \mathbf{r}'|} d\mathbf{r} d\mathbf{r}'. \tag{1.64}$$

Equation (1.64) is important not only because of its accuracy but also because it depends only on the electron density, being the first example of an electron density functional to obtain the energy of a system with many electrons, avoiding the wave function computational problem. After adding up the constraint $\int \rho_e(\mathbf{r}) d\mathbf{r} = N_e$, one can use the variational principle to find out the ground state electron density $\rho_{e,0}(\mathbf{r})$.

In 1951, Slater proposed an approximation for the nonlocal exchange energy of Hartree-Fock, based on the interaction between the charge density with spin s and the Fermi hole with same spin, namely

$$E_X[\rho_e(\mathbf{r},s)] = \frac{1}{2}ke^2 \int \frac{\rho_e(\mathbf{r},s)h_X(\mathbf{r},\mathbf{r}',s)}{|\mathbf{r}-\mathbf{r}'|}d\mathbf{r}d\mathbf{r}', \qquad (1.65)$$

where $h_X(\mathbf{r},\mathbf{r}',s)$ is an exchange hole density assigned to the spin s. Slater supposed that the exchange hole was spherically symmetric and centered at the reference electron at $\mathbf{r}$. Assuming that inside this sphere the exchange hole density is constant and zero outside, the Fermi hole radius is given by

$$r_h(\mathbf{r},s) = (3/4\pi)^{1/3}\rho_e(\mathbf{r},s)^{-1/3}. \qquad (1.66)$$

The radius r_h is sometimes called Wigner-Seitz radius and is a first approximation for the average distance between two electrons in the system. High-electron-density regions are related to small values of r_h. From electrostatics, it is possible to demonstrate that the potential of a uniformly charged sphere of radius r_h is proportional to r_h^{-1} or $\rho_e(\mathbf{r},s)^{1/3}$. Therefore, one finds the following approximation for E_X:

$$E_X[\rho_e(\mathbf{r},s)] \approx C_X \int [\rho_e(\mathbf{r},s)]^{4/3}d\mathbf{r}, \qquad (1.67)$$

where C_X is a numerical constant. In this way, we replace the complicated nonlocal Hartree-Fock energy for a much simpler expression dependent only on the electron density. In order to control the accuracy of this approximation, a semi-empirical adjustable parameter α was introduced instead of C_X, leading to the Slater Xα method:

$$E_{X,\alpha}[\rho_e(\mathbf{r},s)] = -\frac{9}{8}\left(\frac{3}{\pi}\right)^{1/3}ke^2\alpha \int [\rho_e(\mathbf{r},s)]^{4/3}d\mathbf{x}. \qquad (1.68)$$

1.6.2 Hohenberg-Kohn Theorems

In 1964, Hohenberg and Kohn published a seminal paper demonstrating the two fundamental theorems of DFT [5]. The first theorem proved that the ground state electron density is sufficient to determine the Hamiltonian operator of a many-electron system in an external potential and, therefore, each one of its properties.

First Hohenberg-Kohn theorem: *The external potential V is a unique functional (except of a constant) of $\rho_{e,0}(\mathbf{r})$.*

Proof: Let us consider two distinct external potentials $\hat{V}$ and $\hat{V}'$ not differing by a constant (if they differed by a constant, the many electron wave function and, consequently, the electron density would not change) but that produce the same ground state electron density $\rho_{e,0}(\mathbf{r})$. The corresponding Hamiltonians for these potentials are

$$\hat{H} = \hat{T}_e + \hat{V}_{ee} + \hat{V}, \tag{1.69}$$

$$\hat{H}' = \hat{T}_e + \hat{V}_{ee} + \hat{V}'. \tag{1.70}$$

Both Hamiltonians have distinct wave functions for their ground states, $\Phi_{e,0}$ and $\Phi'_{e,0}$, as well as distinct ground state energies $E_{e,0}$ and $E'_{e,0}$. However, both wave functions lead to the same electron density. We can represent this schematically by

$$\hat{V} \Rightarrow \hat{H} \Rightarrow \Phi_{e,0} \Rightarrow \rho_{e,0}(\mathbf{r}) \Leftarrow \Phi'_{e,0} \Leftarrow \hat{H}' \Leftarrow \hat{V}'. \tag{1.71}$$

As $\Phi_{e,0}$ and $\Phi'_{e,0}$ are different, one can use $\Phi'_{e,0}$ as a variational trial function for $\hat{H}$. Therefore, the variational principle implies

$$E_{e,0} < \langle \Phi'_{e,0} | \hat{H} | \Phi'_{e,0} \rangle = \langle \Phi'_{e,0} | \hat{H}' | \Phi'_{e,0} \rangle + \langle \Phi'_{e,0} | \hat{H} - \hat{H}' | \Phi'_{e,0} \rangle, \tag{1.72}$$

or

$$E_{e,0} < E'_{e,0} + \langle \Phi'_{e,0} | \hat{V} - \hat{V}' | \Phi'_{e,0} \rangle = E'_{e,0} + \int \rho_{e,0}(\mathbf{r}) \left(V - V' \right) d\mathbf{r}, \tag{1.73}$$

which implies, by switching the terms with and without the apostrophe,

$$E'_{e,0} < E_{e,0} + \langle \Phi_{e,0} | V' - V | \Phi_{e,0} \rangle = E_{e,0} + \int \rho_{e,0}(\mathbf{r}) \left(V' - V \right) d\mathbf{r}. \tag{1.74}$$

Adding Eqs. (1.73) and (1.74), one finds

$$E_0 + E'_0 < E_0 + E'_0, \tag{1.75}$$

which is evidently a contradiction. Therefore, our premise (two nontrivially distinct external potentials leading to the same ground state electron density) is false, proving the theorem.

As the total energy E of a given system is a functional of the electron density $\rho_e(\mathbf{r})$, all its components must be also functionals of $\rho_e(\mathbf{r})$. For a system of electrons in the external potential $V_{en}[\rho_e(\mathbf{r})]$ of a set of fixed nuclear charges, we have

$$E[\rho_e(\mathbf{r})] = T_e[\rho_e(\mathbf{r})] + V_{ee}[\rho_e(\mathbf{r})] + V_{en}[\rho_e(\mathbf{r})]. \tag{1.76}$$

This expression can be separated into two parts: one that depends on the specific system investigated, defined by the external potential, and the other that is common to all possible many-electron systems, defined by both the electron kinetic and electron–electron interaction potential:

$$E[\rho_e(\mathbf{r})] = \underbrace{T_e[\rho_e(\mathbf{r})] + V_{ee}[\rho_e(\mathbf{r})]}_{\text{Term common to all systems}} + \underbrace{V_{en}[\rho_e(\mathbf{r})]}_{\text{Specific term}}. \tag{1.77}$$

Separating the common term, we define the Hohenberg-Kohn functional, F_{HK},

$$F_{HK}[\rho_e(\mathbf{r})] = T_e[\rho_e(\mathbf{r})] + V_{ee}[\rho_e(\mathbf{r})], \qquad (1.78)$$

yielding

$$E[\rho_e(\mathbf{r})] = F_{HK}[\rho_e(\mathbf{r})] + V_{en}[\rho_e(\mathbf{r})]. \qquad (1.79)$$

In other words, if the Hohenberg-Kohn functional has as input an arbitrary electron density $\rho_e(\mathbf{r})$, it returns the expected value $\langle \Phi_{e,0}| \hat{T}_e + \hat{V}_{ee} |\Phi_{e,0}\rangle$ for the ground state wave function $\Phi_{e,0}$ which corresponds to this electron density. Finding the functional F_{HK} is crucial to ensure that the DFT method works. Its exact form would allow one to solve Schrödinger's equation for an arbitrary number of electrons with minimal computational cost. Even systems as complex as a DNA molecule or a protein could be described accurately at the quantum level. Unfortunately, however, the exact form of F_{HK} is not known. Only approximations can be made.

Looking to the V_{ee} functional, we can separate explicitly the classical Coulomb term $J[\rho_e(\mathbf{r})]$ by writing V_{ee} as

$$V_{ee}[\rho_e(\mathbf{r})] = \frac{1}{2}ke^2 \int \frac{\rho_e(\mathbf{r})\,\rho_e(\mathbf{r}')}{|\mathbf{r} - \mathbf{r}'|}d\mathbf{r}d\mathbf{r}' + E_{NC}[\rho_e(\mathbf{r})]$$

$$= J[\rho_e(\mathbf{r})] + E_{NC}[\rho_e(\mathbf{r})]. \qquad (1.80)$$

Here, $E_{NC}[\rho_e(\mathbf{r})]$ is the nonclassical contribution for the electron–electron interaction, which incorporates all correction effects of self-interaction, exchange, and Coulomb correlation. Finding $E_{NC}[\rho_e(\mathbf{r})]$ and $T_e[\rho_e(\mathbf{r})]$ is the great challenge for DFT.

Second Hohenberg-Kohn theorem: *The ground state energy functional $E_{e,0}[\rho_e(\mathbf{r})]$ is minimized if and only if $\rho_e(\mathbf{r}) = \rho_{e,0}(\mathbf{r})$, the exact electron density of the ground state.*

Proof: From the first theorem, we know that the ground state electron density uniquely determines the external potential V_{en}. It also determines the total number of electrons of the system as $\int \rho_{e,0}(\mathbf{r})d\mathbf{r} = N_e$. So the Hamiltonian of the system is also determined by $\rho_{e,0}(\mathbf{r})$. From the Hamiltonian, in principle, one can find the ground state wave function $\Phi_{e,0}$ associated with $\rho_{e,0}(\mathbf{r})$. If we use an electron density $\rho_e(\mathbf{r})$ distinct from the ground state, it must be associated with a distinct wave function Φ_e, for which we have

$$\langle\Phi_e| \hat{H} |\Phi_e\rangle = T_e[\rho_e(\mathbf{r})] + V_{ee}[\rho_e(\mathbf{r})] + V_{en}[\rho_e(\mathbf{r})]$$

$$= E[\rho_e(\mathbf{r})] > E[\rho_{e,0}(\mathbf{r})] = \langle\Phi_{e,0}|\hat{H}|\Phi_{e,0}\rangle, \qquad (1.81)$$

which is what we wanted to prove, defining the existence of an energy functional of the electron density that, when minimized, allows one to find the total energy and the electron density of a many-electron real system.

However, there is no practical method able to circumvent the necessity of employing some sort of wave function to perform DFT calculations. In particular, there is a subtle relation between the physical and chemical properties of a given material and the electron density, being difficult to discern between a covalent bond and an ionic bond once we know the total electron density.

1.6.3 Kohn-Sham Approach

If the electron density for the ground state $\rho_{e,0}(\mathbf{r})$ is known, the Hohenberg-Kohn theorems imply that it is, in principle, possible to calculate all electronic properties of the ground state from $\rho_{e,0}(\mathbf{r})$ without evaluating the wave function. The Hohenberg-Kohn theorems, however, do not reveal how to calculate $E[\rho_{e,0}(\mathbf{r})]$, since the exact form of the functional $F_{HK}[\rho_e(\mathbf{r})]$ is unknown, and do not say how to find $\rho_{e,0}(\mathbf{r})$ without employing the many-electron quantum state. A crucial step to solve these problems was made in 1965, when Kohn and Sham proposed a strategy that could provide approximate results with good accuracy [6].

Kohn and Sham considered a fictitious reference system of N_e non-interacting (NI) electrons moving in an external potential $V_{NI}(\mathbf{r})$, defined in such a way that its ground state electron density $\rho_{NI,0}(\mathbf{r})$ is equal to the ground state density $\rho_{e,0}(\mathbf{r})$ of a real system with N_e interacting electrons in an external potential $V_{en}(\mathbf{r})$ – i.e, $\rho_{e,0}(\mathbf{r}) = \rho_{NI,0}(\mathbf{r})$. As Hohenberg and Kohn have proved that the ground state electron density specifies the external potential uniquely once $\rho_{NI,0}(\mathbf{r})$ is found, the external potential $V_{NI}(\mathbf{r})$ is well determined, notwithstanding the impossibility of its evaluation.

For the NI system, the Hamiltonian is

$$\hat{H}_{NI} = \sum_{i=1}^{N_e} \left[-\frac{\hbar^2}{2m_e} \nabla_i^2 + V_{NI}(\mathbf{r}_i) \right] \equiv \sum_{i=1}^{N_e} \hat{H}_i^{KS}, \qquad (1.82)$$

where

$$\hat{H}_i^{KS} \equiv -\frac{\hbar^2}{2m_e} \nabla_i^2 + V_{NI}(\mathbf{r}_i). \qquad (1.83)$$

Here, $\hat{H}_i^{KS}$ is the one-electron Kohn-Sham Hamiltonian. As the reference system NI consists of N_e particles, the results already obtained for the Hartree-Fock model show that the wave function for its ground state is a Slater determinant of the lowest energy Kohn-Sham spin orbitals, which are solutions of

$$\hat{H}_i^{KS} u_j^{KS}(\mathbf{r}_i) = \varepsilon_j^{KS} u_j^{KS}(\mathbf{r}_i). \tag{1.84}$$

It is now convenient to introduce the spin component of the wave function, rewriting u_j^{KS} as

$$u_j^{KS}(\mathbf{r}_i, s_i) = \theta_j^{KS}(\mathbf{r}_i)\sigma_j(s_i), \tag{1.85}$$

where s_i can be $\pm 1/2$ and $\sigma_j(s_i)$ is the spin quantum state assigned to the quantum number s_i. So we can change Eq. (1.84) to

$$\hat{H}_i^{KS} \theta_j^{KS}(\mathbf{r}_i) = \varepsilon_j^{KS} \theta_j^{KS}(\mathbf{r}_i), \tag{1.86}$$

where the ε_j^{KS} term is the Kohn-Sham orbital energy for the jth one-electron state.

Let us now define the functional $\Delta T_{e,\text{NI}}$ as

$$\Delta T_{e,\text{NI}}\left[\rho_{e,0}(\mathbf{r})\right] \equiv T_e[\rho_{e,0}(\mathbf{r})] - T_{e,\text{NI}}\left[\rho_{e,0}(\mathbf{r})\right]. \tag{1.87}$$

The value of $\Delta T_{e,\text{NI}}\left[\rho_{e,0}(\mathbf{r})\right]$ is the kinetic energy difference between the real system and the NI system with the same ground state electron density. Let us also define $\Delta V_{ee,\text{NI}}$ as

$$\Delta V_{ee,\text{NI}}\left[\rho_{e,0}(\mathbf{r})\right] \equiv V_{ee}\left[\rho_{e,0}(\mathbf{r})\right] - \frac{1}{2}ke^2 \int \frac{\rho_{e,0}(\mathbf{r})\,\rho_{e,0}(\mathbf{r}')}{|\mathbf{r}-\mathbf{r}'|}d\mathbf{r}d\mathbf{r}'. \tag{1.88}$$

The second term on the right-hand side of Eq. (1.88) is the negative of the electrostatic repulsion between two electrons with their charges spatially spread out according to $\rho_{e,0}(\mathbf{r})$. Using these definitions, the functional that provides the total ground state energy for the many electron system in the external potential V_{en} is

$$E[\rho_{e,0}(\mathbf{r})] = \int V_{en}(\mathbf{r})\rho_{e,0}(\mathbf{r})\,d\mathbf{r} + T_{e,NI}\left[\rho_{e,0}(\mathbf{r})\right] +$$

$$+ \frac{1}{2}ke^2 \int \frac{\rho_{e,0}(\mathbf{r})\,\rho_{e,0}(\mathbf{r}')}{|\mathbf{r}-\mathbf{r}'|}d\mathbf{r}d\mathbf{r}' + E_{XC}\left[\rho_e(\mathbf{r})\right], \tag{1.89}$$

where the exchange-correlation functional E_{XC} is defined as

$$E_{XC}\left[\rho_e(\mathbf{r})\right] \equiv \Delta T_{e,\text{NI}}\left[\rho_e(\mathbf{r})\right] + \Delta V_{ee,\text{NI}}\left[\rho_e(\mathbf{r})\right]. \tag{1.90}$$

All these definitions were made to express the energy functional E as the sum of four quantities: the three first terms of the right-hand side of Eq. (1.89), which are relatively easy to compute from $\rho_e(\mathbf{r})$ and constitute the largest contribution for the ground state energy, and the fourth quantity E_{XC}, which is smaller than the other terms but much more difficult to find out accurately. The problem now is to obtain E_{XC}.

The ground state electron density is given from the Kohn-Sham orbitals as

$$\rho_{e,0}(\mathbf{r}) = \sum_{i=1}^{N_e} \left| \theta_i^{KS}(\mathbf{r}) \right|^2. \tag{1.91}$$

The first two terms term in Eq. (1.89) can be calculated as

$$\int V_{en}(\mathbf{r})\rho_{e,0}(\mathbf{r})d\mathbf{r} = -ke^2 \sum_{j=1}^{N_n} Z_j \int \frac{\rho_{e,0}(\mathbf{r})}{|\mathbf{r} - \mathbf{r}_j|} d\mathbf{r}, \tag{1.92}$$

$$T_{e,\text{NI}}\left[\rho_{e,0}(\mathbf{r})\right] = -\frac{1}{2m_e} \sum_{i=1}^{N_e} \int \theta_i^{KS*}(\mathbf{r})\nabla^2\theta_i^{KS}(\mathbf{r})d\mathbf{r}. \tag{1.93}$$

One can now calculate $E[\rho_{e,0}(\mathbf{r})]$ from $\rho_{e,0}(\mathbf{r})$ if the Kohn-Sham orbitals are determined and if the exchange-correlation functional E_{XC} is known. In order to find the Kohn-Sham orbitals, the variational principle must be applied to Eq. (1.86), selecting the $\theta_i^{KS}(\mathbf{r})$ that minimize $E[\rho_{e,0}(\mathbf{r})]$ with the orthonormality constraint $\langle \theta_i^{KS} | \theta_j^{KS} \rangle = \delta_{ij}$. The final result is a Schrödinger-like equation:

$$\hat{H}^{KS}\theta_i^{KS}(\mathbf{r}) = \left[-\frac{1}{2m_e}\nabla^2 + V_{e,\text{NI}}(\mathbf{r}) \right]\theta_i^{KS}(\mathbf{r}) = \varepsilon_i^{KS}\theta_i^{KS}(\mathbf{r}), \tag{1.94}$$

where the exchange-correlation potential is defined as the variational derivative,

$$V_{XC}(\mathbf{r}) \equiv \frac{\delta E_{XC}[\rho_e(\mathbf{r})]}{\delta \rho_e(\mathbf{r})}, \tag{1.95}$$

and the NI potential is given by

$$V_{e,NI}(\mathbf{r}) = -ke^2 \sum_{j=1}^{N_n} \frac{Z_j}{|\mathbf{r}_j|} + ke^2 \sum_{k=1}^{N_e} \int \frac{\left| \theta_k^{KS}(\mathbf{r}') \right|^2}{|\mathbf{r} - \mathbf{r}'|} d\mathbf{r}' + V_{XC}(\mathbf{r}). \tag{1.96}$$

The Kohn-Sham Hamiltonian $\hat{H}^{KS}$ is very similar to the Fock operator defined in Eq. (1.57), except that the exchange operators are now included in the exchange-correlation potential $V_{XC}(\mathbf{r})$, which also incorporates the electronic correlation effects.

The exchange-correlation energy E_{XC} given by Eq. (1.90) has as components the correlation kinetic energy $\Delta T_{e,\text{NI}}$ (the difference between the real kinetic energy and the kinetic energy of the reference NI system), the exchange energy (due to the antisymmetric wave function), the Coulomb correlation energy (electron–electron repulsion), and a self-interaction correction. The self-interaction term originates from the inclusion of the electron density due to the occupation of the $\theta_i^{KS}(\mathbf{r})$ state in the third term on the left side of Eq. (1.89).

The Kohn-Sham orbitals $\theta_i^{KS}(\mathbf{r})$ are wave functions of the ficticious reference system of NI electrons. Therefore, strictly speaking, these orbitals have no physical meaning beyond their capacity to predict the exact ground state. The total wave function in DFT is not a Slater determinant of "real" spin orbitals. However, calculations of the Kohn-Sham orbitals are very similar to the molecular orbitals obtained from the Hartree-Fock method, and they can be used in qualitative discussions about molecular properties and reactivities. Observe that the Hartree-Fock orbitals are also not "real," as they too refer to a fictitious system where each electron is embedded in a sort of mean field created by the other electrons.

For a closed-shell molecule, the Hartree-Fock energies of the occupied orbitals provide a good approximation for the energy required to remove each electron in them (Koopman's theorem). The same is not true for Kohn-Sham orbitals, with the exception to the energy ε_i^{KS} of the highest occupied molecular orbital (HOMO) which is the negative of the first ionization molecular energy. Yet, the usual approximations for $V_{XC}(\mathbf{r})$ do not allow one to estimate it accurately.

1.7 Exchange-Correlation Energy

There are basically three distinct approximation involved in a DFT calculation. One is conceptual and is related to the interpretation of the Kohn-Sham eigenvalues (energies) and orbitals (wave functions). The second type of approximation is numerical and concerns methods to solve the Kohn-Sham equation given by Eq. (1.94). The third type involves constructing an expression for the unknown functional $E_{XC}[\rho_e(\mathbf{r})]$, given by Eq. (1.90), which contains all many-body aspects of the theory. It is with this type of approximation that we are concerned in this section.

We intend here to discuss three types of functionals, namely

(a) Local density approximation (LDA): Historically, the simplest approximation for the exchange and correlation function $E_{XC}[\rho_e(\mathbf{r})]$. In this approach, the electron density is considered as a uniform electron gas whose electron density is constant throughout the space of the configurations.
(b) Generalized gradient approximation (GGA): An improvement in the LDA approach, incorporating the electron density gradient in the energy expression of the exchange-correlation function. The electronic density ceases to be constant, and its variation is represented by its gradient within the characteristic function of exchange and correlation.
(c) Hybrid functionals: Mix a Hartree-Fock exchange term with different DFT exchange-correlation energies.

Let us discuss all of them in the next sections.

1.7.1 Local Density Approximation (LDA)

Hohenberg and Kohn have shown that if ρ_e (**r**) has a smooth dependence on **r**, the exchange-correlation potential E_{XC} [ρ_e (**r**)] can be expressed as

$$E_{XC}^{LDA} [\rho_e (\mathbf{r})] = \int \rho_e (\mathbf{r}) \, \varepsilon_{XC} (\mathbf{r}) \, d\mathbf{r}, \tag{1.97}$$

where ε_{XC} (**r**) is the exchange-correlation energy per electron in an homogeneous electron gas with electron density ρ_e (**r**). The exchange-correlation functional is obtained from the functional derivative of Eq. (1.95) as

$$V_{XC}^{LDA} (\mathbf{r}) = \frac{\delta E_{XC}^{LDA} [\rho_e (\mathbf{r})]}{\delta \rho_e (\mathbf{r})} = \varepsilon_{XC}(\mathbf{r}) + \rho_e (\mathbf{r}) \frac{\partial \varepsilon_{XC} (\rho_e (\mathbf{r}))}{\partial \rho_e (\mathbf{r})}. \tag{1.98}$$

This exchange-correlation energy per electron can be written as the sum of two parts:

$$\varepsilon_{XC}[\rho_e (\mathbf{r})] = (\varepsilon_r + \varepsilon_C)[\rho_e (\mathbf{r})], \tag{1.99}$$

where ε_{XC}, the exchange energy, is given by

$$\varepsilon_{XC}[\rho_e (\mathbf{r})] = -\frac{3}{4} \left(\frac{3}{\pi} \right)^{1/3} [\rho_e (\mathbf{r})]^{1/3}. \tag{1.100}$$

In a typical DFT calculation, the first step is to select an initial electron density from the superposition of the atomic electron densities in the studied geometry. From it, the exchange-correlation potential is calculated and used to generate the Kohn-Sham orbitals represented in some basis set. Aftward, these Kohn-Sham orbitals are used to generate a new electron density, and the process repeats until there is no significant change in the electron density or in the electron total energy.

In open shell systems, Kohn-Sham spin orbitals replace the spatial orbitals, and the LDA approximation is called local spin density approximation (LSDA), with separate densities $\rho_e^\uparrow$ (**r**) and $\rho_e^\downarrow$ (**r**) for electronic spin up and spin down, respectively. The LDA approximation is surprisingly accurate, considering the simplicity of this model, notwithstanding the fact that it tends to predict too-large interatomic forces and smaller bond lengths. This is a consequence of a cancellation of errors, as the LDA exchange energy is typically overestimated while the LDA correlation energy is typically underestimated [7, 8].

1.7.2 Generalized Gradient Approximations

The LDA and LSDA approximations are based on the homogeneous electron gas model, which is appropriate for a system where the electron density ρ_e (**r**) is a smooth function of **r**. Functionals that go beyond the LSDA method incorporate the electron density gradient:

$$E_{XC}^{GGA}\left[\rho_e^{\uparrow}(\mathbf{r}),\rho_e^{\downarrow}(\mathbf{r})\right] = \int f\left[\rho_e^{\uparrow}(\mathbf{r}),\rho_e^{\downarrow}(\mathbf{r}),\nabla\rho_e^{\uparrow}(\mathbf{r}),\nabla\rho_e^{\downarrow}(\mathbf{r})\right]d\mathbf{r}, \quad (1.101)$$

where f is some function of the electron spin densities and their gradients. Functionals in the form of Eq. (1.101) define the so-called generalized gradient approximation (GGA).

There are many GGA parameterizations for the exchange functional, and they can be separated into two groups: functionals based on some fitting to experimental data and "pure" functionals without any empirical adjustment derived from quantum mechanics. Examples of the first are the Becke88 [9] and Perdew-Wang functionals [10], while the second includes the Perdew86 [11] and the Perdew-Burke-Ernzerhof (PBE) functionals [12]. The empirically fitted exchange functionals predict accurate atomization energies and molecular reaction barriers but are unable to provide a good description of solids, while the functionals without empirical adjustment perform better for solid-state systems.

The most-used GGA correlation functionals are the PBE [12] and LYP (Lee, Yang, and Parr) [13] functionals. In general, GGA methods underestimate the interatomic forces, predicting larger bond lengths in comparison to experiment while, at same time, providing reliable descriptions for ionic, covalent, metallic, and hydrogen bonds. They are, however, unable to take into account dispersive forces such as van der Waals. They also do not improve by much the description of solids or the calculation of ionization potentials and electron affinities in comparison to the LDA approximation. For this reason, meta-GGA [14] and hyper-GGA [15] functionals were proposed, the first incorporating the NI kinetic energy density and the second including the exact exchange energy density (EXX), aiming for chemical accuracy, with errors of about 1 kcal/mol for bonded interactions and 0.1 kcal/mol for nonbonded interactions.

1.7.3 Hybrid Functionals

One can combine exchange and correlation functionals from different authors in order to fit some training set of empirical data, leading to the so-called hybrid functionals. For example, the nonempirical hybrid functional PBE0 [16] uses the GGA–PBE exchange-correlation functional with the exchange term being partially replaced by the exact exchange energy, while the Heyd-Scuseria-Ernzerhof functional [17] mixes the PBE correlation energy with PBE and Hartree-Fock exchange energies modulated by a screened Coulomb potential. These functionals tend to predict much better excitation energies, especially for solids, when compared to the severely underestimated gaps at the LDA and GGA levels.

Meta-hybrid functionals, such as those of the Minnesota family of functionals [18], are adjusted to produce good results in many different situations, including systems where noncovalent forces are significant. They combine a meta-GGA DFT functional with different amounts of exact exchange to achieve good results. Dispersion forces can also be added to "pure" or hybrid functionals by different methods, such as those proposed by Grimme [19, 20] or Tkatchenko and Scheffler [21], being employed to study molecular crystals and biomolecules.

1.8 Molecular Fractionation with Conjugate Caps Method

In order to calculate the DFT interaction energy between a specific molecule and a protein site, only amino acid residues within a certain radius of the ligand can be considered. In this idealized binding region, the ligand and the protein fragments are kept fixed, and a geometry optimization is performed for the hydrogen atoms whose coordinates are not well determined from experimental data.

One method to evaluate the DFT interaction energy between the ligand and the protein site is the molecular fractionation with conjugate caps scheme (MFCC), originally proposed by Zhang and Zhang (2003) [22, 23]. If the protein residue sequence is $R_1 - R_2 - \cdots - R_i - \cdots R_n$ (the $-$ representing the peptide bonds), the DFT interaction energy between the residue R_i and the ligand L, $E_{int}(R_i \cdots L)$ can be obtained by considering only the residue and the ligand. The residue's dangling bonds must be capped with hydrogen atoms or a limited portion of its molecular neighborhood (see Figure 1.6). Labeling the conjugated caps as C^*_{i-1} and C^*_{i+1}, one can write the interaction energy as

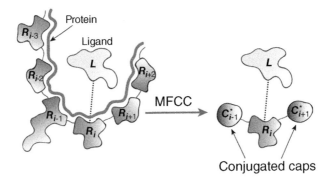

Figure 1.6 MFCC strategy to obtain the interaction energy between a ligand and an amino acid residue in a protein.

$$E_{\text{INT}}(R_i \cdots L) = \left[E(C_{i-1}^* R_i C_{i+1}^* + L) - E(C_{i-1}^* R_i C_{i+1}^* + L, r = \infty) \right] +$$
$$- \left[E(C_{i-1}^* C_{i+1}^* + L) - E(C_{i-1}^* C_{i+1}^* + L, r = \infty) \right] + E_{\text{BSSE}},$$

$$(1.102)$$

where the energy $E(C_{i-1}^* R_i C_{i+1}^* + L)$ is the total DFT energy of the capped residue R_i together with the ligand at the experimental coordinates (obtained from X-ray diffraction, for example). The term $E(C_{i-1}^* R_i C_{i+1}^* + L, r = \infty)$ is the total DFT energy of the capped residue R_i together with the ligand with both being separated by a large distance, typically 100 Å or more, depending on the size of the interacting parts. From the difference of these energies, one must subtract the interaction energy between the ligand and the caps only (with any dangling bonds in them passivated with hydrogen atoms), given by the difference between the total cap-ligand DFT energy $E(C_{i-1}^* C_{i+1}^* + L)$ at experimental coordinates, and the total cap-ligand DFT energy $E(C_{i-1}^* C_{i+1}^* + L, r = \infty)$ with the cap and ligand separated by a large distance. Lastly, an energy correction, E_{BSSE}, due to the different basis sets used to describe the other terms must be included to improve the accuracy of the calculation. The total binding site energy of the ligand L with the protein site S can be obtained by adding up the interaction energies between the ligand and the residues inside it:

$$E_{\text{int}}(S \cdots L) = \sum_{R_i \in S} E_{\text{int}}(R_i \cdots L). \qquad (1.103)$$

The main advantage to obtain $E_{\text{int}}(S \cdots L)$ within the MFCC scheme is that its computational cost grows linearly with the number of residues considered. One must note that in this method, it is important to choose the conjugated caps in such a way that they represent adequately the electronic neighborhood of the residue R_i. All the energies on the right side of Eq. (1.103) can be improved by considering a dielectric solvation background or a distribution of point charges representing distant charged residues [24].

Dispersion-corrected exchange-correlation functionals are essential to provide significant results, as van der Waals forces are not well described by "pure" DFT functionals. By changing the radius of the site S, one can also investigate what the adequate size of the binding site is to ensure good convergence. This fragmentation strategy can be useful to obtain a detailed interaction profile of the ligand, suggesting which amino acid residues are more attractive or repulsive, as well as allowing for the investigation of ligand modifications that can lead to enhanced pharmacological activity, for example.

1.9 Conclusions

Advances in computational power and software combined with theoretical break-
throughs in the modeling of atomic and molecular systems have allowed the simulation of systems, such as nanostructures and proteins, using quantum mechanics. Computational modeling, or in silico experiments, has played a fundamental
role in the development of novel molecules with pharmacological potential and in
the understanding of molecular interactions occurring in biological systems. This
evolution has made feasible the simulation of the physical chemistry properties of
drugs interacting directly with solvated proteins. Parallel processing, enhanced by
the use of graphical processing units (GPUs) capable of performing calculations at
the teraflop scale and at relatively low cost, accelerates significantly the molecular
modeling of systems with thousands or even millions of atoms.

First-principles quantum methods and molecular dynamics are very useful
tools for the investigation of biological receptors and bioactive ligands. DFT
has achieved a huge success in the study of protein-ligand docking from an
atomistic-electronic viewpoint. It provides relevant information on the electronic
structure changes due to the molecular binding, being specific to the point of
identifying which molecular regions contribute more significantly to the drug–
protein interaction. Molecular dynamics, on the other hand, can be employed
in the investigation of a wide range of large scale phenomena, including those
investigated within structural biochemistry, biophysics, enzymology, molecular
biology, pharmaceutical chemistry, and biotechnology. In particular, molecular
dynamics dramatically increases the speed of the screening for new drugs and the
refinement of molecular structures obtained from X-ray diffraction and NMR.

Given the flexibility and competition between different ligand molecules and the
role of weak interactions in the processes of drug incorporation to protein active
sites, accurated systematic methods to estimate their total energy are an absolute
requirement. Besides, protonation and deprotonation processes in molecular chains,
ionic interactions, and formation and breaking of chemical bonds, among other
processes that are not easy to describe from empirical formulations, are relevant to
establish the conformational preference of a drug. The ideal approach would be the
application, under these circumstances, of first-principle quantum methods with a
wider field of applicability. However, high-level quantum methods – such as the
coupled-cluster theory, many-body theory, perturbation theory – are computationally very expensive, being practical only for molecules with a few tens of atoms
at most.

Among the techniques to incorporate quantum effects with acceptable computational cost, DFT is the most promising if one considers the balance between accuracy and efficiency, being already employed for the elucidation of questions related

to proteins. The application of fragmentation strategies, on the other hand, allows the calculation of several properties of the drug–protein complex. The development of exchange-correlation functionals, including dispersion-correction effects, as a matter of fact, has lead to results in good agreement with experimental data. The inclusion of solvation effects can be carried out by dielectric constant modeling or by using an external point charge distribution in which the biomolecule is embedded.

2

Charge Transport in the DNA Molecule

2.1 Introduction

The field of nanotechnology has emerged as one of the most important areas of research in the near future. While scientists have been hardly aspiring to controllably and specifically manipulate structures at the micrometer and nanometer scale, nature has been performing these tasks and assembling structures with great accuracy and high efficiency using specific biological molecules – such as the deoxyribonucleic acid (DNA) molecule, the ribonucleic acid (RNA) molecule, amino acids, and proteins. As a consequence, there has been a tremendous interest in recent years to develop concepts and approaches for self-assembled systems and to find their electronic and optical applications. Biology can provide models and mechanisms for advancing this approach, but there is no straightforward way to apply them to electronics since biological molecules are essentially electrically insulating [25].

However, exquisite molecular recognition of various natural biological materials can be used to form a complex network of potentially useful systems for a variety of optical, electronic, and sensing applications. For instance, investigations of electrical junctions, in which single molecules or small molecular assemblies operate as conductors connecting traditional electrical components such as metal or semiconductor contacts, constitute a major part of what is nowadays known as molecular electronics. Their diversity, versatility, and amenability to control and manipulation make them potentially important components in nanoelectronic devices [26].

For scientists, this continuing progress and the consequent need for further size miniaturization makes the DNA molecule, the basic building block of living species and the carrier responsible of the genetic code [27], the best candidate to fulfill this place. Arguably one of the main challenging research subjects of current science, human DNA, is around 6 mm long, has about 2×10^8 nucleotides, and is tightly packed in a volume equal to 500 μm^3. If a set of three nucleotides can be assumed to be analogous to a byte, then these numbers represent either 1 KB μm^{-1} (linear density) or 1.2 MB μm^{-3} (volume density) – an appreciation of how densely information can be stored in DNA.

A complete DNA molecule is a chromosome, with protein components present as structural support. The DNA of each gene carries a chemical message that signals to the cell how to assemble the amino acids in the correct sequence to produce the protein for which that gene is responsible. The information is contained in the sequence of the monomers called nucleotides, which make up the DNA molecule and whose structure consists of a base together with a backbone of alternating sugar molecules and phosphate ions. There are four different nucleotides in DNA, differing by their chemical components, linked together forming a backbone of alternating sugar–phosphate residues with the bases that carry the information of the gene. For practical reasons, these nucleotides can be symbolized by a sequence of four letters – namely guanine (G), adenine (A), cytosine (C), and thymine (T) – whose repeated stacks are formed by either AT (TA) or GC (CG) pairs coupled by hydrogen bonds, the so-called Chargaff's rule [28], and held in a double-helix structure by a sugar–phosphate backbone (see the schematic drawing depicted in Figure 2.1). The specificity of this base pairing and the ability to ensure that it occurs in this fashion (and not some other way) is a key factor to use DNA in materials applications.

The double-helical arrangement of the two molecules leads to a linear helix axis – linear not in the geometrical sense of being a straight line but in the

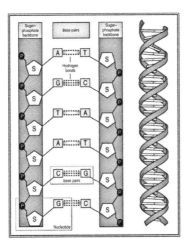

Figure 2.1 Schematic illustration of the chemical composition of the DNA molecule, showing its four bases guanine (G), adenine (A), cytosine (C) and thymine (T) and the sugar–phosphate backbone (S and P).

topological sense of being unbranched. This robust, although malleable, one-dimensional structure of DNA is unique and may be used to design functional nanostructures, and its charge transport capability in the appropriate energy regime can be quite good. Besides, DNA molecules encode all the information needed to build every cell, tissue, and membrane of a living organism and, consequently, occupy an outstanding position in life sciences for DNA's crucial role as carrier of the genetic code.

Numerous algorithms have been introduced to characterize and graphically represent the genetic information stored in the DNA nucleotide sequence. The goal of these methods is to generate a representative pattern for certain sequences or groups of sequences.

Notwithstanding, the design of DNA-based devices for molecular nanoelectronics is not yet an easy task since they are crucially dependent upon elucidation of the mechanism and dynamics of electrons and hole transport in them. Besides, unlike proteins, DNA is not primarily an electron/hole-transfer system, and its suitability as a potential building block for molecular devices may not depend only on long-distance transfer of electrons and holes through the molecule. However, the discovery that DNA, like proteins, can conduct an electrical current has made it an interesting candidate for nanoelectronic devices, which could help to overcome the limitations that classical silicon-based electronics is facing presently. Indeed, individual DNA molecules are very suitable for producing a new range of devices that are much smaller, faster, and more energy efficient than the present semiconductor-based ones.

In fact, DNA offers a solution to many of the hurdles that need to be overcome, since it has the capacity of self-assembly and self-replication, making it possible to produce nanostructures with a precision that is not achievable with classical silicon-based technologies. The DNA molecule is usually described as a short-ranged correlated random ladder, but nothing prevents the DNA chain from being artificially grown following quasiperiodic sequences, such as the Fibonacci (FB) and Rudin-Shapiro (RS) sequences, presenting long-range correlations. These structures exhibit interesting properties, namely:

(a) They have a complex fractal spectra of energy, which can be considered to be their indelible mark.
(b) They also exhibit collective properties that are not shared by their constituents.

These collective properties are due to the presence of long-range correlations, which are expected to be reflected somehow in their various spectra (as in light propagation, electronic transmission, density of states, etc.) defining another description of disorder (for up-to-date reviews, see References [29, 30]). Besides, the introduction of long-range correlations in aperiodic or genomic DNA sequences markedly changes their physical properties and can play a crucial role in their charge transfer efficiency, making a strong impact on their engineering biological processes like gene regulation and cell division. Moreover, the nature of this long-range correlation has been the subject of intense investigation, whose possible applications on electronic delocalization in the one-dimensional Anderson model have been recently discussed [31].

The influence of the sugar–phosphate backbone is also important because it promotes the emergence of a bandgap of the order of the hopping integral. Recent results showed that the hybridization of the overlapping *p*-orbital in the base-pair stack coupled to the backbone is sufficient to predict the existence of a gap in the nonequilibrium current-voltage characteristics, with a minimal number of parameters [32].

On the other hand, advances in novel DNA constructions, with the creation of 3-D DNA topological structures, are opening up a new world for charge transport in DNA junctions and lattices, a scientific advance bridging the molecular world to the world where we live [33]. As a matter of fact, the DNA base-pair recognition system can be used to direct the assembly of highly structured materials into a series of 3-D triangle-like motifs with specific nanoscale features, as well as in DNA computation to process complex information, as explicitly stated by Seeman [34]. The combination of synthetic stable branched DNA and sticky-ended cohesion (small cohesive sequences on each end of the motif), which attach to other molecules and place them in a set order and orientation, has led to the development of structural DNA nanotechnology over the last 30 years.

The early topological constructs built from DNA led also to the development of specific single-stranded DNA topologies. The action of staple strands was later explored to fold DNA strands into a variety of shapes – a technique named DNA origami, which uses a few hundred short DNA strands to direct a very long DNA strand to form structures to any desired shape, serving as their assembly line's framework [35]. Nowadays, these several nanometric structures, with great potential biotechnological applications, have evolved to the ability to produce 2-D and 3-D DNA crystals with linear dimensions of the order of 1 mm. To explore the complex charge transport in these 3-D connected DNA, wire fragments give rise to novel DNA-based circuits and/or devices, opening up a powerful new direction in the field of integrated nanoelectronic-biological structures.

Within this context, the purpose of this chapter is to present a comprehensive and up-to-date account of the main electronics properties of the DNA molecule, mainly within the context of quasiperiodicity of the bases arrangement, and the role played by short- and long-range correlation effects in the search for nano-size devices [36]. The chapter is organized so that we start with the basic properties of the quasiperiodic structures, stressing those that have already been grown by experimentalists, namely the Fibonacci and Rudin-Shapiro structures (Section 2.2). Then, in Sections 2.3 and 2.4, we discuss the important issue of charge transport in the DNA molecule, based on an effective tight-binding model that describes an electron moving in a chain with a single orbital per site and nearest-neighbor interactions, together with a transfer-matrix approach to simplify the algebra, which can be otherwise quite complex. Several different DNA topologies are taken into account.

2.2 Quasiperiodic Structures

The subject of quasicrystals first achieved prominence in 1984, following the report by Schechtman et al. [37] of metallic Al-Mn alloys, showing amazing and interesting electron diffraction data. They mixed Al and Mn in a roughly six-to-one proportion and heated the mixture until it melted. The mixture was then rapidly cooled back to the solid state by dropping the liquid into a cold spinning wheel, a process known as melt spinning.

When the solidified alloy was examined with an electron microscope, a novel structure was revealed. It exhibited fivefold symmetry, which is forbidden in ideal crystals, and a long-range order, which is lacking in amorphous solids. Its order, therefore, was neither truly amorphous nor crystalline. Subsequent measurements, using X-ray scattering at much higher resolution, led to electron diffraction patterns showing not only fivefold but also icosahedral symmetries, forbidden by the rules of crystallography (for a review, see Reference [38]). These achievements led to the award of the 2011 Nobel Prize in Chemistry to D. Schechtman, *for the discovery of quasicrystals*.

Theoretical studies developed by Levine and Steinhardt [39] explained these types of symmetry through the aperiodic 2-D and 3-D Penrose tilings in their diffraction patterns. Tiling is the geometrical operation that results in filling a space with an arrangement of regular polyhedra. Their predictions were, indeed, qualitatively similar to those observed by Schechtman et al. [37]. In addition to further experimental studies, the subsequent challenge has been the development of theoretical models to characterize these artificial structures.

Although the term quasicrystal is more appropriate when applied to natural compounds or artificial alloys, in 1-D there is no difference between quasicrystals and the quasiperiodic structures formed by the incommensurate arrangement of periodic unit cells. The particular mathematical sequences (Fibonacci and Rudin-Shapiro, among others) that define the type of quasiperiodic structure will be discussed in the following subsections. An appealing motivation for studying such structures is that they exhibit a highly fragmented energy spectrum displaying a self-similar pattern [40–42]. Indeed, from a strictly mathematical perspective, it has been proven that their spectra are Cantor sets in the thermodynamic limit.

A fascinating feature of these quasiperiodic structures is that they exhibit distinct physical properties, found neither in periodic arrangements nor in their constituent parts, giving rise to a novel description of disorder. Indeed, theoretical transfer-matrix treatments can be used to show that these spectra are fractals, defining intermediate systems between periodic crystals and random amorphous solids [43, 44]. This is one of the features that make them of particular interest to study.

The presence and nature of long-range correlations in such systems preclude using canonical approaches, like perturbation theory, where one first separates a small localized piece of the system, treating the rest as a perturbation a posteriori. This approach typically does not work for the cases under consideration here because the behavior of the overall macroscopic system is quite distinct from the behavior of its separate small pieces, due to the long-range correlations. Fortunately, the presence of long-range correlations itself gives the key to circumvent this difficulty, namely that these systems are normally robust to wide modifications on a microscopic scale.

An important consequence of this robustness, where many systems that are distinct within a microscopic scale may exhibit the same critical behavior, is that one can classify the various systems in a few universality classes. As an analogy, we may consider the topic of continuous phase transitions: the critical behavior is known to depend only upon global properties, namely the geometric dimension of the system and the symmetries of its order parameter, being insensitive to the details of the microscopic interactions between the atoms or molecules.

The spectra of many types of elementary excitations in quasiperiodic structures have been extensively studied by numerous research groups. In all cases, the spectra were found to be Cantor-like with critical eigenfunctions [45]. For electronic systems, exact eigenfunctions were found only at the special null energy value. However, there are infinitely many eigenvalues in the energy spectrum, although they are rare for the electron chaotic orbits.

An important issue is to understand the wave functions corresponding to these chaotic orbits. We note that it does not necessarily follow that the wave functions themselves are chaotic, because the orbits represent only selected points on the lattice. In addition, there may be a discrete set of extended states and a quite complex fractal energy spectrum, a common feature of these systems that can be considered their basic signature. Several different mathematical techniques – including renormalization group theory [46], the transfer-matrix method [47], and chaotic Hamiltonian systems [48], to mention just a few – have been successfully applied to describe quasiperiodic structures, leading to remarkable results.

Another important motivation for studying quasiperiodic structures comes from the recognition that the localization of electronic states, one of the most active fields in condensed matter physics, could occur not only in disordered systems but also in the deterministic quasiperiodic systems [49]. Localization of electronic states in quasiperiodic structures was studied using a tight-binding Schrödinger equation by several research groups (see [50] and the references therein).

The quasiperiodic structures considered here are of the type generally known as substitutional sequences; they have been studied in several areas of mathematics, computer science, and cryptography. The sequences are characterized by the nature of their Fourier spectrum, which can be dense pure points (as for the Fibonacci sequence) or singular continuous ones (as for the Rudin-Shapiro sequence).

We start with some general mathematical considerations and terminology. First, we give the definition of a substitutional sequence of the type used here. Consider a finite set ξ, where $\xi = \{A, B\}$, for example, with A and B being two different building blocks, called an alphabet, and denote by ξ^* the set of all words of finite length (such as $AABAB$) that can be written in this alphabet. Now let us define ζ as a map from ξ to ξ^* by specifying that ζ acts on a word by substituting each letter (e.g., A) of this word by its corresponding image, denoted by $\zeta(A)$. A sequence is then called a substitutional sequence if it is a fixed point of ζ – i.e., if it remains invariant when each letter in the sequence is replaced by its image under ζ.

These substitutional sequences are described in terms of a series of generations that obey particular inflation rules. Let $a_1, a_2, \ldots, a_g$ be g basic units, and define this pattern as stage n of the sequence. Then, the next stage of the sequence, $n + 1$, is obtained inductively from stage n by the inflation rule $\vec{a} \to \bar{M}\vec{a}$, where $\vec{a}$ represents the column vector $(a_1, a_2, \ldots, a_g)^t$, with t denoting the transpose. Also, $\bar{M} = (m_{ij})$ is a $g \times g$ matrix with nonnegative integer matrix elements. The matrix $\bar{M}$ and its successive applications fully determine the sequence. At each stage, a_i is replaced by $m_{i1}a_1$, followed by $m_{i2}a_2, \ldots$, etc., for $i = 1, 2, \ldots, g$. For example, for the case of the Fibonacci lattice to be discussed shortly, we have $g = 2$, and it turns out that we operate with the 2×2 substitution matrix $\bar{M}$

$$\bar{M} = \begin{pmatrix} 1 & 1 \\ 1 & 0 \end{pmatrix}, \tag{2.1}$$

on the vector $(a_1, a_2)^t$ at each stage. This gives, in terms of the building blocks A and B, the substitution rules $A \to AB$, $B \to A$, which will then generate the whole sequence, provided we start with AB as the first compound word of the sequence. Similar procedures can be identified to generate other quasiperiodic sequences. In the following, we proceed to give explicit definitions of the main substitutional sequences to be used here.

2.2.1 Fibonacci

The Fibonacci (FB) sequence is the oldest example of an aperiodic chain of numbers. It was conceived by Leonardo de Pisa (whose nickname was Fibonacci) in 1202 as a result of his investigation on the growth of a population of rabbits. The successive Fibonacci numbers are generated by adding together the two previous numbers in the sequence, after specifying suitable initial conditions.

For our purposes, a Fibonacci structure can be realized experimentally by juxtaposing the two basic building blocks A and B in such a way that the N_{FB}th stage of the process $S_{N_{FB}}$ is given by the recursive rule $S_{N_{FB}} = S_{N_{FB}-1} S_{N_{FB}-2}$, for $N_{FB} \geq 2$, starting with $S_0 = B$ and $S_1 = A$. It has the property of being invariant under the transformations $A \rightarrow AB$ and $B \rightarrow A$.

The Fibonacci generations are (see Figure 2.2a)

$$S_0 = B; \quad S_1 = A; \quad S_2 = AB; \quad S_3 = ABA; \quad S_4 = ABAAB; \quad \text{etc.} \qquad (2.2)$$

In this case, the number of building blocks increases in accordance with the Fibonacci number $F_{N_{FB}}$ defined by the rule $F_{N_{FB}} = F_{N_{FB}-1} + F_{N_{FB}-2}$ (with $F_0 = F_1 = 1$). Also, the ratio between the number of the building blocks A and the number of the building blocks B in the sequence tends to the golden mean number $\tau_{FB} = (1 + \sqrt{5})/2 \simeq 1.62$ for large generation number N_{FB}. This particular irrational number is related to fivefold symmetries (e.g., it is twice the ratio of the distance between the center vertex and the center mid-edge in a

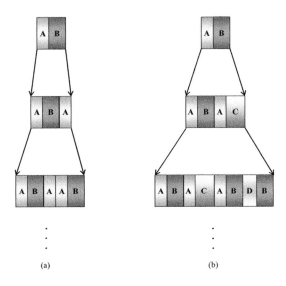

(a)   (b)

Figure 2.2 Schematic illustration of the two quasiperiodic structures considered in this work: (a) Fibonacci; (b) Rudin-Shapiro. After Albuquerque et al. [36].

pentagon). It is interesting to note that all the Fibonacci numbers can be generated from the golden mean number through the relation

$$F_{N_{FB}} = \left[(\tau_{FB})^{N_{FB}} - (-\tau_{FB})^{-N_{FB}} \right] \Big/ \sqrt{5}. \tag{2.3}$$

This means that a sequence of rational numbers (namely the integer-valued Fibonacci numbers) can be obtained from powers of irrational numbers.

There are variations of the preceding sequence that lead to generalized Fibonacci (GFB) structures involving different relationships between the number of the building blocks A and the number of the building blocks B (thus also generalizing the golden mean number).

In these cases, the Nth stage of the structure $S_{N_{GFB}}$ is taken to be generated by the sequence given recursively as

$$S_{N_{GFB}+1} = S_{N_{GFB}}^{p} S_{N_{GFB}-1}^{q}, \tag{2.4}$$

with, as before, $S_0 = B$ and $S_1 = A$. Here the indexes p and q are arbitrary positive integer numbers and $N_{GFB} \geq 1$.

This above notation means that $S_{N_{GFB}}^{p}$ represents p adjacent repetitions of the stack $S_{N_{GFB}}$. This type of inheritance is normal in iterative processes and frequently produces self-similar structures that are the basis of fractal configurations. When $p = q = 1$ (the simplest possible case), we have just the well-known Fibonacci sequence, discussed previously. Equivalently, generalized Fibonacci sequences can also be generated by the substitutional relation

$$B \to A, \quad A \to A^p B^q, \tag{2.5}$$

where A^p (or B^q) represents a string of p A-blocks (or q B-blocks). The total number of blocks in $S_{N_{GFB}}$ is equal to the generalized Fibonacci number denoted by $F_{N_{GFB}}$, given now by the recurrence relation

$$F_{N_{GFB}} = p F_{N_{GFB}-1} + q F_{N_{GFB}-2}, \tag{2.6}$$

with initial values $F_0 = F_1 = 1$.

The characteristic value $\tau(p,q)$, defined as being the ratio of $F_{N_{GFB}}$ to $F_{N_{GFB}-1}$ in the limit of $N_{GFB} \to \infty$, must satisfy the quadratic equation

$$\tau(p,q)^2 - p\tau(p,q) - q = 0. \tag{2.7}$$

Solving for the positive root gives explicitly

$$\tau(p,q) = \lim_{N_{GFB} \to \infty} F_{N_{GFB}}/F_{N_{GFB}-1} = \frac{p + \sqrt{p^2 + 4q}}{2}. \tag{2.8}$$

This expression generalizes the previous golden mean result and introduces other types of means, depending on the values of p and q. For instance, for $p = q = 1$

we have $\tau(1,1) \equiv \tau_{FB} \simeq 1.62$, the well-known golden mean number. Similarly, $\tau(2,1) \simeq 2.41$ is the silver mean number, $\tau(3,1) \simeq 3.30$ is the bronze mean number, and $\tau(1,3) \simeq 2.30$ is the nickel mean number.

It is worth to mention briefly here a mathematical aspect that has interesting implications for the physical properties of a quasiperiodic system. We may note that the expression for $\tau(p,q)$ in Eq. (2.8) is formally equivalent to a result that arises when determining the eigenvalues of the substitution matrix $\bar{M}$ introduced earlier in this section. This was exemplified by Grimm and Baake [51] in treating a quantum spin chain with quasiperiodic pair interactions. Essentially, they were able to classify the different substitutional sequences based on the irrationality of $\tau^-(p,q)$, which denotes the negative root of Eq. (2.8). They found that if $|\tau^-(p,q)| < 1$, it is a so-called Pisot-Vijayraghavan (PV) irrational number, and the fluctuations of the physical properties associated with the sequence are relatively well behaved and stable.

On the other hand, if $|\tau^-(p,q)| > 1$, the fluctuations of the physical properties are almost chaotic. For the examples of generalized Fibonacci cases mentioned earlier, only the nickel mean sequence ($p = 1$ and $q = 3$) is not a PV type, and, therefore, we expect its physical properties to behave more chaotically (as found in the specific heat calculations discussed in [52]).

2.2.2 Rudin-Shapiro

To set up a quasiperiodic chain of the Rudin-Shapiro (RS) type, we consider the juxtaposition of four basic building blocks A, B, C, and D, in such a way that they have the property of being invariant under the transformations $A \rightarrow AB$, $B \rightarrow AC$, $C \rightarrow DB$, and $D \rightarrow DC$.

The RS sequence belongs to the family of the so-called substitutional sequences, which are characterized by the nature of their Fourier spectrum. It exhibits an absolutely continuous Fourier measure, a property that it shares with the random sequences. The total number of building blocks in the unit cell increases with $2^{N_{RS}+1}$, N_{RS} being the Rudin-Shapiro generation number. The generations of this quasiperiodic structure are

$$S_0 = [AB]; \quad S_1 = [ABAC]; \quad S_2 = [ABACABDB]; \quad \text{etc.,} \qquad (2.9)$$

as depicted in Figure 2.2b.

2.3 Tight-Binding Hamiltonian

In condensed matter physics, the tight-binding model (or TB model) is an approach to calculate the electronic band structure using an approximate set of wave functions based upon their superposition for isolated atoms located at each atomic site. The method is closely related to the linear combination of atomic orbitals (LCAO) method used in quantum chemistry.

Tight-binding models are applied to a wide variety of systems, including organic materials such as the DNA molecule, to set up their electronic density of states. Though the tight-binding model is a one-electron model, it also provides a basis for more advanced calculations (like that of surface states) and applications to various kinds of many-body problems and quasiparticle calculations. Generally, it gives good qualitative results in many cases, and their parameters can be calculated by quantum models such as in the DFT case [5, 6].

The name tight-binding refers to the properties of tightly bound electrons in solids. The electrons in this model should be tightly bound to the atom to which they belong, and they should have limited interaction with states and potentials on the surrounding atoms of the solid. As a result, the wave function of the electron will be rather similar to the atomic orbital of the free atom it belongs to. The energy of each electron will also be rather close to its corresponding ionization energy in the free atom or ion because the interaction with potentials and states on neighboring atoms is limited.

Many interesting theoretical results concerning the electronic properties of one-dimensional chains have been obtained by using the Schrödinger equation in the tight-binding approximation. A considerable amount of work has been devoted to the study of this equation, for both random and quasiperiodic sequences of the on-site potential ϵ_n and/or the hopping potential t_{nm} between the quantum states $\langle n|$ and $|m\rangle$. The main achievements are the following:

(a) If the Hamiltonian parameters are independent random variables, the system exhibits Anderson localization – i.e., the eigenstates are exponentially localized – and the energy spectrum itself is a regular object with at most a finite number of bands. In the case of a binary potential distribution, the spectrum has one or two bands.

(b) If the hopping potentials t_{nm} are a binary sequence arranged in a pure Fibonacci or generalized Fibonacci way, the energy spectrum is a Cantor set of zero (Lebesgue) measure – i.e., there is an infinite number of gaps – and the total bandwidth vanishes.

Specifically for the pure Fibonacci case, the eigenstates are neither extended nor localized, but exhibit an intermediate behavior. For the generalized Fibonacci case, the eigenstates are extended [53]. In higher-dimensional cases, the energy spectra can be band-like with finite measure, fractal-like with zero band-width, or a mixture of partly band-like and partly fractal-like character.

In the next sections, we intend to investigate the charge transport properties of a DNA molecule using the tight-binding Halmitonian model described earlier, considering several types of DNA topologies.

2.4 Charge Transport in DNA

Charge transport in DNA molecules attracts considerable interest among the physics, chemistry, and biology communities, not only because of its relevance as the carrier of genetic code of all living organisms but also as a promising candidate for molecular electronics.

In fact, the use of molecules as electronic components is a powerful new direction in the science and technology of nanometer-scale systems due to their scientific and engineering applications [54]. Besides, charge mobility in DNA has its own importance based on its biological context, as well as on its technological applications (e.g., the use of DNA in electrochemical sensors and in future nanotechnologies). No doubt, the electronic conduction in DNA molecules is a research frontier in molecular electronics because of their potential use in nanoelectronic devices, both as a template for assembling nanocircuits and as a circuit element. Processes that possibly use charge transfer include, among others, the action of DNA damage response enzymes, transcription factors, and polymerase cofactors, all of them playing important roles in the cell. Indeed, it was proven that damaged regions have significantly different behavior than healthy regions in DNA after the passage of an electric current [55].

Although the use of DNA molecules in nanoelectronic circuits is very promising due to their self-assembly and molecular recognition abilities, their conductivity properties are not yet properly recognized. Different conclusions are obtained from several experiments. On the theoretical side, both ab initio calculations and model-based Hamiltonians are extensively adopted to interpret the diversity of the experimental results and to ascertain the underlying charge transport mechanisms. The former can provide a detailed description but is currently limited to relatively short molecules. The latter is much less detailed although it allows addressing systems of more realistic length. However, the model-based approach can play an additionally important complementary role because it grasps usually the underlying physics.

Earlier models of the electronic transport in DNA molecules assumed that the transmission channels are along their longitudinal axis. A π-stacked array of the DNA nucleobases – namely guanine (G), adenine (A), cytosine (C), and thymine (T) – provides the way to promote long-range charge migration, which in turn gives important clues to mechanisms and biological functions of charge transport.

The increasing diversification of applications requiring materials with specific electronic properties at nanoscale size and the consequent need for further

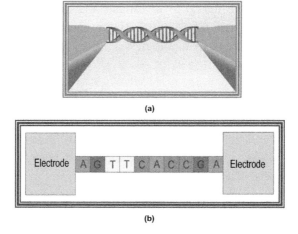

(a)

(b)

Figure 2.3 Schematic illustration of the the DNA molecule sandwiched between two electrodes. (a) pictorial view; (b) the single-strand case. After Albuquerque et al. [36]

miniaturization make the DNA molecule an excellent candidate for molecular electronics. This immediately leads to the question of what the possible mechanisms governing charge migration through DNA are, and what parameter ranges (length, temperature, geometrical arrangements, etc.) may support an electric current.

2.4.1 Single and Double Strand Geometries

Now consider a model in which a finite DNA molecule is sandwiched between two electrodes, as depicted in Figure 2.3a. For a single-strand DNA chain (see Figure 2.3b), the tight-binding Hamiltonian is written in terms of a localized basis as [56]

$$H = \sum_n \epsilon_n |n\rangle\langle n| + \sum_{n,m} t_{nm} |n\rangle\langle m|, \qquad (2.10)$$

where ϵ_n represents the energy (in units of $\hbar$) of the site n, and t_{nm} is the hopping potential. The sum over m is limited to the nearest neighbors.

The Dyson equation is [57]

$$G(\epsilon) = \epsilon^{-1}[I + HG(\epsilon)], \qquad (2.11)$$

where I is the identity matrix and H is the Hamiltonian given by Eq. (2.10).

Starting with the guanine (G) nucleotide as seed, let us consider that the DNA can be modeled by a Fibonacci quasiperiodic sequence in such a way that it can be built through the inflation rules G → GC and C → G. For the first generation of the FB sequence, with only a guanine linked to the electrodes, its Green function can be found by applying the tight-binding Hamiltonian Eq. (2.10) to the Dyson equation, Eq. (2.11), to obtain [58]

$$G_{nn}^{-1}(\epsilon) = \epsilon - \epsilon_G + 2\gamma(1), \tag{2.12}$$

where

$$\gamma(1) = -\frac{t_{GS}^2}{\epsilon - \epsilon_S + t_{SS}T(\epsilon)}. \tag{2.13}$$

Here, t_{GS} (t_{SS}) is the hopping term between the guanine G and the substrate S (within the substrate S). Also, $T(\epsilon)$ is the transfer function given by

$$T(\epsilon) = -(2t_{SS})^{-1}\left[(\epsilon - \epsilon_S) \pm \sqrt{(\epsilon - \epsilon_S)^2 - 4t_{SS}^2}\right]. \tag{2.14}$$

Repeating the procedure for any Fibonacci generation, we get

$$G_{nn}^{-1}(\epsilon) = \epsilon - \epsilon_G + \gamma(1) + \kappa(N), \tag{2.15}$$

where

$$\kappa(N) = -\frac{t_{n,n\pm1}^2}{\epsilon - \epsilon_n + \kappa(N-1)}, \tag{2.16}$$

with $\kappa(1) = \gamma(1)$, provided we replaced the hopping term t_{GS} by t_{CS}, the hopping term between the cytosine C and the substrate S. Also, N is the number of nucleotides in the strand. The symmetry $t_{n,n\pm1} = t_{n\pm1,n}$ holds.

Let us now contrast these results (Fibonacci sequence) with a Rudin-Shapiro (RS) sequence modeling the single-strand DNA molecule. Starting again from a G (guanine) nucleotide as seed, the quasiperiodic RS sequence is then built through the inflation rules G → GC, C → GA, A → TC, and T → TA. The RS sequence starts to deviate from the FB sequence in the third generation, when the sequence has four nucleotides GCGA connected to the electrodes.

Using a procedure similar to the quasiperiodic Fibonacci case, we can get for any RS generation number the same expression as for the Fibonacci case provided that, in Eq. (2.16), we replace $\kappa(N)$ with $\gamma(N)$, given by

$$\gamma(N) = -\frac{t_{n,n\pm1}^2}{\epsilon - \epsilon_n + \gamma(N-1)}. \tag{2.17}$$

Differently from the FB sequence, we note that, for the Rudin-Shapiro sequence, $t_{n,n\pm1}$ represents four distinct values of hopping potentials, namely t_{CT}, t_{GC}, t_{GA}, and t_{TA}, where we have assumed that $t_{n,n\pm1} = t_{n\pm1,n}$ in both cases.

The electronic density of state (DOS) – i.e., the number of electronic states per interval of energy at each energy level that are available to be occupied by the electrons – is given by:

$$\rho(\epsilon) = -(1/\pi)\text{Im}\left[\text{Tr}\langle n|G(\epsilon)|n\rangle\right] \qquad (2.18)$$

where Im (Tr) means the imaginary part (trace) of the argument shown between brackets. The energies ϵ_n are chosen from the ionization potentials of the respective nucleotides.

In the following, we will use as representative values $\epsilon_G = 7.75$ eV (guanine), $\epsilon_A = 8.24$ eV (adenine), $\epsilon_C = 8.87$ eV (cytosine), and $\epsilon_T = 9.14$ eV (thymine) [59]. All the hopping terms t_{nm} among the bases were taken equal to 1.0 eV, considering that theoretical calculations using ab initio methods yield for this potential values in the range 0.4–1.0 eV [59]. The potential at the DNA-electrode interface (here considered as a platinum metal) is considered to be the energy difference between the Fermi's level of the platinum and the HOMO's (highest occupied molecular orbital) isolate guanine (cytosine) state, giving us $t_{GS} = 2.39$ ($t_{CS} = 2.52$) eV.

We are aware that the HOMO state of the guanine (cytosine) may significantly change in the presence of the electrode, yielding a different potential at the interface DNA-electrode interface. Although we do not expect any relevant change in the DOS main features, the actual electron's localization length may be influenced, especially at the band edges. The hopping term inside the electrode (t_{SS}) is 12 eV [60]. Further, the on-site energy for the electrode (platinum) is $\epsilon_S = 5.36$ eV, which is related to the work function of this metal [61].

Figure 2.4 (left) depicts the density of states for a DNA quasiperiodic chain corresponding to the 11th (bottom), 13th (middle), and 15th (top) FB sequence generation number, respectively. Here, N_{FB} means the sequence generation number, while n_{FB} corresponds to the number of nucleotides in a given sequence generation. From there, we can infer the following main properties:

(a) The parity of the FB generation is not important for the DOS spectrum.
(b) Although the DOS for each generation as a whole does not show any symmetry, there are two very well-defined and symmetrical regions lying in the energy's intervals (in units of eV) $5.75 < \epsilon < 9.30$ (we call it region I), and $9.30 < \epsilon < 10.30$ (region II).
(c) Region II, which appears as a sort of anomaly in the DOS spectrum, is due to the presence of the cytosine nucleotide in the quasiperiodic chain. We can also notice that this region represents a kind of region I profile inverted in a smaller scale.
(d) Each region defines a clearly auto-similar spectrum for different generations. The auto-similarity holds also for the whole spectrum (regions I+II).

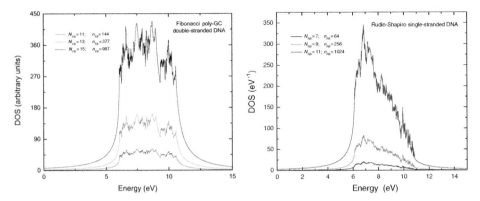

Figure 2.4 Left: Electronic density of states (DOS) spectra for the Fibonacci (FB) poly (GC) DNA single-strand model corresponding to the 11th (bottom), 13th (middle), and 15th (top) FB sequence generation number. Right: same as before but for the Rudin-Shapiro (RS) DNA single-strand model corresponding to the 7th (bottom), 9th (middle), and 11th (top) RS sequence generation number. After Albuquerque et al. [36]

(e) The central peak for region I is close to the guanine's ionization energy $\epsilon_G = 7.75$ eV, while the central valley in region II corresponds to $\epsilon_G + 2t_{nm} = 9.75$eV.

(f) The ratio among the distances of consecutive generations tends to the golden mean number, $\tau_{FB} = (1+\sqrt{5})/2$, a number intrinsically linked to the Fibonacci sequence, as we saw in Section 2.2.1.

The density of states for a DNA quasiperiodic chain following a Rudin-Shapiro quasiperiodic sequence are shown in Figure 2.4 (right), corresponding to its 7th (bottom), 9th (middle), and 11th (top) sequence generation number. Although some similarities with the Fibonacci case persist (for instance, the asymmetry of the spectra and the fact that again the parity of the quasiperiodic generation is not important), they are completely different, indicating how important is the model considered to simulate the DNA structure. As their main features, their central peaks, which are sequence independent, lie around 6.8 eV, which is about $\epsilon_C - t_{nm}$, with the bandwidth approximately given by $\epsilon_G \pm 4t_{nm}$.

Consider now the so-called double-strand model to describe an infinite DNA molecule, as depicted in Figure 2.5. The tight-binding Hamiltonian describing one electron moving in this ladder geometry composed of two interconnected chains of sites, side by side, with a single orbital per site and nearest-neighbor interactions, can be given by ($\hbar = 1$):

$$t(\psi_{n+1}^{\alpha} + \psi_{n-1}^{\alpha}) + w\psi_n^{\beta} = (E - \epsilon_n^{\alpha})\psi_n^{\alpha},$$
$$t(\psi_{n+1}^{\beta} + \psi_{n-1}^{\beta}) + w\psi_n^{\alpha} = (E - \epsilon_n^{\beta})\psi_n^{\beta}. \tag{2.19}$$

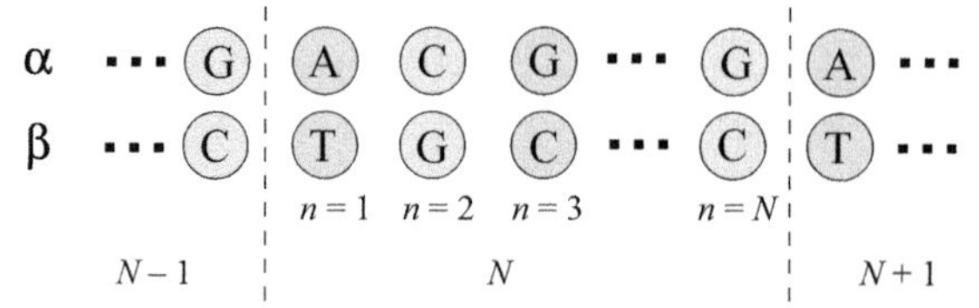

Figure 2.5 Schematic representation of an infinite double-strand DNA molecule, considering the intrachain (t) and the interchain (w) first-neighbor hopping terms. The Nth cell is also depicted.

Here, ϵ_n^α (ϵ_n^β) is the single energy of the orbital ψ_n^α (ψ_n^β) (the upper index refers to the chain, while the lower index refers to the site position in each chain – see Figure 2.5). Also, t and w are the intrachain and the interchain first-neighbor electronic overlaps (hopping amplitude), respectively.

Within this framework, the (discrete) Schrödinger equation can be written as

$$
\begin{pmatrix} \psi_{n+1}^\alpha \\ \psi_{n+1}^\beta \\ \psi_n^\alpha \\ \psi_n^\beta \end{pmatrix} = M(n) \begin{pmatrix} \psi_n^\alpha \\ \psi_n^\beta \\ \psi_{n-1}^\alpha \\ \psi_{n-1}^\beta \end{pmatrix},
\tag{2.20}
$$

where $M(n)$ is the transfer matrix

$$
M(n) = \begin{pmatrix} (E-\epsilon_n^\alpha)/t & -w/t & -1 & 0 \\ -w/t & (E-\epsilon_n^\beta)/t & 0 & -1 \\ 1 & 0 & 0 & 0 \\ 0 & 1 & 0 & 0 \end{pmatrix}.
\tag{2.21}
$$

After successive applications of the transfer matrix $M(n)$, we have

$$
\begin{pmatrix} \psi_{n+1}^\alpha \\ \psi_{n+1}^\beta \\ \psi_n^\alpha \\ \psi_n^\beta \end{pmatrix} = M(n)M(n-1)\cdots M(2)M(1) \begin{pmatrix} \psi_1^\alpha \\ \psi_1^\beta \\ \psi_0^\alpha \\ \psi_0^\beta \end{pmatrix}.
\tag{2.22}
$$

In this way, we have the wave function at an arbitrary site. Calculating this product of transfer matrices is completely equivalent to solve the Schrödinger equation for the system.

Defining the ket formed by the orbitals of the Nth unitary cell, i.e.,

$$
|\psi^{(N)}\rangle = \begin{pmatrix} \psi_{N+1}^\alpha \\ \psi_{N+1}^\beta \\ \psi_N^\alpha \\ \psi_N^\beta \end{pmatrix},
\tag{2.23}
$$

and taking into account that, in our model, each generated sequence is an unitary cell whose repetition builds up the entire infinite DNA molecule. Bloch's ansatz for each chain yields

$$|\psi^{(N+1)}\rangle = T_n |\psi^{(N)}\rangle = \exp{(i Q_i L)}|\psi^{(N)}\rangle, \qquad (2.24)$$

where $T_n = M(n)M(n-1) \cdots M(2)M(1)$, Q_i is the Bloch's wavevector and L is the periodic distance. Therefore,

$$[T_n - \exp{(i Q_i L)}I]|\psi^{(N)}\rangle = 0, \qquad (2.25)$$

where I is the identity matrix. Since T_n is an unimodular matrix (det $T_n = 1$), its eigenvalue should satisfy $\lambda_1 \lambda_2 \lambda_3 \lambda_4 = 1$, i.e., $\lambda_2 = \lambda_1^{-1}$ and $\lambda_4 = \lambda_3^{-1}$, implying the existence of only two independent eigenvalues.

Therefore, Bloch's wavevector should satisfy

$$\exp{(i Q_i L)} = \lambda_i, \quad i = 1, 2. \qquad (2.26)$$

The secular equation is then

$$\lambda_i^4 + \Xi \lambda_i^3 + \Gamma \lambda_i^2 + \Xi \lambda_i + 1 = 0, \qquad (2.27)$$

where $\Xi = -Tr[T_n]$ (the trace of the matrix T_n), and

$$\Gamma = (T_{11} + T_{22})(T_{33} + T_{44}) - T_{34}T_{43} - T_{12}T_{21} - T_{13}T_{31} \qquad (2.28)$$
$$- T_{14}T_{41} - T_{23}T_{32} - T_{24}T_{42} + T_{11}T_{22} + T_{33}T_{44}.$$

Here, T_{ij} are the elements of the transfer matrix T_n. Rearranging Eq. (2.27), we have

$$\gamma_i^2 + \Xi \gamma_i + \Gamma - 2 = 0. \qquad (2.29)$$

Here, $\gamma_i = (\lambda_i + \lambda_i^{-1})$ are the roots of the second-order degree equation, each one corresponding to one of the independent eigenvalues of the transfer matrix T_n. Its explicit form is

$$\gamma_{1,2} = \frac{-\Xi \pm \sqrt{\Xi^2 - 4(\Gamma - 2)}}{2}. \qquad (2.30)$$

For the DNA ordering of the first sequenced human chromosome 22 (Ch22), entitled NT_{011520}, the number of letters of this sequence is about 3.4×10^6 nucleotides. This sequence was retrieved from the website of the National Center of Biotechnology Information. The energies ϵ_n are again chosen from the ionization potential of the respective bases. The hopping term t among the bases was also taken equal to 1.0 eV, while the hopping potential w due to the hydrogen bonds linking the two strands is considered to be 0.1 eV.

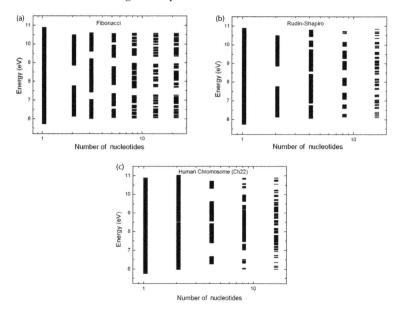

Figure 2.6 Energy spectra for (a) Fibonacci infinite DNA chain; (b) Rudin-Shapiro infinite DNA chain; (c) Ch22 DNA chain. After Albuquerque et al. [36]

With the intention of comparing the quasiperiodic sequences FB and RS with the genomic sequence, we assume also that the energies ϵ_n take the values ϵ_G, ϵ_A, ϵ_C, and ϵ_T, as in the DNA genomic sequence, with the same numerical values.

Figure 2.6 shows the electron energy spectra, as measured by their equivalent bandwidth Δ (the sum of all allowed energy regions in the band structures), for the Fibonacci (Figure 2.6a) and the Rudin-Shapiro (Figure 2.6b) quasiperiodic sequences, as well as for the genomic DNA Ch22 (Figure 2.6c), respectively, up to the number of nucleotides n equal to 93 in each unit cell N. This is nothing but the Lebesgue measure of the energy spectrum. From there, one can infer the forbidden and allowed energies as a function of the number of nucleotides n. Notice that, as expected, as n increases, the allowed band regions get narrower and narrower, as an indication of more localized modes.

2.4.2 Influence of the Sugar–Phosphate Backbone

Now we consider the charge transport in a poly (dG)-poly (dC) DNA finite segment, taking into account its double-strand geometry, including the sugar–phosphate backbone. Our theoretical model is again based on a tight-binding Hamiltonian, within a Dyson's framework, together with a transfer-matrix technique employed to simplify the algebra.

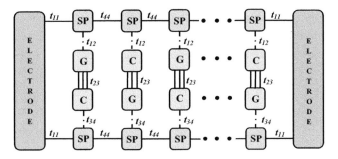

Figure 2.7 Schematic representation of a double-strand DNA molecule, including the sugar–phosphate (SP) contribution.

The electronic density of states are calculated considering that the DNA molecule is sandwiched between two electrodes. Besides, the poly (dG)-poly (dC) geometry follows a FB quasiperiodic structure. The spectra are then compared with those found from a genomic DNA sequence, considering again a finite segment of the first sequenced human chromosome Ch 22.

Our model for the double-strand poly (dG)-poly (dC) DNA, including the contribution of the sugar–phosphate (SP) backbone, is depicted in Figure 2.7. The tight-binding Hamiltonian is written in terms of a localized basis as [62]

$$
H = \sum_n \left[\epsilon_{SP}^n |n, 1><n, 1| + \epsilon_{\alpha}^n |n, 2><n, 2| + \right.
$$

$$
+ \epsilon_{\beta}^n |n, 3><n, 3| + \epsilon_{SP}^n |n, 4><n, 4| \Big]
$$

$$
+ \sum_n \Big[w_{12}[|n, 1><n, 2| + |n, 2><n, 1|] \Big]
$$

$$
+ \sum_n \Big[w_{23}[|n, 2><n, 3| + |n, 3><n, 2|] \Big]
$$

$$
+ \sum_n \Big[w_{34}[|n, 3><n, 4| + |n, 4><n, 3|] + t_{SS}(|n, S><n, S|) \Big]
$$

$$
+ \sum_n \Big[t_{11}(|n, 1><n - 1, 1|) + t_{44}(|n, 4><n \pm 1, 4|) \Big], \qquad (2.31)
$$

where ϵ_{SP}^n represents the single energy, in units of $\hbar$, at site n of the sugar–phosphate orbital, with $\epsilon_{\alpha,\beta}^n$ ($\alpha, \beta = $ C or G) being the ionization energy of the respective base α, β. Also, $w_{12} = w_{12}(\alpha \rightarrow$ SP), $w_{23} = w_{23}(\alpha \rightarrow \beta)$, and $w_{34} = w_{34}(\beta \rightarrow$ SP) are the interchain first-neighbor electronic overlaps (hopping amplitude), with $\alpha, \beta = $ C, or G, while t_{SS} is the hopping term in the electrodes. Furthermore, $t_{11} = t_{11}($SP $\rightarrow$ S$) = t_S$ and $t_{44} = t_{44}($SP $\rightarrow$ SP$) = t_{SP}$ are the intrachain hopping amplitudes. Here, the letter S means the electrode (considered, as before, a platinum metal), while SP means the sugar–phosphate backbone.

To model a DNA segment, we consider a quasiperiodic chain of Fibonacci type, starting with a G (guanine) base as seed. It can now be built in a similar way, as described earlier, through the inflation rules G → GC and C → G for the first strand. For the second strand, we use Chargaff's rule [28], leading to a GC or a CG base pair.

For the first generation of the FB sequence, in which only a guanine base is linked to the electrodes, the Dyson equation leads to

$$G(\epsilon)^{-1} = K_G + 2\Gamma(1), \tag{2.32}$$

where

$$\Gamma(1) = -L_{SPS}[K_S + L_S T]^{-1} L_{SPS}. \tag{2.33}$$

Here, T is a transfer matrix linking the Green functions of two next-neighbors sites. Also, $L_S = -t_{SS} I$ – I being a 4×4 identity matrix – and K_G, K_S, L_{SPS} are matrices given by

$$K_G = \begin{bmatrix} \epsilon - \epsilon_{SP} & -v_{12} & 0 & 0 \\ -v_{12} & \epsilon - \epsilon_G & -v_{23} & 0 \\ 0 & -v_{23} & \epsilon - \epsilon_C & -v_{34} \\ 0 & 0 & -v_{34} & \epsilon - \epsilon_{SP} \end{bmatrix} \tag{2.34}$$

where $v_{12} = w_{12}(G \to SP)$, $v_{23} = w_{23}(G \to C)$, and $v_{34} = w_{34}(C \to SP)$.

$$K_S = \begin{bmatrix} \epsilon - \epsilon_S & -t_{SS} & 0 & -t_{SS} \\ -t_{SS} & \epsilon - \epsilon_S & -t_{SS} & 0 \\ 0 & -t_{SS} & \epsilon - \epsilon_S & -t_{SS} \\ -t_{SS} & 0 & -t_{SS} & \epsilon - \epsilon_S \end{bmatrix}, \tag{2.35}$$

$$L_{SPS} = \begin{bmatrix} -t_S & 0 & 0 & 0 \\ 0 & 0 & 0 & 0 \\ 0 & 0 & 0 & 0 \\ 0 & 0 & 0 & -t_S \end{bmatrix}. \tag{2.36}$$

Repeating the procedure for any Fibonacci generation, we get

$$G(\epsilon)^{-1} = K_G + \Gamma(1) + \Lambda(n_{FB}), \tag{2.37}$$

where n_{FB} is the number of nucleotides in the strand. Here,

$$\Lambda(n_{FB}) = -L_{SPSP}[K_i + \Lambda(n_{FB} - 1)]^{-1} L_{SPSP}, \tag{2.38}$$

whose initial condition is

$$\Lambda(1) = -L_{SPS}[K_S + L_S T]^{-1} L_{SPS}. \tag{2.39}$$

We will now turn our discussion to the determination of the electronic density of states (DOS), as given by Eq. (2.18). The energies $\epsilon_{\alpha,\beta}$ are again chosen from the ionization potential of the respective bases – i.e., $\epsilon_G = 7.77$ eV (guanine), and $\epsilon_C = 8.87$ eV (cytosine). Also, we use the energy of the sugar–phosphate backbone as $\epsilon_{SP} = 12.27$ eV, while the hopping term between the base pair is $w_{23}(G \rightarrow C) = 0.90$ eV [63]. The hopping potentials between the base (G or C) and the sugar–phosphate (SP) backbone is $w_{12}(G \rightarrow SP) = w_{34}(C \rightarrow SP) = 1.5$ eV [60]. Finally, the hopping potential between the sugar–phosphate backbones is $t_{SP} = 0.02$ eV [60].

The electronic DOS spectrum, considering the DNA molecule modeled by the quasiperiodic Fibonacci sequence, is shown in Figure 2.8 as a function of the energy (in eV), for a Fibonacci's generation numbers $N_{FB} = 12$ and $n_{FB} = 610$. For comparison, we also consider segments of natural DNA, as part of the human chromosome Ch22 (dashed lines), whose spectra depict a strike agreement with those modeled by the Fibonacci sequence.

Although the DOS for each generation as a whole does not show any symmetry, it presents for the spectrum depicted here (Fibonacci and Ch22) two symmetrical regions, located around the energies 7.03 eV (peak I), and 15.35 eV (peak III), respectively, besides an asymmetrical one around 9.03 eV (peak II). Despite the

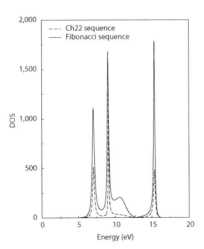

Figure 2.8 Electronic DOS spectra, as a function of the energy (in eV), for the 14th-generation number Fibonacci poly (dG)-poly (dC) DNA double-strand model, corresponding to the number of nucleotides $n_{FB} = 610$ (full line). For comparison we are also showing the DOS of a segment of natural DNA, a part of the human chromosome Ch22 (dashed lines). After Albuquerque et al. [36].

fact that peak I does not have a direct correlation with the ionization energies of the bases (guanine and cytosine), its energy value is near the hopping term in the interface DNA-electrode, suggesting an important influence for the choice of the electrodes on the DOS properties of the DNA molecule.

Peak II has a strong correlation with the ionization energy of cytosine, which is around 9 eV. This result is interesting because shows that while the amount of cytosine in the FB generations as well as in the first sequenced human chromosome Ch22 is less than the guanine one, its bigger ionization energy makes a difference regarding the electronic DOS of the whole system.

The third peak (peak III) occurs to an amount approximately equal to twice the ionization energy of guanine. We can also observe an anomaly in the spectrum around 10.6 eV, which is twice the value of the ionization energy of the electrode ϵ_S. Moreover, the DOS in each level of energy increases with the Fibonacci generation approximately for the intervals (in units of eV) $5.36 < \epsilon < 15.98$. Note that this interval is comprised between the ionization energy of the electrode ϵ_S and twice the ionization energy of the guanine ϵ_G. However, around 12.5 eV, we observed that the DOS is almost null for both FB and Ch22 cases. This result may be related to the hopping term of the electrode and/or to the ionization energy of the sugar–phosphate, which are both around 12 eV.

2.4.3 Renormalization Approach

So far, we have assumed the electronic transport in DNA molecules throughout a transmission channel along their longitudinal axis through a π-stacked array of DNA nucleobases, formed by the four nucleotides A,T, G, and C. Further improvements in this model include the backbone structure of the DNA molecule explicitly, which reduces the DNA base-pair architecture into a single site per pair, the so-called fish bone model. Later, Klotsa et al. [64] generalized the fishbone DNA model by considering each base as a distinct site, weakly coupled by hydrogen bonds. As a consequence, two central branches are thus obtained, whose interconnected sites represent the DNA base pairs; they are coupled to upper and lower disconnected backbone sites, giving rise to the so-called dangling backbone ladder (DBL)-DNA model.

Taking into account these latest developments, in this subsection, we now use a model Hamiltonian within a one-step renormalization approach to describe the charge transport properties of a DBL-DNA molecule (see Figure 2.9). Our description of the DNA molecule considers the contributions of the nucleobase system as well as the sugar–phosphate backbone molecules.

We consider a DBL-DNA model following a Fibonacci (FB) and a Rudin-Shapiro (RS) quasiperiodic nucleobase arrangement, as well as the DNA sequence

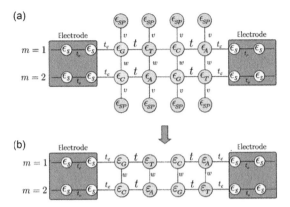

Figure 2.9 Sketch illustrating the renormalization process mapping the DBL-DNA chain model into a linear diatomic lattice. Three different hopping integrals are considered: the intrastrand term (t), the interstrand term (w), and the coupling between the sugar–phosphate backbone and the basepairs (v), respectively. (a) starting effective tight-binding model for the Fibonacci and Rudin-Shapiro sequence for a DBL-DNA model; (b) renormalized model of the DBL-DNA molecule after the first decimation step. After Albuquerque et al. [36].

of the first sequenced human chromosome Ch22 for the sake of comparison. The resulting variations of the charge transport efficiency are analyzed in these sequences by numerically computing the main features of their electron transmittance coefficients.

The tight-binding Hamiltonian for a DBL-DNA molecule describes a single electron moving in a geometry composed by two interconnected chains of sites sandwiched between two metallic electrodes (donor DN and acceptor AC) with a single orbital per site and nearest-neighbor interactions, yielding

$$H_{total} = H_{DNA} + H_{electrode} + H_{coupling}. \tag{2.40}$$

In order to get a simple mathematical description of the DBL-DNA molecule, keeping most of its relevant physical information, we use now a one-step renormalization process to map the DBL-DNA chain into a linear diatomic lattice (see Figure 2.9). This model allows us to incorporate the sugar–phosphate backbone contribution into an energy-dependent on-site ionization potential in the main DNA's base pairs, whose renormalized site energies are given by [64]

$$\varepsilon_{\alpha,\beta}^n = \epsilon_{\alpha,\beta}^n + t_{\alpha,SP}(\alpha \rightarrow SP)^2/(E - \epsilon_{SP}^n). \tag{2.41}$$

Here, $\epsilon_{\alpha,\beta}^n$ ($\alpha, \beta = $ C, G, A or T) is the ionization energy (in units of $\hbar$) of the respective base α, β; $t_{\alpha,SP}(\alpha \rightarrow SP)$ is the hopping potentials between the base α (G, C, A, or T) and the sugar–phosphate (SP) backbone; finally, ϵ_{SP}^n represents the single energy at site n of the sugar–phosphate orbital, taking into account the nature of the neighborhood nucleobase as well as the presence of water molecules and/or counterions attached to the backbone.

Considering the renormalization procedure, the first term of the Hamiltonian given by Eq. (2.40) can be read as [65]

$$
\begin{aligned}
H_{DNA} = & \sum_n [\epsilon_\alpha^n |n,1><n,1| + \epsilon_\beta^n |n,2><n,2|] \\
& + \sum_n w(\alpha \to \beta)[|n,1><n,2| + |n,2><n,1|] \\
& + \sum_n t(\alpha \to \alpha)[|n,1><n\pm 1,1|] \\
& + \sum_n t(\beta \to \beta)[|n,2><n\pm 1,2|].
\end{aligned}
\tag{2.42}
$$

The second term, related to the two semi-infinite metallic electrodes, reads

$$
\begin{aligned}
H_{electrode} = & \sum_{n=-\infty}^{0} \sum_{m=1}^{2} [\epsilon_S^n |n,m><n,m| + t_0 |n,m><n\pm 1,m|] \\
& + \sum_{n=N+1}^{\infty} \sum_{m=1}^{2} [\epsilon_S^n |n,m><n,m| + t_0 |n,m><n\pm 1,m|].
\end{aligned}
\tag{2.43}
$$

Our DNA molecule is coupled to the electrodes by the tunneling Hamiltonian

$$
H_{coupling} = \sum_{m=1}^{2} t_c [|0,m><1,m| + |n,m><n+1,m|],
\tag{2.44}
$$

where $t_c = \sqrt{tt_0}$ represents the hopping amplitude between the AC (DC) electrode and the beginning (end) of the DNA base-pair structure.

The electronic DOS spectrum (in arbitrary units) is depicted in Figure 2.10 as a function of the energy in units of eV. We have considered the four nucleotides arranged in a quasiperiodic fashion following a Fibonacci sequence (full line with $n_{FB} = 34$ nucleotides). In the inset, we consider the Rudin-Shapiro one (full line with $n_{RS} = 32$ nucleotides), as well as the spectrum for the human chromosome Ch22 (dotted line with $n_{Ch22} = 32$ nucleotides).

It is relevant to stress that the presence of long-range correlations in the disorder distribution is a possible mechanism to induce delocalization in low-dimensional systems. However, the actual correlations in our model (hopping mechanism) are not strong enough to produce this correlation-induced transition, and the stationary states remain all localized. Nevertheless, the presence of long-range correlations

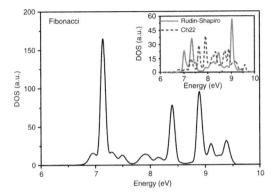

Figure 2.10 The eletronic DOS in arbitrary units (a.u.) plotted against the energy E (in eV) for the DBL-DNA model considering the Fibonacci quasiperiodic sequences with the number of nucleotides $n_{FB} = 34$ (full line). In the inset, we show also the eletronic density of states for the Rudin-Shapiro quasiperiodic sequences, with $n_{RS} = 32$ (full line), and a segment of natural DNA, as part of the human chromosome Ch22, whose number of nucleotides is $n_{Ch22} = 32$ (dotted line). After Albuquerque et al. [36].

enhances the localization length, and, therefore, the DOS spectra, as shown in Figure 2.10, survive in larger segments as compared with a non-correlated random sequence.

2.5 Conclusions

Nanobiotechnology has advanced to the stage that a large variety of nanoelectronic devices are suitable for integration with biological structures. Among them, in the last few years, many promising developments have occurred by using DNA-based nanostructures.

No doubt, precision DNA-mediated self-assembly devices have the potential to profoundly impact this field, representing major advances in the use of human-made bioelectronic technology. The reason for that is the ability to select the sequence of DNA's nucleotides and, hence, provide addressability during the assembly phase.

In order to account for some of these achievements, we have provided in this chapter the state of the art of the DNA's unique electronic properties by considering its several topologies, each of them having their own particularities. The understanding of DNA's charge transport properties, as discussed here, is therefore crucial in this process, giving rise to a vast array of possible uses such as medical monitoring devices, drug delivery systems, and patient monitoring systems, to cite just a few.

3

Electronic Transmission Spectra of the DNA Molecule

3.1 Introduction

As discussed in the previous chapter, due to their potential applications in nanoelectronics, there has been a growing interest in the synthesis, characterization, and electronic properties of DNA-based molecules with periodic nucleotide sequences. Using a full range of physical and biochemical methods, studies have now established that double-helical DNA is a medium for the efficient transport of electrons, triggering a series of experimental and theoretical investigations (see [66] and the references therein). The earliest studies involved physical measurements of current flow in DNA fibers, leading their conductivity properties to a mixture of conclusions.

Controversial reports consider that DNA may be a good linear conductor, while others have found that it is somewhat more effective than proteins, even when DNA has perfectly ordered base pairs [67]. Recent measurements of electrical transport through individual short DNA molecules indicated that it has a wide-bandgap semiconductor behavior [68]. Besides, strongly deformed DNA molecules deposited on a substrate and connected to metallic electrodes can behave as an insulator or a conductor depending on, among other things, the ratio between the thickness of the substrate and molecule. In fact, one experiment even suggested that DNA can behave as a superconductor [69].

These physical studies have not yet been reconciled with one another. Their discrepancies probably depend heavily on the connections between the DNA and the electrodes used, as well as on the integrity of the DNA itself in the absence of water and when exposed to very high voltages. Besides, these experiments have not only underscored that the DNA base pair stack can mediate hole and electron transport chemistry but have also shown the exquisite sensitivity of the charge transport through the DNA structure.

These seemingly contradictory theoretical and experimental results are caused mainly by three factors:

(a) Native DNA consists of a double helix with an aperiodic sequence, sugar–phosphate side chains and water as well as ions surrounding it.
(b) The topology of the double helix, which is not a rigid object, with the different constituents of DNA moving relative to each other.
(c) The works so far have been performed by using quite different theoretical methods and experimental techniques.

In addition to its own structural complexity, measuring charge transport in a DNA chain is strongly biased by the invasive role of contacts, interaction with some inorganic substrate, and temperature/atmosphere experimental conditions. As a result, focus has been now shifted from asking whether DNA can mediate long-range charge transport to questioning how it works. Specifically, we are seeking answers to how DNA structure and sequence affect this reaction and how important is DNA-mediated charge transport.

To this end, several experimental groups have reported measurements of the current–voltage ($I \times V$) characteristics obtained from electrical transport measurements throughout individual or small numbers of DNA-like molecules captured between two metal nanoelectrodes. The traditional molecular view of electron transfer between donor and acceptor species gives rise to a novel view of the molecule as a current-carrying conductor, and observables such as electron-transfer rates are replaced by the conductivities or, more generally, by current–voltage relationships in molecular junctions.

Of primary importance is the need to understand the interrelationship between the molecular structure of such junctions and their function – i.e., their transmission and conduction properties. Such investigations, in which single molecules or small molecular assemblies operate as conductors connecting "traditional" electrical components such as metal or semiconductor contacts, constitute a major part of what has become the active field of molecular electronics. Their diversity, versatility, and amenability to control and manipulation make molecules and molecular assemblies potentially important components in nanoelectronic devices. A standard electron transfer system thus containing a donor, an acceptor, and a molecular bridge connecting them.

Our target is the electron transfer between the two conducting electrodes (donor and acceptor) through a molecular medium, which bears strong similarity to the more conventional systems that involve at least one molecular species in the donor/acceptor pair. These systems have attracted much attention from the semiconductor industry, and there is a great interest from an applied point of view to model and understand the capabilities of molecular conductors. At the same time, this is also a topic of great interest from the point of view of fundamental physics. Three possible charge transfer mechanisms might be considered:

(a) *Superexchange*: The charge tunnels from the donor (DN) electrode to the acceptor (AC) through the DNA segment in a nonadiabatic process. An exponential decrease in the rate of charge transport with increasing length of the DNA segment is predicted.

(b) *Hopping*: Charge occupies the DNA segment in traveling from the DN electrode to the AC one by hopping between the DNA's discrete molecular orbitals. If the rate of charge migration is faster than trapping processes, the charge should be able to migrate over long distances.

(c) *Domain hopping*: Charge occupies the DNA segment by delocalizing over several bases or a domain. This domain hops along the DNA segment to travel from the DN electrode to the AC one. As in a pure hopping mechanism, the charge should be able to travel long distances before getting trapped.

The DNA molecule found in cells is a double helix consisting of two antiparallel strands held together by specific hydrogen-bonded base pairs. The double-helical arrangement leads to a linear helical axis, linear not in the geometrical sense of being a straight line but in the topological sense of being unbranched.

Unlike proteins, a stacked array of DNA base pairs derived from these nucleotides can provide the way to promote long-range charge migration, which in turn gives important clues to mechanisms and biological functions of transport. So far, numerous algorithms have been introduced to characterize and graphically represent the genetic information stored in the DNA nucleotide sequence. The goal of these methods is to generate patterns for certain sequences or groups of sequences [63].

With this aim in mind, we report in this chapter an analytical as well as numerical investigation of the electronic transmission spectra of DNA segments, considering a model in which the DNA molecule is sandwiched by two electrodes – donnor (DN) and acceptor (AC), respectively – following Fibonacci (FB) and Rudin-Shapiro (RS) quasiperiodic patterns. Afterward, we compare them, as we have done in the previous chapter, to the DNA sequence of the first sequenced human chromosome 22 (Ch22).

Initially, we investigate the conductivity of the DNA molecule models through their electron transmittance coefficient. Furthermore, by solving numerically a time-independent Schrödinger equation, we also compute some basic properties of the current–voltage (I×V) characteristics, for all DNA models considered here, following a Landauer-Büttiker formulation [70, 71]. The investigation of twisted geometries, methylated states, and diluted base pairing follow.

3.2 Electrical Conductivity

Electron transmission conductivity through molecules (such as DNA) and molecular interfaces is currently a subject of intensive research. This is due to the scientific/technological interest in the electron-transfer phenomena underlying the operation of the scanning tunneling microscope on one hand and the transmission properties of molecular bridges between conducting leads on the other.

In view of that, the traditional molecular assumption of electron transfer – in which rates depend on the donor and acceptor electrodes, properties, the solvent, and the electronic coupling between the states involved – gives rise to a novel approach of the molecule as a current-carrying conductor whose observables, such as electron-transfer rates and yields, are being replaced by their conductivities or, more generally, by their current–voltage ($I\times V$) profiles. Such investigations of electrical junctions, in which single molecules or small molecular assemblies operate as conductors, constitute a major part of the new molecular electronics field.

Electron conductivity properties though DNA are still controversial, as we pointed out in the previous chapter, mainly due to the tremendous difficulties in setting up the proper experimental environment and the DNA molecule itself. Despite the lack of a consistent picture, many theoretical explanations for the charge transport phenomena have been suggested so far on the basis of the standard solid-state physics approach such as polarons, solitons, and the hole hopping model on guanine sites, to cite just a few, but the situation has been still far from unifying the theoretical scheme. For instance, recently, the electric conductance of DNA molecules was studied using a tight-binding small polaron model, and the length dependence of the electric current was derived [72]. The study's main conjecture was that the drift of polarons states may lead to a rapid motion of charges introduced on DNA.

From the experimental side, one of the main problems is how to attach proper electrodes to the single DNA molecule so that any recorded conductivity comes from the molecule itself and not from some residual conductivity in the surrounding medium. Therefore, it is mandatory to provide reliable electrical contacts to the DNA molecule that neither allow any electron transfer reactions through the ionic medium surrounding the molecule nor propagate noise from other means.

3.2.1 Electronic Transmission Spectra

Consider the dangling backbone ladder (DBA)-DNA model, described in Section 2.4, further assumed being connected to two semi-infinite electrodes, whose energies ϵ_m are adjusted to simulate a resonance with the guanine highest occupied molecular orbital (G-HOMO) energy level – i.e., $\epsilon_m = \epsilon_G$.

For this system, the transmission coefficient $T_N(E)$, which gives the transmission rate through the chain and is related with the Landauer resistance, is defined by [73]

$$T_N(E) = [|T_1|^2 + |T_2|^2]/2, \tag{3.1}$$

where T_1 (T_2), is given by

$$T_1(T_2) = N_1(N_2)/D. \tag{3.2}$$

Here,

$$
\begin{aligned}
N_1 = (\nabla_{33}\nabla_{11}\nabla_{22} &- \nabla_{33}\nabla_{12}\nabla_{21} + \nabla_{34}\nabla_{11}\nabla_{22} - \nabla_{34}\nabla_{12}\nabla_{22} \\
&- \nabla_{31}\nabla_{22}\nabla_{13} + \nabla_{31}\nabla_{12}\nabla_{23} - \nabla_{31}\nabla_{22}\nabla_{14} + \nabla_{31}\nabla_{12}\nabla_{24} \\
&+ \nabla_{32}\nabla_{21}\nabla_{13} - \nabla_{32}\nabla_{11}\nabla_{23} + \nabla_{32}\nabla_{21}\nabla_{14} - \nabla_{32}\nabla_{11}\nabla_{24}),
\end{aligned} \tag{3.3}
$$

$$
\begin{aligned}
N_2 = (\nabla_{43}\nabla_{11}\nabla_{22} &- \nabla_{43}\nabla_{12}\nabla_{21} + \nabla_{44}\nabla_{11}\nabla_{22} - \nabla_{44}\nabla_{12}\nabla_{21} \\
&- \nabla_{41}\nabla_{22}\nabla_{13} + \nabla_{41}\nabla_{12}\nabla_{23} - \nabla_{41}\nabla_{22}\nabla_{14} + \nabla_{41}\nabla_{12}\nabla_{24} \\
&+ \nabla_{42}\nabla_{21}\nabla_{13} - \nabla_{42}\nabla_{11}\nabla_{23} + \nabla_{42}\nabla_{21}\nabla_{14} - \nabla_{42}\nabla_{11}\nabla_{24}),
\end{aligned} \tag{3.4}
$$

$$D = (\nabla_{11}\nabla_{22}) - (\nabla_{12}\nabla_{21}). \tag{3.5}$$

In these equations, ∇_{ij} are the components of the 4x4 matrix ∇, defined as $\nabla = \Theta^{-1}S^{-1}PS$, with

$$
\Theta = \begin{pmatrix}
e^{-ikNa} & 0 & 0 & 0 \\
0 & e^{-ikNa} & 0 & 0 \\
0 & 0 & e^{ikNa} & 0 \\
0 & 0 & 0 & e^{ikNa}
\end{pmatrix}, \tag{3.6}
$$

$$
S = \begin{pmatrix}
e^{-ika} & 0 & e^{ika} & 0 \\
0 & e^{-ika} & 0 & e^{ika} \\
1 & 0 & 1 & 0 \\
0 & 1 & 0 & 1
\end{pmatrix}, \tag{3.7}
$$

where k is given by

$$k = \cos^{-1}[(E - \epsilon_S)/2t_0]. \tag{3.8}$$

Also, $P = M_R\left(\prod_{n=N}^{1} M_n\right)M_L$, where the M's matrices are given by

$$
M_n = \begin{pmatrix}
(E - \varepsilon_\alpha^n)/t & -w/t & -1 & 0 \\
-w/t & (E - \varepsilon_\beta^n)/t & 0 & -1 \\
1 & 0 & 0 & 0 \\
0 & 1 & 0 & 0
\end{pmatrix}, \tag{3.9}
$$

$$M_L = \begin{pmatrix} (E - \epsilon_S)/t_c & 0 & -t_0/t_c & 0 \\ 0 & (E - \epsilon_S)/t_c & 0 & -t_0/t_c \\ 1 & 0 & 0 & 0 \\ 0 & 1 & 0 & 0 \end{pmatrix}, \tag{3.10}$$

$$M_R = \begin{pmatrix} (E - \epsilon_S)/t_0 & 0 & -t_c/t_0 & 0 \\ 0 & (E - \epsilon_S)/t_0 & 0 & -t_c/t_0 \\ 1 & 0 & 0 & 0 \\ 0 & 1 & 0 & 0 \end{pmatrix}. \tag{3.11}$$

The transmission coefficients $T_N(E)$, given by Eq. (3.1), are depicted in Figure 3.1 as a function of the energy (in units of eV). We have considered the four nucleotides arranged in a quasiperiodic fashion, either following a Fibonacci sequence (with $n_{FB} = 34$ nucleotides) or a Rudin-Shapiro one (with $n_{RS} = 32$ nucleotides), respectively, both showing a long-range pair correlation. For comparison, we also show the spectrum for the human chromosome Ch22 (with $n_{Ch22} = 32$ nucleotides).

The transmission bands in the spectrum are fragmented, which is related to the localized nature of the electrons eigenstates in disordered chains and reflects the number of passbands in each structure. Although the presence of long-range correlations in the disorder distribution is a possible mechanism to induce delocalization, the actual correlations in our model (hopping mechanism) are not strong enough

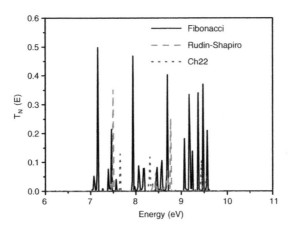

Figure 3.1 Transmittance coefficient $T_N(E)$ as a function of the energy E (in units of eV) for the DBL-DNA model considering the Fibonacci and Rudin-Shapiro quasiperiodic sequences, with number of nucleotides $n_{FB} = 34$ (full line) and $n_{RS} = 32$ (dashed line), respectively. For comparison, we show a segment of natural DNA, as part of the human chromosome Ch22, whose number of nucleotides $n_{Ch22} = 32$ (dotted line). After Albuquerque et al. [36]

to produce this correlation-induced transition, and the stationary states remain all localized.

Nevertheless, the presence of long-range correlations enhances the localization length, and, therefore, the transmission resonances, as shown in Figure 3.1, survive in larger segments as compared to a non-correlated random sequence. Observe also that the transmission coefficient for long-range correlated Rudin-Shapiro sequences depicts a trend similar to the one observed for the genomic Ch22 sequence.

3.2.2 Current–Voltage Characteristic Curves

The transmission coefficient is a useful quantity to describe the transport efficiency in quantum systems. However, $T_n(E)$, as discussed in the previous section, is usually difficult to be directly measured experimentally. One appropriated alternative is to consider the profiles of their current–voltage (I×V) characteristic curves.

With the tight-binding Hamiltonian describing a double-strand DNA model, as given by Eq. (2.19), one can evaluate the I×V characteristics by applying the Landauer-Büttiker [70, 71] formulation:

$$I(V) = (2e/h) \int_{-\infty}^{+\infty} T_N(E)[f_{DN}(E) - f_{AC}(E)]dE, \tag{3.12}$$

where the Fermi-Dirac distribution is

$$f_{DN(AC)} = \left[\exp[(E - \mu_{DN(AC)})/k_B T] + 1 \right]^{-1}. \tag{3.13}$$

Also, $\mu_{DN(AC)}$ is the electrochemical potential of the two leads – donor (DN) and acceptor (AC) – fixed by the applied bias voltage V as

$$|\mu_{DN} - \mu_{AC}| = eV. \tag{3.14}$$

The current onset is crucially dependent on the electrochemical potentials of the electrodes, which can be altered by the coupling to molecules. For simplicity, before bias voltage is applied, the electrochemical potential of the whole system is taken to be zero. It is important to emphasize that the transmittance $T_N(E)$ should be calculated in the forward and backward applied electric field directions.

For the double-strand DNA sequences, as described previously, the current-voltage characteristics are plotted in Figure 3.2 for Fibonacci (Figure 3.2a), Rudin-Shapiro (Figure 3.2b), the random case (Figure 3.2c), and the human chromosome Ch22 (Figure 3.2d), respectively [74]. We assume a linear voltage drop across the DNA molecules by means of the usual expression, numerically computed near zero temperature, as given by Eq. (3.14). To reproduce the potential mismatch at zero bias, the energy difference between the guanine HOMO energy level and the

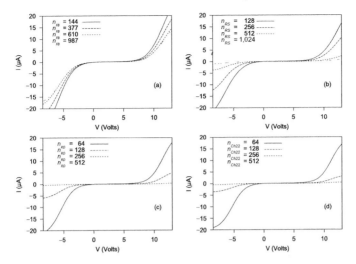

Figure 3.2 Current–voltage (I×V) characteristics of a double-strand DNA finite segment considering different numbers of nucleotides for (a) Fibonacci sequence (FB); (b) Rudin-Shapiro sequence (RS); (c) the random case (RD); (d) the human chromosome Ch22, respectively. After Albuquerque et al. [36]

metallic Fermi level of the electrode is set to 1.2 eV [75]. As the voltage drop is switched on, the transmission coefficient $T_N(E)$ becomes voltage-dependent, resulting in transmission band shifts (shown in Figure 3.2 for all cases studied here), which in turn lead to a voltage threshold modulation.

To extract the main features of the tunneling currents in DNA chains, let us compare the behavior of the genomic Ch22 (Figure 3.2d) with those characterizing the quasiperiodic and random structures (Figure 3.2a, b, and c) under the resonance condition given by the hopping term $t_{nm} = 1$ eV. In this case, if the potential barrier between the metallic contacts and the DNA is set to zero, a staircase in the plot I×V is found.

As soon as a potential barrier between the DNA and the metals is introduced (1.2 eV), the I×V characteristic curves show the profiles depicted in Figure 3.2. The current threshold at a given voltage scale is not sensitive in respect to the different structures considered here, mainly due to the electronic correlations presented by the structures. However, such correlations shall depend strongly on the intrachain coupling, and further studies considering more realistic model parameters would be needed in order to infer the actual relevance of this threshold enhancement in DNA molecules.

Observe the striking agreement between the I×V characteristic curves for the random and the genomic Ch22 case. Such agreement can be accounted for by the short-range pair correlations shared by them, suggesting that the inclusion of

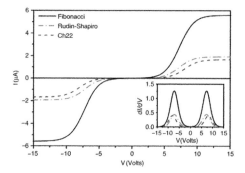

Figure 3.3 Current–voltage ($I \times V$) characteristics of a DBL-DNA sequences for (a) Fibonacci sequence (full line); (b) Rudin-Shapiro sequence (dashed line); (c) the human chromosome Ch22 (dotted line), respectively. The inset shows the differential conductance dI/dV versus the voltage V of the devices. After Albuquerque et al. [36]

just first-neighbors intrastrand pair correlations on the nucleotide distribution may provide an adequate description of the DNA electronic properties.

On the other hand, the current–voltage characteristics of the DBL-DNA model (see Section 2.4) are plotted in Figure 3.3 for the Fibonacci (full line), Rudin-Shapiro (dashed line), and the human chromosome Ch22 (dotted line) simulations, respectively. In this case, there are a characteristic ohmic region for $-5.0 \leq V_{bias} \leq +5.0$ eV and nonlinear regions indicating transitions toward current saturation for $V_{bias} < -5.0$ eV and $V_{bias} > +5.0$ eV. The inset in Figure 3.3 shows the transconductances $dI/dV \times V$ of the devices, which are highly nonlinear. All of them have semiconductor characteristics, as in the double-strand case depicted in Figure 3.2.

3.3 Twisted Geometry

Usually, charge transfer studies for a wide variety of systems, including organic materials such as the DNA molecule, are based on the Schrödinger equation in the tight-binding approximation for both random and quasiperiodic sequences of the on-site potential ϵ_n and/or the hopping potential t_{nm} between the quantum states $|n\rangle$ and $\langle m|$ – the former (latter) leading to Anderson localization (Cantor set of zero Lebesgue measure) [54]. Focusing on the DNA case, the model considers a simplified vision of the Watson-Crick pairs attached to the sugar–phosphate backbone condensed into a single nucleotide site.

Although this model was successfully employed to describe numerous experimental data [36], it has some critical assessment. One key point is the "one-orbital-per-site" picture, insufficient in some aspects to characterize the quantum state. A reasonable solution is the adoption of nucleotide pairs, instead of the separate nucleotides to define the quantum states to be investigated by using more sophisticated quantum chemistry models such as the DFT approach, whose accuracy is achieved at a greater computational cost when contrasted to a much simpler tight-binding model.

Other relevant consideration is the topology of the double helix, which is not a rigid object, with the different constituents of DNA moving relative to each other, depicting a twisted geometry as well as presenting linear deformations of its structure in response to the charge arrival at this particular site (polaronic effects). The presence of water molecules and counterions interacting with the nucleobases and the backbone can be taken into account by considering a realistic value of the backbone ionization energy.

In this context, we use a model Hamiltonian within a two-step renormalization approach to describe the charge transport properties of a twisted DNA molecule following the model described in [63, 65]. Our description of the DNA molecule takes into account the contributions of the nucleobase system, as well as the sugar–phosphate backbone molecules on a tight-binding Hamiltonian model. The DNA's helicoidal structure is considered by means of not only the longitudinal intrachain hopping term but also through the twist angle $\theta_{n,n\pm1}$ between two adjacent base pairs $(n, n \pm 1)$ attached along the molecule backbone [76]. The resulting variations of the charge transport efficiency are analyzed by numerically computing the main features of their transmittance and current–voltage (I×V) characteristic curves.

The tight-binding model Hamiltonian describing the twisted DNA molecule sandwiched between two platinum electrodes (a platinum metal) is, as discussed in the previous chapter,

$$H_{total} = H_{DNA} + H_{electrode} + H_{coupling}. \tag{3.15}$$

Here, we have considered the base pairing between the complementary strands and the stacking interaction between nearest-neighbor bases, as well as the sugar–phosphate backbone. A relevant new feature in the DNA's Hamiltonian H_{DNA} is the inclusion of torsional effects, which are responsible for the helicoidal DNA structure [77]. These effects are quite important since, in physiological conditions, the DNA's double-helix structure exhibits a full-fledged three-dimensional geometry.

As a consequence, every two consecutive nucleobases are twisted by a certain angle $\theta_{n,n\pm1}$ (in equilibrium conditions, $\theta_{n,n\pm1} = \theta_0 = \pi/5$), and, therefore, the orbital overlapping is substantially reduced, yielding smaller values for the hopping integral values [78].

In addition, at physiological temperatures, the relative orientation of neighboring bases becomes a function of time, thereby modifying their mutual overlapping in an oscillatory way (dynamical effect). The twisted DNA model was also considered by [79] using a path integral formalism to study the thermodynamics of a short fragment of heterogeneous DNA interacting with a stabilizing solvent on the temperature range in which denaturation takes place.

Considering the nucleobases as identical point masses, helically arranged and mutually connected by means of elastic rods, which describe the sugar–phosphate backbone, the position of the nth nucleobase can be expressed in cylindrical coordinates as

$$x_n = r_n \cos \phi_n, \quad y_n = r_n \sin \phi_n, \quad z_n = r_0 \phi_n, \tag{3.16}$$

where n labels the considered base pair along the DNA double strand, r_n and ϕ_n are the usual cylindrical coordinates, and $r_0 = h_0/\theta_0$, $h_0 \approx 0.34$ nm are the equilibrium separation between two successive base-pair planes (B-DNA form). Thus, the geometrical distance between two neighboring nucleobases can be expressed as [77]

$$d_{n,n\pm1} = \left[r_0^2 \theta_{n,n\pm1}^2 + r_n^2 + r_{n\pm1}^2 - 2r_n r_{n\pm1} \cos \theta_{n,n\pm1} \right]^{1/2}, \tag{3.17}$$

where $\theta_{n,n\pm1} = \pm\phi_{n+1} \mp \phi_n$. In equilibrium conditions,

$$d_{n,n\pm1} = l_0 = \left[h_0^2 + 4r_n^2 \sin^2(\theta_0/2) \right]^{1/2}. \tag{3.18}$$

Considering that the atomic orbitals are orthogonal to each other, in the equilibrium condition ($d_{n,n\pm1} = l_0$), the full 3-D description of the helix geometry yields, for the longitudinal intrachain hopping term,

$$t_L(\theta_{n,n\pm1}) = t_L(\theta_t) = t_L(0) \left[1 - \eta \left(\frac{2r_0}{l_0} \sin \frac{\theta_{n,n-1}}{2} \right)^2 \right]. \tag{3.19}$$

Here,

$$\eta = 1 + |\eta_{pp\pi}|/\eta_{pp\sigma}, \tag{3.20}$$

where $\eta_{pp\pi}$ and $\eta_{pp\sigma}$ describe the hybridization matrix elements between the orbitals of neighboring bases p_z orbitals [78]. Further simplification can be done if one considers the propagation of low-frequency twist oscillations (acoustic modes), leading to a small (although non-zero) fixed twist angle θ_t, i.e.,

$$t_L(\theta_t) = t_L(0)(1 - \chi\theta_t^2). \tag{3.21}$$

Here, χ is a dimensionless coupling strength between the charge and the lattice system. For small twists, $\chi = 2.92$ [77].

In order to get a simple mathematical description of the DNA molecule, we use now a two-step renormalization process similar to that described in the previous chapter (Section 2.4). In the first step, we map the full DNA chain into a linear

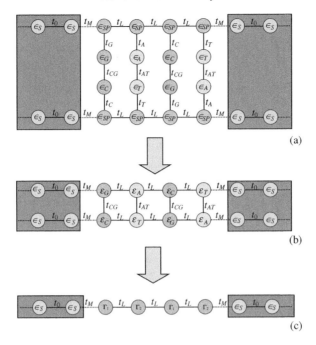

Figure 3.4 (a) Sketch illustrating the renormalization process mapping the DNA chain model; (b) first into a linear diatomic lattice; (c) and then a linear monoatomic lattice. In this way we reduce the eleven physical parameters, namely, the ionized energies ϵ_{SP} (sugar–phosphate backbone) and ϵ_α (α being the four nitrogen bases A,T, C, and G); the hopping terms t_α (between the nitrogen bases and the sugar–phosphate backbone) and $t_{\alpha\beta}$ (the transverse hopping between the nitrogen bases), into three variables only, the hopping term t_L (the longitudinal hopping between the nitrogen bases) and the ionized energies $\Gamma_j(E)$ ($j = 1,2$). After de Almeida et al. [76].

diatomic lattice (see Figure 3.4b). Afterward, we renormalize the linear diatomic lattice into a 1-D lattice as shown in Figure 3.4c, simplifying the model with 11 physical parameters, namely

(a) the ionized energies ϵ_{SP} (sugar–phosphate backbone) and ϵ_α (α being the four nitrogen bases A,T, C, and G),

(b) the hopping terms t_α (between the nitrogen bases and the sugar–phosphate backbone) and $t_{\alpha\beta}$ (the transverse hopping between the nitrogen bases),

into three variables only: the hopping term t_L (the longitudinal hopping between the nitrogen bases) and the ionized energies $\Gamma_j(E)$ ($j = 1,2$), defined by

$$\Gamma_1(E) = t_{CG} + \sum_{\alpha=C,G} \frac{\tau_\alpha^2(E)}{E - \epsilon_{SP}}, \qquad (3.22)$$

with

$$\tau_\alpha(E) = t_\alpha + \frac{\epsilon_\alpha(E - \epsilon_{SP})}{t_\alpha}, \quad \alpha = C, G. \tag{3.23}$$

The term $\Gamma_2(E)$ can be obtained from Eqs. (3.22) and (3.23), provided we replace C, G by A, T.

The energies ϵ_α are chosen from the ionization potential of their respective bases. Taking into account explicitly the contribution of water molecules, their experimental values are [80] $\epsilon_A = 7.7$ eV (adenine), $\epsilon_T = 8.1$ eV (thymine), $\epsilon_G = 7.4$ eV (guanine), and $\epsilon_C = 8.1$ eV (cytosine). We use the energy of the electrode (platinum) $\epsilon_M = 5.36$ eV, which is related to the work function of this metal [61]. The energy of the sugar–phosphate backbone is $\epsilon_{SP} = 12.27$ eV, justified by the presence of a number of counterions and water molecules, located along the DNA backbone structure, interacting with the nucleobases and the backbone itself by means of hydration, solvation, and charge transfer processes [63]. Furthermore, we take the hopping potentials between the nucleobases (G, C, A, or T) and the sugar–phosphate (SP) backbone as $t_G = t_C = t_A = t_T = 1.0$ eV, while the hopping between the base pair intrachain term (interchain terms) is $t_L(0) = 0.15$ eV (are $t_{GC} = 0.9$ eV and $t_{AT} = 0.3$ eV) [63], which are within the range of values obtained by quantum chemistry calculations [81]. Finally, the hopping term in the electrode is $t_0 = 12.0$ eV.

Now, considering the renormalization procedure depicted schematically in Figure 3.4, the first term of the Hamiltonian (Eq. (3.15)) is described by

$$H_{DNA} = \sum_{j=1,2} \sum_n \Gamma_j(E)|n><n| + \sum_n t_L(\theta_t)|n><n \pm 1|. \tag{3.24}$$

The second term, related to the two semi-infinite metallic electrodes, reads

$$H_{electrode} = \sum_{n=-\infty}^{0} \left[\epsilon_M|n><n| + t_0|n><n \pm 1| \right]$$

$$+ \sum_{n=N+1}^{\infty} \left[\epsilon_M|n><n| + t_0|n><n \pm 1| \right], \tag{3.25}$$

where ϵ_M (t_0) is the ionized energy (hopping term) of the electrode. Our DNA molecule is coupled to the electrodes by the tunneling Hamiltonian

$$H_{coupling} = t_M \left[|0><1| + |N><N + 1| \right], \tag{3.26}$$

where $t_M = 0.63$ eV represents the hopping amplitude between the source (drain) electrode and the beginner (end) of the DNA base-pair structure, and N represents the number of nucleotides in the structure under consideration.

Considering the tight-binding Hamiltonian, the transmission coefficient $T_N(E)$, is defined by (single-strand model) [56]

$$T_N(E) = \frac{4 - X^2(E)}{[-X^2(E)(\mathcal{P}_{12}\mathcal{P}_{21} + 1) + X(E)(\mathcal{P}_{11} - \mathcal{P}_{22})(\mathcal{P}_{12} - \mathcal{P}_{21}) + Y + 2]} \tag{3.27}$$

where $Y = \sum_{i,j=1,2} \mathcal{P}_{ij}^2$ and

$$X(E) = \left[E - \sum_{j=1,2} \Gamma_j(E) \right] \bigg/ t_L(\theta_t). \tag{3.28}$$

Also, $\mathcal{P}_{ij}$ are elements of the transfer matrix $\mathcal{P} = M(N)M(N-1)\cdots M(1)$, with [53]

$$M(j) = \begin{pmatrix} X(E) & -1 \\ 1 & 0 \end{pmatrix}. \tag{3.29}$$

For a given energy E, $T_N(E)$ measures the level of backscattering events in the electrons (or hole) transport through the chain.

Figure 3.5 depicts the transmittance spectra as a function of the energy (minus the Fermi energy) in units of eV for the poly (GC) sequence, in which $\Gamma_1(E) \neq 0$ and $\Gamma_2(E) = 0$ (Figure 3.5a); the poly (AT) sequence, in which $\Gamma_1(E) = 0$ and $\Gamma_2(E) \neq 0$ (Figure 3.5b); and the poly (GCAT) sequence, in which $\Gamma_1(E)$ and $\Gamma_2(E)$ are different of zero (Figure 3.5c). We have considered the Fermi energy equal to guanine's ionization energy for the poly (GC) and poly (GCAT) structures, and adenine's for the poly (AT) structure. Also, N, the number of nucleotides, is equal to 24, and the torsion angles are $\theta_t = 0$ (black full line) and $\pi/10$ (gray full line).

The transmittance spectra show several energies with high transmission resonances ($T_N(E) = 1$), besides a striking symmetry around the energies 1.71 (poly (GC), see Figure 3.5a); 1.46 (poly (AT), see Figure 3.5b); and 1.21 (poly (GCAT), see Figure 3.5c). All units are in eV and independent of the value of the torsion angle θ. The reason for the former (symmetrical spectra) is due to the periodicity of the DNA structures considered here, as depicted in Figure 3.4, since, for quasiperiodic DNA structures the energy spectra exhibit distinct physical properties, leading to a novel description of disorder.

The independence on the torsion angle θ depends on the role played by the sugar–phosphate backbone ionization energy ϵ_{SP}. Indeed, the expected two transmission bands around the ionization energies of the guanine-ϵ_G and cytosine-ϵ_C (adenine-ϵ_A and thymine-ϵ_C) bases for the poly (GC) (poly (AT)) structure, formed when the backbone ionization energy ϵ_{SP} is equal to zero, progressively approach each other as ϵ_{SP} increases, leading to the one-band structure depicted in

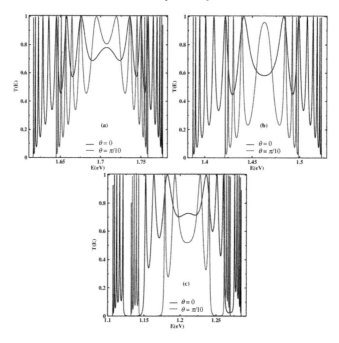

Figure 3.5 Transmittance spectra as a function of the energy, minus the Fermi energy, which is equal to the guanine's ionization energy for the poly (GC) and poly (GCAT) structures, and adenine's for the poly (AT) one, in units of eV considering N (the number of nucleotides) equal to 24, and a torsion angle $\theta_t = 0$ (black full line) and $\pi/10$ (gray full line). (a) DNA poly (GC); (b) DNA poly (AT); (c) DNA poly (GCAT). After de Almeida et al. [76]

Figures 3.5a, b, and c for the realistic value considered here. This profile is insensitive to the adopted value for the hydrogen bonding strength ($t_{GC} = 0.9$ eV and $t_{AT} = 0.3$ eV, respectively), which is responsible for the binding of the complementary DNA strands, because its energetic value is much less than the backbone ionization energy $\epsilon_{SP} = 12.27$ eV, which is justified by the presence of a number of counterions and water molecules.

It is important to mention that stationary electron transmission spectra through finite DNA chains are usually investigated throughout a dynamical map, as it is quite common in quasiperiodic systems. According to that, the electronic transmission properties are affected by the localization of the electronic wave functions, despite the fact that long-range correlations in DNA finite segments could be a possible mechanism to induce delocalization. However, the actual correlations are not strong enough, and the stationary states remain all localized [82].

For inhomogeneous random sequences, the scenario is worst since almost all states are strongly localized and the electronic transport is dominated by dissipative

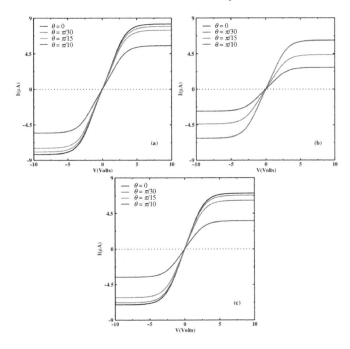

Figure 3.6 I×V characteristic curves for torsion angles $\theta_t = 0$ and $\pi/30$ (bottom full lines), $\theta_t = \pi/15$ (middle full line) and $\theta_t = \pi/10$ (top full line). (a) DNA poly (GC); (b) DNA poly (AT); (c) DNA poly (GCAT). After de Almeida et al. [76].

processes. Nevertheless, as in our case, the presence of long-range correlations due to the periodicity of the renormalized structure might enhance the localization length and, therefore, the transmission resonances persist in larger segments as compared with a non-correlated random sequence.

The I×V characteristics curves can be obtained by applying the Landauer-Büttiker formulation described previously, and are depicted in Figure 3.6 for the DNA poly (GC) (Figure 3.6a), for the DNA poly (AT) (Figure 3.6b), and for the DNA poly (GCAT) (Figure 3.6c). As one can see, the current–voltage profiles for torsion angles $\theta_t = 0$ and $\pi/30$, the full lines at the bottom, are very close with a noticeable overlap, mainly for the DNA poly (AT) case. The middle (top) current–voltage curve represents a torsion angle $\theta_t = \pi/15$ ($\pi/10$). All of them present an ohmic region -2.5 V $\leq V \leq 2.5$ V, and symmetric regions otherwise, indicating a semiconductor behavior that can be attributed to the resonant dipoles of the DNA segment leading to an overall depletion region effect. This behavior can be explained by the tunneling of electrons under an energy barrier between adjacent localized states of the basis so that electrons can travel through the molecule mainly by the hopping mechanism.

The localized states may be located in the vicinity of the Fermi energy of the electrodes, so that when a voltage bias is applied, the Fermi energy aligns with a localized state, and the charge electron transport is initiated. No rectifying behavior is observed, indicating that the charge transport in the highest occupied molecular orbital (HOMO) of the DNA poly (GC), located at the guanine base, is of the same magnitude of its lowest unoccupied molecular orbital (LUMO), located at the cytosine base, when the voltage bias is applied to the sample.

It is worth mentioning that the observed symmetry might be contaminated by a residual Schottky effect from the contacts, as measured by the contribution of $H_{electrode}$ depicted in Eq. (3.25). The maximum value of the current, obtained when there is no torsion angle, is 8.26 µA. Observe that as the torsion angle θ_t increases, the absolute value of the current decreases, indicating a transition toward an insulator phase, reaching a critical value at θ_t approximately equal to $\pi/7.5$ (poly (GC)), $\pi/7$ (poly (AT)), and $\pi/6.2$ (poly (GCAT)) – all of them below the torsion angle $\theta_t = \pi/5$ between the base pairs of the B-form of DNA.

3.4 Methylated States

Characterization of electronic processes in biomolecules, with a consequent impact in the development of medical technologies and therapies, is becoming a standard procedure nowadays. For instance, investigations of the compatibility of the interface between electronic devices and biological systems are much encouraged, since the quality of bio-interfaces can play a central role in the prospective success and impact of technologies such as thin film electronics, in vitro cell culture models, and medical devices that make use of organic materials in place of conventional semiconductors [83].

Among these biological systems, DNA methylation plays an important role, being a process by which methyl groups (CH_3) are added to the fifth carbon atom of the cytosine base or the sixth nitrogen atom of the adenine base. Besides, it can be associated with heritable and functionally relevant changes in gene activity, although not modifying the DNA biochemical structure [84].

Usually, genes are silenced when methylation occurs in their promoter region, where the transcription process is initiated. However, when the methylation occurs in the gene body, they may have a positive correlation to their expression (transcription/translation) process. For example, the 5-methylcytosine-based DNA methylation occurs through the covalent addition of a methyl group at the 5-carbon of the cytosine heterocyclic aromatic ring by an enzyme called DNA methyltransferase, resulting in the formation of 5-methylcytosine. This is a typical vertebrate DNA methylation pattern that occurs predominantly in CpG islands, dinucleotide-rich

Figure 3.7 Schematic representation of the methylated DNA strand sandwiched between two electrodes. Observe the methylation of the cytosine nitrogened base, the so-called 5-methylcytosine, expressed by its ionization energy ε_M.

regions that possess high relative densities of CpG and are positioned at the 5' ends of many human genes.

Most cell types display relatively stable DNA methylation patterns, with 70%–80% of all CpGs being methylated. This event is associated with a number of key processes including embryonic development, chromosome stability, genomic imprinting, X chromosome inactivation, suppression of repetitive elements, and carcinogenesis [85].

In this section, we suggest a nanoelectronic device to investigate the electron transport properties of a DNA strand by taking into account the contributions of the 5-methylcytosine. Our main purpose is to observe how methylation of cytosine in a DNA strand modifies its charge transport properties in comparison to a non-methylated DNA strand. Both molecules are set in a linear geometry covalently linked to two platinum electrodes (see Figure 3.7).

We model the non-methylated/methylated DNA strands by adopting a tight-binding Hamiltonian constructed from ab initio parameters with a single orbital per site and nearest-neighbor interactions and by neglecting any environment or complex contacts, as described previously [86]:

$$H_{total} = H_{DNAm} + H_{electrode} + H_{coupling}. \qquad (3.30)$$

The first term of Eq. (3.30) includes the intrastrand charge propagation through the methylated DNA, being given by

$$H_{DNAm} = \sum_{n=1}^{n_M} \left[\varepsilon_{\alpha}^{n} |n, 1\rangle \langle n, 1| + t_{n,n\pm1} |n, 1\rangle \langle n \pm 1, 1| \right], \qquad (3.31)$$

where n_M is the number of nucleotides in the methylated DNA strand, and ε_{α}^{n} is the on-site ionization energy of the respective base α at the nth site. Here, α = adenine (A), cytosine (C), guanine (G), thymine (T), and 5-methylcytosine (M). Also, $t_{n,n\pm1}$ are the hopping parameters between the adjacent sites $n, n \pm 1$ of the methylated DNA strand.

The second term, related to the two semi-infinite metallic electrodes, is given by

$$H_{electrode} = \sum_{n=-\infty}^{0} \left[\varepsilon_E^n |n, 1\rangle \langle n, 1| + t_E |n, 1\rangle \langle n \pm 1, 1| \right]$$

$$+ \sum_{n=N+1}^{\infty} \left[\varepsilon_E^n |n, 1\rangle \langle n, 1| + t_E |n, 1\rangle \langle n \pm 1, 1| \right]. \tag{3.32}$$

Here, ε_E^n (t_E) is the ionization energy (hopping parameter) of the electrode.

Finally, the third term describes the coupling between the methylated DNA strand and the semi-infinite metallic electrodes, yielding

$$H_{coupling} = t_c \left[|0, 1\rangle \langle 1, 1| + |N, 1\rangle \langle N + 1, 1| \right], \tag{3.33}$$

where $t_c = 0.63$ eV is the hopping amplitude between the source (drain) electrodes and the ends of the methylated DNA strand.

The vertical ionization energy and hopping parameters for methylated cytosine and the base pairs, respectively, were obtained from first-principles calculations using the Gaussian 09 code within the DFT framework. For these calculations, we used the Becke's half-and-half hybrid exchange-correlation (BH and HLYP) functional [87] and a Dunning's correlation consistent polarized valence double ζ basis set (cc-pVDZ) [88] to optimize the structures of the base pairs, as well as to calculate their occupied (HOMO) orbital energies. The BH and HLYP method is strongly recommended for the study of small peptides and other similar biomolecular systems with hydrogen bonding and charge transfer interactions [89], especially when the objective is to calculate HOMO eigenvalues, as this hybrid functional includes a fraction of the exact orbital exchange.

Most biomolecular interactions take place in an aqueous environment, making it, therefore, important to consider the effects of the aqueous solvent. In particular, the DNA molecule displays considerable sensitivity to its ionic surroundings during its various structural transitions and charge transfer states, mainly in the base-pairing effect of its double-stranded (native) topology [90]. However, in its single-stranded (denatured) form, as considered in this section, its structure and intermolecular interactions are much less affected with a minimal exposure of its hydrophobic groups to the aqueous solvent. In view of that, and to avoid the increase of the computational cost, all calculations were carried out only for molecules in the gaseous phase (no solvation effects were taken into account).

We used the relation $IE = E(N - 1) - E(N)$ to evaluate the ionization energy for 5-methylcytosine (7.02 eV), where the ionized (non-ionized) nucleotide with one missing electron was characterized by the total energy $E(N - 1)$ ($E(N)$) [91]. Taking advantage of the published ab initio ionization potential calculation

Table 3.1 *Hopping parameters (in eV) between adjacent bases in a DNA strand. After de Almeida et al. [86].*

X|Y	A	T	C	G	M
A	0.260	0.179	0.223	0.220	0.227
T	0.173	0.278	0.167	0.245	0.225
C	0.163	0.190	0.198	0.282	0.248
G	0.252	0.221	0.144	0.210	0.201
M	0.369	0.450	0.507	0.145	0.430

of stacked bases, the values for the other nucleotides are as follows: guanine, 7.75 eV; cytosine, 8.87 eV; adenine, 8.24 eV; and tymine, 9.14 eV [59].

The hopping parameters are estimated from [92]

$$t_{n,n\pm1} = 0.5\left(E_{n,n\pm1}^{HOMO} - E_{n,n\pm1}^{HOMO-1}\right), \tag{3.34}$$

where $E_{n,n\pm1}^{HOMO}$ and $E_{n,n\pm1}^{HOMO-1}$ are, respectively, the first- and second-highest occupied molecular orbital energies for the base pairs formed by n and $n \pm 1$ residues. The hopping parameters between adjacent bases in the methylated DNA strand are listed in Table 3.1.

In order to perform a comparative analysis of the electronic conductance of the standard and methylated DNA forms, current–voltage characteristics of the DNA strands with 16 bases are shown in Figure 3.8. When the potential barrier between the metallic contacts and the DNA is set to zero, a steplike feature in the plot $I \times V$ is found, roughly depicting its semiconducting characteristics. The inset illustrates the transconductance dI/dV versus V of the devices, which is highly nonlinear.

The approximately linear $I \times V$ curves observed for $5 < |V| < 10$ V indicate a semiconductor–metallic transition (SMT), allowing for the possibility to interpret the variation of experimental results found in the standard DNA form. The zero-conductance plateau (current slope) in the resulting $I \times V$ characteristics in the region $|V| < 5$ V ($5 < |V| < 10$ V) is due to the gap (band) width of the electronic band structures of the DNA molecule. From the $I \times V$ data, no rectifying behavior was observed, indicating that the charge transport features in HOMO states are similar to those found in the LUMO.

Figure 3.8a deals with the DNA sequence GAGCTGACGTTCACGG retrieved from the first sequenced Ch22, which contains the basic elements relevant for the analysis of the methylation effect on the electronic transport. Figure 3.8b-d exhibits the results for some methylated forms, differing in the amount of 5-methylcytosine, characterized by one (Figure 3.8b), two (Figure 3.8c), and three (Figure 3.8d) methylated cytosines, respectively. Amazingly, one can see that even

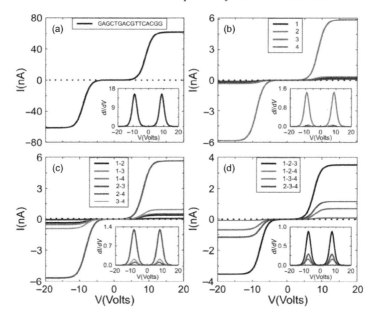

Figure 3.8 Current–voltage I (in nA) against V (in volts) curves for a non-methylated/methylated DNA strands. The insets show the transconductance (dI/dV) versus V of the devices. (a) GAGCTGACGTTCACGG chain with four (1,2,3,4, in sequence) non-methylated cytosine bases. (b) Methylation of only one cytosine in the DNA strand (1, 2, 3, or 4). Observe the overlap of the curves 1, 2, and 4 near the horizontal axis. (c) Methylation of two cytosines in the DNA strand (pairs 1–2, 1–3, 1–4, 2–3, 2–4, and 3–4). (d) Methylation of three cytosines in the DNA strand (labeled by the triplets 1–2–3, 1–2–4, 1–3–4, and 2–3–4). After de Almeida et al. [86]

a single methylation site in a DNA strand (Figure 3.8b) reduces the current–voltage curve by one order of magnitude, being a very convenient and cheap way to probe their particular characteristics.

The overall reduction of the saturation current by the presence of methylated cytosine sites is directly associated with the fact that these act as additional impurity centers, thus enhancing the Anderson localization of the electronic states around their closest vicinity. The sensitivity of the saturation current on the position of the methylated cytosine is related to a secondary phenomenon, namely the impact of cytosine methylation on the hopping amplitudes to the neighboring nucleotides.

The data shown in Table 3.1 reveals that the hopping amplitude connecting a cytosine to a guanine is reduced upon methylation. In contrast, the hopping amplitudes between cytosine and adenine as well as between cytosine and thymine are enhanced. Hence, electronic transport has an additional suppression when methylation occurs in a cytosine directly connected to a guanine because this leads to a smaller average hopping amplitude. This feature is clearly illustrated in Figure 3.8b which shows that the saturation current is much smaller when methylation occurs in the cytosines located at position 1, 2, and 4, which are directly connected to a guanine base, as compared to the corresponding saturation current when methylation occurs at the cytosine at position 3, which is connected to adenine and thymine.

When methylation takes place at two cytosines, the suppression of the saturation current is smaller when the second methylation site corresponds to a cytosine that, although having one neighboring guanine, is connected to an adenine. This is so because it has the largest increase of the hopping amplitude. Note, however, that cytosines located at positions 2 and 4 have similar neighborhoods (one guanine and one adenine). Figure 3.8c shows that the saturation current is larger when the second methylated site (beside the one at position 3) is at position 2, corresponding to well-separated scattering centers.

Finally, when methylation occurs at all cytosine sites except for one, the non-methylated base acts as a defect. As such, the suppression of the saturation current will be larger when the defect (non-methylation) is in cytosine 3, which is the only one that has both hopping amplitudes to the neighboring nucleotides enhanced by methylation, as illustrated in Figure 3.8d. The preceding physical mechanisms influencing the electronic transport are expected to hold for quite general methylated DNA sequences. Note also the opposite change in the current when one goes from the methylated sites 1–2 to the methylated sites 1–2–3 (increasing in current) as compared to the change when site 4 is methylated from the methylated sites 2–3 (decreasing in current). This feature is related to the distinct influence produced by methylation on the hopping amplitudes of neighboring sites. In fact, the methylation of site 3 enhances the hopping amplitude to both nearest nucleotides, namely adenine and thymine (see Table 3.1), thus favoring charge transport.

On the other hand, the methylation of site 4 enhances (reduces) the hopping to the adenine (guanine) site, leading to a larger barrier for the charge transport. This strongly suggests the feasibility of using I×V curve measurements to develop biosensors for the diagnosis of human diseases related to aberrant gene expression caused by DNA methylation.

3.5 Diluted Base-Pairing

A very instructive model that unveils the special role played by correlations in the electronic properties of DNA-based structures incorporates diluted base pairing. In this model, we consider DNA poly (CG) and DNA poly (CT) segments at which guanine bases (G) are attached laterally at a fraction of the cytosine (C) bases.

Within a tight-binding description, the density of states and eigenfunctions of the one-electron states can be mapped onto that of the Anderson chain with diluted disorder. As such, the influence of the effective disorder on the nature of the

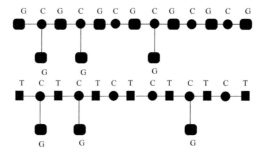

Figure 3.9 Schematic representation of the single-strand DNA molecule showing the main periodic chains of alternate bases (CG and CT sequences) with diluted base pairing. Guanine (G) bases are laterally attached at random to a fraction p of the cytosine (C) sites. After Albuquerque et al. [36]

one-electron states, as well as on the wave-packet dynamics, can be explored. In particular, base-pairing dilution indeed leads to a complete exponential localization of all one-electron states in segments formed with complementary units (as in poly (CG)).

On the other hand, a resonant state is not affected by disorder and remains extended in chains with noncomplementary units (as in DNA poly [CT]). In the presence of such resonant state, the wave packet develops a diffusive dynamics.

The theoretical framework makes use of an effective tight-binding Hamiltonian describing one electron moving in a geometry composed of a periodic chain of alternate bases (CG or CT sequences) [93]. The model assumes that G bases are laterally attached to C sites at random with probability p (see Figure 3.9), taking into account just a single orbital per site and nearest-neighbor transfer integrals t (along the main chain) and w (among paired bases). The corresponding time-independent Schrödinger equation for a DNA poly (CG) sequence is given by

$$E\psi_j^G = t(\psi_{j-1}^C + \psi_{j+1}^C) + \epsilon_G\psi_j^G, \text{ for odd } j, \tag{3.35}$$

$$E\psi_j^C = t(\psi_{j-1}^G + \psi_{j+1}^G) + w\beta_j\psi_j^G + \epsilon_C\psi_j^C, \text{ for even } j. \tag{3.36}$$

For a DNA poly (CT), G is replaced with T. Here, ϵ_α ($\alpha =$ G, T, or C) represents the on-site potential at the bases G, T, or C, and ψ_j^α is the wave-function coefficient in the single orbital basis, defined by

$$|\Psi\rangle = \sum_{(j,\alpha)} \psi_j^\alpha |j,\alpha\rangle, \tag{3.37}$$

where (j,α) runs over all base units. Also, $\beta_j = 1$ with probability p and $\beta_j = 0$ with probability $1 - p$, where p is the concentration of G sites attached to the single-stranded main periodic chain. At the sites where $\beta_j = 1$, there is an additional equation:

$$E\psi_j^G = w\psi_j^C + \epsilon_G\psi_j^G, \tag{3.38}$$

for even j.

A clear picture of the nature of the electronic states on the preceding model is achieved after performing a decimation procedure of the attached base units. The preceding tight-binding model for a DNA-based molecule is mapped onto an effective 1-D diluted Anderson model [94]. This model contains a diagonal disorder diluted by an underlying periodicity. The resulting sequence is composed of two interpenetrating sublattices, one composed of random potentials (Anderson chain) while the other has nonrandom segments.

The degrees of freedom associated with the lateral DNA bases appearing in the preceding equations are removed by substituting

$$\psi_j^G = [w/(E - \epsilon_G)]\psi_j^C, \text{ for even } j, \tag{3.39}$$

into the equation for the coefficients ψ_j^C, yielding

$$E\psi_j^C = \epsilon_C^*\psi_j^C + t(\psi_{j-1}^G + \psi_{n+1}^G), \tag{3.40}$$

where

$$\epsilon_C^* = \epsilon_C + [w^2/(E - \epsilon_G)] \tag{3.41}$$

is the renormalized potential at the cytosine sites at which the G bases are laterally attached. For those cytosine bases with no lateral attachment, the potential remains the bare one.

After eliminating the coefficients associated with the lateral G bases, the remaining set of equations expresses an alternate sequence of CG (or CT) nucleotides. The C sites have two possible values for the on-site potential, namely ϵ_C^* with probability p or ϵ_C with probability $(1 - p)$, respectively. The remaining bases of the periodic sequence have all the same potential: ϵ_G for poly (CG) or ϵ_T for poly (CT).

The random character of the diluted base pairing is reflected in a random sequence for the effective on-site energies of the cytosine sites. This kind of sequence is similar to the structure so-called diluted Anderson model. It consists of two interpenetrating sequences: a periodic sequence containing the guanine or thymine sites for DNA poly (CG) or DNA poly (CT), respectively, and a random sequence containing bare and renormalized cytosine sites.

Due to the periodicity of the nonrandom sublattice, a special resonance energy E_0 appears with vanishing wave-function amplitudes on the random sublattice.

Therefore, this mode is mainly insensitive to the presence of disorder and may lead to a possible mechanism to induce conductance in such DNA-based molecules. For the DNA poly (CT) molecule, the resonance energy is $E_0 = \epsilon_T$. At this energy, the renormalized cytosine potential remains finite, leading to a divergence of the localization length of the one-electron eigenmodes, as the resonance energy is approached.

On the other hand, the resonance energy for DNA poly (CG) molecules is $E_0 = \epsilon_G$, in which the renormalized cytosine potential diverges. This case corresponds to an effectively infinite disorder that counteracts the delocalization effect. As a consequence, diluted base pairing induces a stronger localization of the one-electron eigenfunctions in DNA poly (CG) than in DNA poly (CT) structures.

The spectrum of the Lyapunov exponent $\gamma(E)$ (which is the inverse of the localization length) of long DNA segments nicely illustrates the previously described features. The Green's function recursion method based on Dyson's equation provides

$$G_{n+1,n+1}^{n+1} = \left[E - H_{n+1,n+1}^0 - t_{n+1,n} G_{n,n}^n t_{n,n+1} \right]^{-1}, \tag{3.42}$$

with

$$G_{1,n+1}^{n+1} = G_{1,n}^n t_{n,n+1} G_{n+1,n+1}^{n+1}, \tag{3.43}$$

where $G_{1,n+1}^{n+1}$ denotes the $M \times M$ Green's function operator between the first and the $(n + 1)$th base pairs. Also, $G_{n+1,n+1}^{n+1}$ and $H_{n+1,n+1}^0$ are the Green's function operator and the free Hamiltonian for the isolated $(n + 1)$ base pair, $t_{n,n+1}$ is the diagonal $M \times M$ matrix coupling the base pairs at position n and $n + 1$, and E is the diagonal $M \times M$ matrix for the electron energy.

In the recursive equation, $G_{1,1}^1 = I$ (identity matrix) and $G_{0,0}^0 = 0$ (null matrix). We stress that, for a single-strand DNA-like segment, $M = 1$.

The Lyapunov exponent for a DNA segment is given by

$$\gamma(E) = (1/2N) \ln \left[Tr \left| G_{1,N+1}^{N+1} \right|^2 \right]. \tag{3.44}$$

For extended states, $\gamma(E)$ vanishes in the thermodynamic limit.

The resonance effect is robust with respect to distinct transfer integral values and is more clearly analyzed without considering the additional energy scale associated to distinct intrastrand (t) and interstrand (w) hopping integrals.

In the following illustration, it was considered $t = w = 1$ eV, which is somewhat larger than previously reported estimates of intrastrand transfer integrals [95], and a typical set for the ionization energies values, namely $\epsilon_C = 8.87$ eV (cytosine), $\epsilon_G = 7.75$ eV (guanine), and $\epsilon_T = 9.14$ eV (thymine), all units in eV. An exact

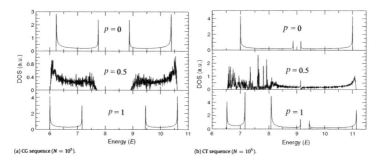

Figure 3.10 Plot of the electronic density of states (DOS) in arbitrary units (a.u.) versus the reduced energy E (in units of eV), for the particular cases of the hopping term $t = 1$ eV. (a) poly (CG)-based DNA sequences: the bandgap persists for poly (CG) chains with diluted base-pairing, and all van Hove singularities are rounded off; (b) poly (CT)-based DNA sequences: the bandgaps coalesce for base-pair diluted poly (CT) before splitting into three bands. Disorder does not affect the van Hove singularity at $E = \epsilon_T$. The gap-less band structure, together with the nonlocalization of the resonance state, favors the electronic transport in this case. After Albuquerque et al. [36]

diagonalization of the complete tight-binding Hamiltonian, given by Eqs. (3.35) and (3.36), provides the participation number of all eigenstates.

The electronic density of states (DOS) is obtained directly from the recursive Dean's method. Figure 3.10 shows the DOS for three representative values of the concentration of paired cytosine bases, namely

(a) $p = 0$, corresponding to pure DNA poly (CG) and DNA poly (CT) chains
(b) $p = 1$, describing the DNA poly (CG) and DNA poly (CT) chains with guanine bases laterally attached to all cytosine bases
(c) $p = 0.5$, representing a typical sequence of diluted base pairing.

The DOS for the DNA poly (CG) sequences is shown in Figure 3.10a. The electronic density of states has two main bands, which is typical of binary sequences, with the gap for $p = 1$ being larger than for $p = 0$. Such enhancement of the energy gap is a direct consequence of the base pairing. For $p = 0.5$, all van Hove singularities at the band edges are rounded off by the presence of disorder. The fluctuations in the DOS have been exploited in the literature to identify the nature of the states. The variance in the number of states in a given energy window scales linearly with the system size for localized states, while having just a slow logarithmic scaling for extended states. These two regimes reflect the distinct level spacing statistics of localized and extended states. As a result, much smaller fluctuations are attained in the normalized DOS when extended states are present as compared to the fluctuations observed in the energy range corresponding to localized states. These fluctuations are of the same magnitude in both bands, which indicate that these bands are equally affected by disorder.

The DOS curves for DNA poly (CT)-based chains are depicted in Figure 3.10b. For these sequences, a series of relevant features is absent in the previous case. Firstly, the two-band structure of the binary $p = 0$ case evolves to a three-band structure at $p = 1$, as expected for a periodic structure with three distinct sites in the unit cell. The bottom of the upper band at $p = 0$ coincides with the top of the middle band at $p = 1$. This energy corresponds exactly to the resonance energy $E_0 = \epsilon_T$. When the concentration of the attached guanine bases increases, the two-band structure first coalesces into a single band before splitting into three bands, as shown for the particular case, $p = 0.5$. Further, the van Hove singularities are rounded off, except for the one located at E_0, which corresponds to the resonance state insensitive to disorder. Therefore, diluted base pairing produces a gap-less band structure while keeping the states around E_0 extended, an ideal scenario for electronic transport. Additionally, the DOS exhibits stronger fluctuations at the bottom than at the top of the energy band, pointing out that the low-energy states are more localized than the high-energy ones.

The Lyapunov exponent γ directly probes the disorder effect on the nature of the electronic eigenstates. In Figure 3.11 (upper panel), the spectrum of the Lyapunov exponent for the base-pair-diluted DNA poly (CG) molecule with $p = 0.5$ is shown. For the hopping amplitude $t = 1$ eV, the Lyapunov exponent achieves a minimum value of the order of 10^{-2} in both energy bands. Therefore, the maximum localization length in this chain is of the order of 100 sites – i.e., no delocalized mode survives to diluted base pairing in binary periodic DNA sequences of corresponding bases, such as DNA poly (CG). The average localization length scales as $1/t^2$. Further, this kind of disorder affects both bands in a similar way, as already pointed out through the analysis of the DOS fluctuations.

Figure 3.11 (lower panel) shows the corresponding Lyapunov exponent spectrum for a DNA poly (CT) chain with $p = 0.5$ diluted base pairing. The presence of two singularities are evident. The first one is at $E = \epsilon_G$, which corresponds to the energy at which the renormalized ϵ_C^* diverges, thus leading to an effective infinite disorder. The one-electron mode at this energy is strongly localized. The second singularity is at $E = E_0 = \epsilon_T$. This corresponds to the energy mode not affected by the disorder and has a Bloch-like character. The low-energy modes are more localized than the high-energy ones, in agreement with the observation that these regions depict distinct DOS fluctuations.

The spectra of the Lyapunov exponents consider chains with weaker hopping amplitudes ($t = 0.5$ and 0.25 eV). They also display the same resonances, thus corroborating the robustness of the resonance mechanism with regard to distinct energy scales of the transfer integrals. The quite distinct effects caused by diluted

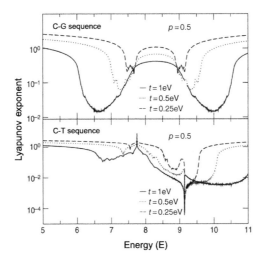

Figure 3.11 (Upper panel): Lyapunov exponent $\gamma(E)$ versus the energy E (in units of eV) for a DNA poly (CG) sequence with $p = 0.5$ diluted base pairing. All states are exponentially localized, with the maximum localization length being of the order of 10^2 sites for the hopping term $t = 1$ eV, decreasing as $1/t^2$. (Lower panel): Same as in upper panel but for a DNA poly (CT) sequence. For $E = \epsilon_G$, one observes a strong localization as the renormalized energy ϵ_C^* diverges. At the resonance energy $E = E_0 = \epsilon_T$, the mode is not affected by disorder and keeps its Bloch-like character. Data from different values of the hopping term t confirm that the resonance mechanism is robust with respect to this energy scale. After Albuquerque et al. [36]

base pairing in DNA poly (CG) and DNA poly (CT) may have a significant impact on the electronic transport.

3.6 Conclusions

In conclusion, by using an effective tight-binding Hamiltonian model, we have investigated the transport properties of a molecular device made up of a DNA double-helix molecule directly coupled to two platinum electrodes. As the double-strand model does not account for variability of the hopping amplitudes and their dependence on the electron energy, we explicitly also take into account the transport along the sugar–phosphate side chains. The electronic transmittance and I×V characteristics were discussed in terms of its on-site ionization and electrode energies, as well as its different hopping parameters.

The I×V characteristic curve seems to be accounted for by the short-range pair correlations, suggesting that the inclusion of just first-neighbors intrastrand pair correlations on the nucleotide distribution provides an adequate description of the DNA's electronic properties. However, as the electron transmissivity depends strongly on the intrachain coupling, more realistic model parameters were included, such as a twist angle θ_t between two adjacent base pair attached along the molecule backbone.

In order to get a simple mathematical description of the DNA molecule, keeping most of its relevant physical information, we used also a two-step renormalization process to map it into a linear monoatomic lattice, allowing us to incorporate the sugar–phosphate backbone contribution into an energy-dependent on-site ionization potential on the main DNA's base pairs.

Methylated DNA forms were also discussed in order to identify and describe their impact on charge transport. For a single methylation defect, the saturation current is strongly suppressed when cytosine is connected to guanine. If two methylation defects are allowed in a strand, current suppression is smaller, but very specific I×V curves were found for strands with three methylation defects. This strongly suggests the feasibility of using I×V curve measurements to develop biosensors for the diagnosis of human diseases related to aberrant gene expression caused by DNA methylation.

Future works in this field should point to the possibility of developing new sophisticated nanodevices integrating human-made nanostructures with different biomolecules, as well as the previously mentioned DNA base pair stack within the cell. In doing so, care should be taken with the DNA spontaneous point mutations, whose plausible cause is the so-called rare tautomers arising from the proton transfer (PT) reactions between the base pairs' double-helix architecture of DNA proposed by Watson and Crick. Nanobiomolecular sensors may be developed for the detection of these point structural defects in DNA associated with the formation of the mismatches of the canonical A–T and G–C Watson-Crick base pairs. Our I×V characteristic curves may shed some light on this biologically important question.

Other important mechanisms, besides the biological environment, are the vibrational modes and the formation of polarons (the bound state of an electron with a lattice distortion). Surely, they might open up the possibility of monitoring and controlling critical biological functions and processes in unprecedented ways, giving rise to a vast array of potential technological achievements.

We hope that the present findings described here may stimulate further developments in the synthesis and characterization of new DNA-based molecules with potential applications in nanobioelectronics.

4

Thermodynamic Properties of the DNA Molecule

4.1 Introduction

One important issue worthy of attention and, so far, little explored is the connection between the scale invariance of quasiperiodic structures' energy spectra and their thermodynamic properties. The first steps in this direction were made by considering simplified fractals structures based on the Cantor sequence [96] as well as on the critical attractor of the logistic and circle maps at the onset of chaos [97]. From their energy spectrum, their thermodynamic behavior was derived. Self-similar spectra displayed some anomalous features; the most prominent one was related to the emergence of log-periodic oscillations in the low-temperature behavior of the specific heat. Besides, the average low-temperature specific heat was intimately connected with some underlying fractal dimension characterizing the energy spectrum [98].

In order to fill this gap, it is our intention in this chapter to explore the thermodynamics properties of the DNA molecule, as described by quasiperiodic systems, in the classical (Maxwell-Boltzmann), quantum (Fermi-Dirac), and the so-called nonextensive (beyond Boltzmann-Gibbs) statistics states. One of the main reason, for that lies in the fact that the unique structure of the DNA molecule allows various alterations of its material properties – which could modify its electrical, optical, and thermodynamic properties, revealing additional features.

Early theoretical and experimental works on the low-temperature heat capacity of DNA primarily took into account the phonon contributions – specifically, the redundant low-energy density of the vibrational states. In doing that, the main conclusion was that the low-energy of the DNA molecule is not unique among biopolymers, and its specific heat possesses a combination of the properties similar to those of glasses and other disordered materials [99].

Another important issue concerns the relationship between the low-temperature thermodynamic properties and the multifractal character of the energy spectra of a sequence dependent finite segment of a DNA molecule, as already discussed in Chapters 2 and 3. More specifically, what happens to the specific heat spectra profile in these cases? Does it present log-periodic oscillations as a function of the temperature T in the low-temperature region, around a mean value given by a characteristic dimension of the energy spectrum?

Besides, one of the most intriguing properties of the DNA macromolecule rises when a solution is heated up. As a consequence, the base pairs of the double helix break up and dissociate from each other to form two separated random coils. This phenomenon is referred to as DNA denaturation or thermal DNA melting, leading to interesting temperature profiles of some DNA's thermodynamical functions, such as the stretching of the hydrogen bonds, specific heat, and entropy [100].

The answers for the preceding questions are the main purposes of this chapter. Our main intent is to compare the different spectra profiles, seeking possible differences and similarities among them, with the objective to establish some kind of standard behavior.

4.2 Classical Statistics: The Single-Strand DNA Structure

Consider the effective tight-binding Hamiltonian with a single orbital per site ϵ_n at the orbital ψ_n and nearest-neighbor interactions through a hopping potential t, describing a single-strand DNA structure, as discussed in Chapter 2. This (discrete) Schrödinger equation can be also written as [101]

$$\begin{pmatrix} \psi_{n+1} \\ \psi_n \end{pmatrix} = M(n) \begin{pmatrix} \psi_n \\ \psi_{n-1} \end{pmatrix}, \tag{4.1}$$

where $M(n)$ is the transfer matrix

$$M(n) = \begin{pmatrix} t - \epsilon_n & -1 \\ 1 & 0 \end{pmatrix}. \tag{4.2}$$

After successive applications of the transfer matrices, we have

$$\begin{pmatrix} \psi_{n+1} \\ \psi_n \end{pmatrix} = M(n)M(n-1)\cdots M(2)M(1) \begin{pmatrix} \psi_1 \\ \psi_0 \end{pmatrix}, \tag{4.3}$$

yielding the wave function ψ_n at an arbitrary site n. As before, calculating the product of transfer matrices is completely equivalent to solve the Schrödinger equation for the system. The criterium for allowed energy is when $(1/2)Tr$ $[T_n] < 1$, with $Tr[T_n]$ meaning the trace of the transfer matrix $T_n = M(n)$ $M(n-1)\cdots M(2)M(1)$.

Figure 4.1 shows the electron energy spectra, as measured by their equivalent bandwidth Δ (the sum of all allowed energy regions in the band structures), as a

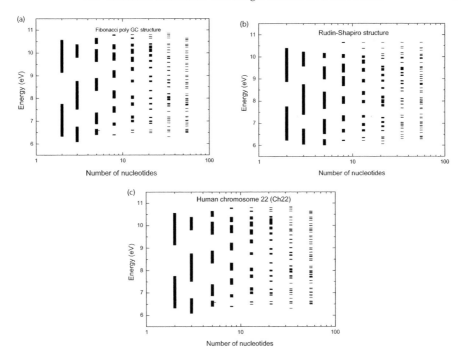

Figure 4.1 Energy spectrum for finite segment of a single-strand DNA molecule following (top left) Fibonacci sequence; (top right) Rudin-Shapiro sequence; (bottom) DNA sequenced human chromosome 22 (Ch 22). After Albuquerque et al. [36]

function of the number of nucleotides N, for the quasiperiodic sequences of (a) Fibonacci, (b) Rudin-Shapiro, and (c) the genomic Ch22 DNA structure.

The bandwidth Δ_i for the Nth number of nucleotides is given by

$$\Delta_1 = E_2 - E_1 \implies E_2 = E_1 + \Delta_1, \tag{4.4}$$

$$\Delta_2 = E_4 - E_3 \implies E_4 = E_3 + \Delta_2, \tag{4.5}$$

$$\vdots$$

$$\Delta_i = E_{2i} - E_{2i-1} \implies E_{2i} = E_{2i-1} + \Delta_i, \tag{4.6}$$

where E_1 and E_2 are the energy values of the bottom and the top of the first energy band (counting from its smallest to its largest values). Also, E_3 and E_4 are the energy value of the bottom and the top of the second energy band, and so on, for increasing N. We take the level density inside each band to be constant, and the same for all bands in a given hierarchy. In this case, a fractal or multifractal structure emerges at the $N \to \infty$ limit. Obviously, the number of bands depends on the number of nucleotides, as it is shown in Figure 4.1. In what follows, we consider a normalization in the frequency spectrum in such a way that the bands stay within the limits 0 and 1.

Within a classical Maxwell-Boltzmann statistics, the partition function Z_{MB} is given by

$$Z_{MB} = \int_0^\infty \rho(E) \exp(-\beta E_i) dE, \qquad (4.7)$$

where we have considered a unit Boltzmann's constant – i.e., β equal to $1/T$. Also, we take the DOS $\rho(E) = 1$.

We justify the use of the classical Maxwell-Boltzmann statistics because the electrons here behave as Boltzmann particles once the gaps in their energy spectra become smaller than the Fermi temperature T_F (which is our case). The quantum Fermi-Dirac statistics case will be the topic of Sections 4.4 and 4.5.

After some calculations, we find the partition function Z_{MB} as

$$Z_{MB} = \frac{1}{\beta} \sum_{i=1,3,\ldots}^{2N-1} (1 - e^{-\beta \Delta_i}) \exp(-\beta E_i). \qquad (4.8)$$

Note that it is only necessary to know the distribution of the energy spectrum of a given multifractal system to calculate the partition function [98]. Once we know the partition function, it is possible to calculate the specific heat using

$$C(T) = \frac{\partial}{\partial T} \left[T^2 \frac{\partial \ln Z_{MB}}{\partial T} \right], \qquad (4.9)$$

which can be written as

$$C(T) = 1 + \frac{\beta f_N}{Z_{MB}} - \frac{g_N^2}{Z_{MB}^2}. \qquad (4.10)$$

Here,

$$f_N = \sum_{i=1,3,\ldots}^{2N-1} [E_i^2 e^{-\beta E_i} - E_{i+1}^2 e^{-\beta E_{i+1}}], \qquad (4.11)$$

and

$$g_N = \sum_{i=1,3,\ldots}^{2N-1} [E_i e^{-\beta E_i} - E_{i+1} e^{-\beta E_{i+1}}]. \qquad (4.12)$$

Therefore, once we know the electronic energy spectra of a given DNA chain, we can determine the associated specific heat by using Eq. (4.10).

Figure 4.2a shows the electrons' specific heat $C(T)$ (in units of the Boltzmann's constant k_B) as a function of the temperature T (in units of Δ, the sum of all allowed energy regions in the band structures) for the Fibonacci poly (CG) single-stranded

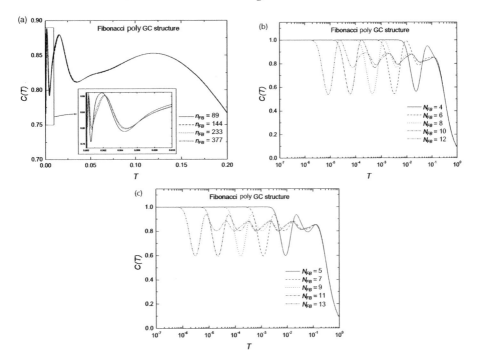

Figure 4.2 (a) Specific heat spectra $C(T)$ (in units of the Boltzmann's constant k_B) versus temperature (T) (in units of Δ, the sum of all allowed energy regions in the band structures) corresponding to the 10th, 11th, 12th, and 13th (89, 144, 233, and 377) Fibonacci poly (GC) sequence generation (number of nucleotides) DNA single-stranded model. The inset shows the rather interesting oscillatory behavior of the specific heat spectra. (b) Log-periodic spectra for even Fibonacci generation number ($N_{FB} = 4$, 6, 8, 10, and 12). (c) Log-periodic spectra for odd Fibonacci generation number ($N_{FB} = 5, 7, 9, 11$, and 13). After Albuquerque et al. [36]

DNA structure. It is possible to see that these spectra are almost independent of the Fibonacci's generation number N_{FB}. More important, the inset of this figure shows the oscillatory behavior of the specific heat for low temperatures with two classes of oscillations, one for the *even* and the other for the *odd* generation numbers of the sequence, the amplitude of the even oscillations being slight bigger than the amplitude of the odd one. The number of times that the specific heat oscillates for a given generation number N_{FB} is $(N_{FB} - 2)/2$ for N_{FB} even, and $(N_{FB} - 1)/2$ for N_{FB} odd.

These behaviors are illustrated in Figure 4.2b and c, where *log* plots of the specific heat against the temperature are depicted, corresponding to the even and odd generation numbers of the Fibonacci sequence, respectively. This peculiar behavior is a kind of signature with no counterpart in the other quasiperiodic structures and should be connected to the properties found for other Fibonacci spectra [98].

The log-periodic behavior of the specific heat shows a mean value d; around it, $C(T)$ oscillates log-periodically. This value is not related to the fractal dimension of the Fibonacci quasiperiodic structure because our specific heat spectra are not strictly invariant under changes of scales. Instead, this mean value d can be given approximately by the so-called spectral dimension (the exponent of a power law fit of the integrated DOS) – which, in this case, is approximately equal to 0.8. It is also associated to the minimum singularity exponent α_{min} in the Fibonacci multifractal $f(\alpha)$ spectrum.

Furthermore, the self-similarity of the specific heat spectra increases for sequences with a difference of two in the generation process. Note that the specific heat properties in log scale are basically controlled by the behavior of the low-energy region at the scale considered (i.e., each oscillation can be considered as a change of scale in the spectrum). In this sense, at a high scale defined by the Fibonacci's generation number N_{FB}, the low-energy region would be controlled by the generation number $(N_{FB} - 2)$; at a smaller scale, the low-energy region would be controlled by the generation number $(N_{FB} - 2) - 2 = N_{FB} - 4$ and so on.

Figure 4.3 (left) shows the specific heat spectrum against the temperature for a poly (CG) single-stranded DNA structure modeled by the Rudin-Shapiro sequence. In this case, there are oscillations with amplitude much larger than those found in the Fibonacci case. Moreover, the number of oscillations is not directly proportional to the number of generations, and there is no well-defined parity behavior, as in the Fibonacci model. The inset of this figure clearly reveals this.

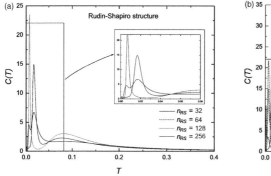
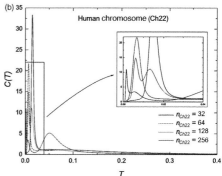

Figure 4.3 Left: Specific heat for a poly (CG) single-stranded DNA structure modeled by the Rudin-Shapiro sequence as a $C(T)$ versus temperature (T) plot for n_{RS}=32, 64, 128, and 256 nucleotides. The inset shows the oscillatory behavior of the specific heat spectra. Right: same as before but for the single-strand human chromosome 22 (Ch22) DNA mode. After Albuquerque et al. [36]

A similar spectra was found for the genomic single-strand Ch22 DNA structure, as depicted in Figure 4.3 (right), which shows random oscillations with amplitudes slightly larger than those obtained for the Rudin-Shapiro chain.

In all situations studied here, for high temperatures ($T \rightarrow \infty$), the specific heat for all generation numbers N_{FB} converges and decays as T^{-2}. This asymptotic behavior is mainly due to the fact that we have considered our system bounded. On the other hand, for low temperatures ($T \rightarrow 0$), the specific heat displays an oscillatory profile, no matter the model considered.

4.3 Classical Statistics: The Double-Strand DNA Structure

Consider the tight-binding Hamiltonian appropriated to the double-strand DNA molecule described in Chapter 2, whose energy spectrum, as measured by their equivalent bandwidth Δ, is depicted in Figure 2.7a (Fibonacci case), 2.7b (Rudin-Shapiro case), and 2.7c (Ch22 human chromossome) [102].

Following the lines of the previous section, the specific heat spectra can be obtained after the determination of their partition function using a Maxwell-Boltzmann statistics, as yielded by Eq. (4.10).

Figure 4.4(a) shows the electronic specific heat (ESH) spectra for the Fibonacci double-strand DNA chains, corresponding to its 10th (89), 11th (144), 12th (233), and 13th (377) generation numbers (number of nucleotides n_{FB}), as a function of the temperature. As in the previous case, for the high-temperature limit ($T \rightarrow \infty$), the specific heat for all cases converges and decays as T^{-2} as a consequence of the existence of a maximum energy value in the spectrum (once the spectrum is bounded). As the temperature decreases, the specific heat increases up to a maximum value. The corresponding temperature for this maximum value depends on the number of nucleotides n_{FB}, although one can see a clear tendency for a common temperature value as n_{FB} increases.

After the maximum value, the specific heat falls into the low-temperature region, where it starts to present a nonharmonic small oscillation behavior, as shown in the inset of Figure 4.4a. This can be interpreted as a superposition of Schottky anomalies corresponding to the scales of the energy spectrum. Furthermore, the profiles of these oscillations also define two oscillation classes, as far as the parity (even or odd) of the generation number of the Fibonacci sequence is concerned, as in the single-strand case. These behaviors are better illustrated in Figure 4.4b and c, where log plots of the specific heat against the temperature are depicted, showing clearly a log-periodic behavior – i.e., $C(T) = AC(aT)$, where A is a constant, and a is an arbitrary number.

The mean value d, around which $C(T)$ oscillates log-periodically, can be given approximately, as in the previous case, by the so-called spectral dimension

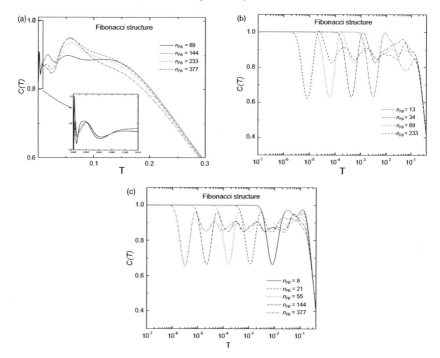

Figure 4.4 (a) Specific heat (in units of the Boltzmann's constant k_B) versus temperature (in units of Δ, the sum of all allowed energy regions in the band structures) for Fibonacci double-strand DNA chains. The inset shows the low-temperature behavior of the specific heat. (b) Log-periodic behavior of the specific heat for the even (6th, 8th, 10th, and 12th generation, respectively) Fibonacci DNA chain. (c) Log-periodic behavior of the specific heat for the odd (5th, 7th, 9th, 11th, and 13th generation, respectively). After Albuquerque et al. [36]

associated to the minimum singularity exponent α in the multifractal curve $f(\alpha)$, namely $\alpha_{min} = 0.835$. Of course, the number of oscillations observed in the specific heat spectra is related to the number of nucleotides n_{FB}, as n_{FB} as on the hierarchical generation of the Fibonacci sequence (more oscillations appear as n_{FB} increases).

A different scenario occurs when one considers the other quasiperiodic structure studied here (i.e., modeling the DNA molecule by the RS sequence), which is depicted in Figure 4.5 (left). Similarly to the Fibonacci case, in the limit when $T \rightarrow \infty$, the specific heat goes to zero as T^{-2} for all values of n_{RS}. Also, there are oscillations in the region near $T \rightarrow 0$ (which are better shown by the inset of the figure).

Although these oscillations can be interpreted as Schottky anomalies, as in the Fibonacci case, they do not have the same standard of behavior – i.e., two groups

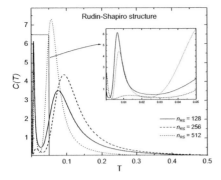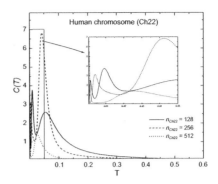

Figure 4.5 Left: Specific heat (in units of the Boltzmann's constant k_B) versus temperature (in units of Δ) for the Rudin-Shapiro double-strand DNA chain, corresponding to its 8th (128), 9th (256), and 10th (512) generation number (number of nucleotides n_{RS}). The inset shows the low-temperature behavior of the specific heat. Right: same as before but for the single-strand Ch22 DNA mode. After Albuquerque et al. [36]

of oscillations corresponding to even and odd generation numbers of the sequence. Additional differences should be pointed out. In this case, there are oscillations with amplitude very superior to those found in the Fibonacci case. More important, the log plot does not show a log-periodic behavior. Instead, it shows an erratic profile, which can be attributed to the more disordered structure of the Rudin-Shapiro sequence. Therefore, apart from the common asymptotic behavior of the specific heat when $T \to \infty$ and $T \to 0$, there is no other connection between the Fibonacci and Rudin-Shapiro DNA chains, regarding their specific heat spectra, considered here.

Finally, for comparison purposes, we present in Figure 4.5 (right) the specific heat behavior of the double-stranded human chromosome Ch22 DNA chain. As in the two previous cases, in the limit when $T \to \infty$, the specific heat goes to zero as T^{-2}, and there are also oscillations in the low-temperature region due to Schottky anomalies (which are better shown by the inset of the figure). One can see clearly that the overall behavior of the specific heat of Ch22 DNA chains is very close to the specific heat of the Rudin-Shapiro structure, in contrast with the Fibonacci case. For example, Ch22 and Rudin-Shapiro specific heats both present similar amplitude of oscillations, as well as erratic behavior in their log plots, instead of the log-periodic behavior found in the Fibonacci case.

Before concluding, let us comment on a possible connection between these present results with those of Mrevlishvili [103]. Their experimental data show oscillations of the specific heat at low temperatures, which are qualitatively similar to our present numerical theoretical results. They attribute their results to the non-crystalline order of the DNA sample – which may be modeled, as we have shown here, by quasiperiodic systems.

4.4 Quantum Statistics: The Single-Strand DNA Structure

The multifractal energy spectrum of the FB and RS sequences for both single- and double-strand DNA-like sequences was obtained in previous sections, and it will be taken into account to determine the specific heat spectra by using quantum mechanical Fermi-Dirac statistics. Since the spin degree of freedom will not be considered here, each occupied quantum state can support only one particle. According to the Fermi-Dirac statistics, the average occupation number of each state is given by

$$\langle n_i \rangle = \frac{1}{1 + \exp[\beta(E_i - \mu)]}, \tag{4.13}$$

where μ is the chemical potential, which can be computed as a function of temperature and band filling from

$$N_e = \sum_{i=1}^{N} \langle n_i \rangle, \tag{4.14}$$

from which $\mu(N_e/N, T)$ can be extracted numerically. Here, N_e is the number of non-interacting Fermi particles (electrons), while N is the total number of one-particle accessible states (electrons and holes).

The average internal energy is obtained from

$$U(N_e/N, T) = \sum_{i=1}^{N} E_i \langle n_i \rangle, \tag{4.15}$$

where the temperature dependence of the chemical potential $\mu(N_e/N, T)$ is explicitly taken into account.

We compute the specific heat at constant volume by differentiating the internal energy U with respect to the temperature T, i.e., $C_V = dU/dT$. It is then straightforward to calculate the fermionic specific heat, yielding [104]

$$C_V = \frac{1}{4T^2} \left[\sum_i E_i^2 \cosh^{-2}(\delta_i) - \frac{\left[\sum_i E_i \cosh^{-2}(\delta_i) \right]^2}{\sum_i \cosh^{-2}(\delta_i)} \right], \tag{4.16}$$

where $\delta_i = [(E_i - \mu)/2T]$. Figure 4.6a shows a log–log plot of the ESH spectra at constant volume (in units of N_e, the number of non-interacting Fermi particles) versus the temperature T for the 14th generation of the single-strand Fibonacci DNA chain, corresponding to $n_{FB} = 610$ nucleotides. Several values of the band fillings N_e/N are considered and indicated in the figure.

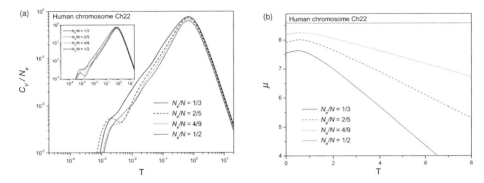

Figure 4.6 (a) Log–log scale of the fermionic specific heat at constant volume (in units of N_e) versus the temperature T for the 14th generation of the single-strand Fibonacci DNA chain, corresponding to $n_{FB} = 610$ nucleotides. Four different concentrations are analyzed. The inset in Figure 4.8a depicts the case for the 15th generation of the single-strand Fibonacci DNA chain, corresponding to $n_{FB} = 987$ nucleotides. (b) Chemical potential μ versus the temperature T. After Albuquerque et al. [36]

For the high temperature limit ($T \to \infty$), the specific heat for all cases converges and decays with T^{-2}. As the temperature decreases, the specific heat increases up to a maximum value, with the corresponding temperature for this maximum value depending on the number of band fillings N_e/N, although one can see a clear tendency for a common temperature value as N_e/N increases, independently of the occupation number ratio. After the maximum value, the specific heat falls into the low-temperature region and starts, due to the fractality of the energy spectrum, a complex pattern of log-periodic oscillations which signals the discrete scale invariance of the spectrum at the vicinity of the Fermi energy.

These oscillations occur around a linear trend (in log–log scale), whose power-law behavior is $C_V \propto T^{\phi_{FB}}$, with $\phi_{FB} = 0.74$, lasting until the temperature reaches a value around 10^{-3}. At this point, a phase transition (in the sense of an oscillatory regime) occurs, and the specific heat falls again linearly with T. The inset of this figure considers the 15th generation of the Fibonacci DNA chain, corresponding to $n_{FB} = 987$ nucleotides. From there, we can see a larger number of oscillations of the specific heat for low T (in general, the number of oscillations of the specific heat for fractal spectra increases with the order of the generation of the fractal). Besides, the oscillatory regime disappears at a lower temperatures, when compared to the 14th generation of the Fibonacci DNA chain.

In Figure 4.6b, the profile of the chemical potential $\mu(T, N_e/N)$ against the temperature T is presented for a single-stranded Fibonacci DNA fragment, considering the occupation ratios $N_e/N = 1/2, 4/9, 2/5$, and $1/3$. For lower values of T, there is a transient period, on which the chemical potential reaches a maximum value,

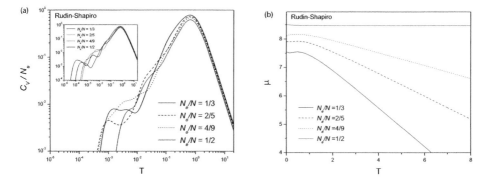

Figure 4.7 Log–log scale of the fermionic specific heat at constant volume (in units of N_e) versus the temperature T for the 10th generation of the single-stranded Rudin-Shapiro DNA chain, corresponding to $n_{RS} = 512$ nucleotides. The inset in Figure 4.9a depicts the case for the 11th generation of the Rudin-Shapiro DNA chain, corresponding to $n_{RS} = 1,024$ nucleotides. (b) Chemical potential μ versus the temperature T. After Albuquerque et al. [36]

and then starts to decrease (in all cases but $N_e/N = 0.5$) linearly as the ratio N_e/N decreases. For $N_e/N = 0.5$, the chemical potential has a constant value. This is an expected feature, since the chemical potential is a measure of the energy per particle for a given entropy.

A similar scenario is observed when one considers the other quasiperiodic structure studied, modeling the single-strand DNA molecule by the Rudin-Shapiro sequence, whose log–log plot of the specific heat at constant volume (in units of N_e) is depicted in Figure 4.7a for its 10th generation (which means 512 nucleotides). Its profile is very similar to those of Figure 4.6 when one consider the same band fillings, N_e/N. There is a transient region where the C_V oscillates nonharmonically around an inclined straight line, $C_V \propto T^{\phi_{RS}}$, with $\phi_{RS} = 1.01$, and then suddenly it falls to zero, linearly with T. However, now this decrease depends more strongly on the band fillings N_e/N considered. The inset shows the case for the Rudin-Shapiro's 11th generation, corresponding to $n_{RS} = 1,024$ nucleotides. The chemical potential for this sequence is shown in Figure 4.7b, with qualitative behavior similar to the FB case.

Now, let us compare our results with a real system, the single-strand human Ch22 chromosome. For this purpose, the log–log plot of the specific heat at constant volume (also in units of N_e), analyzed through a Fermi-Dirac statistics, is depicted in Figure 4.8a. Again, the specific heat falls to zero when $T \to \infty$, but now in a slightly higher ratio. Also, after the maximum value of C_V is reached, in the low-temperature region, the specific heat falls roughly linearly with T, and at $T = 0.5 \times 10^{-2}$ it falls more rapidly with T. Note that in this case there are less oscillations

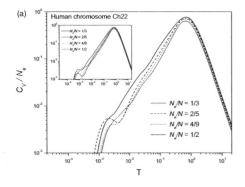
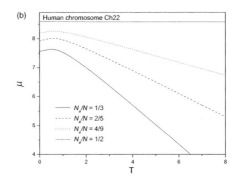

Figure 4.8 Log–log scale of the fermionic specific heat at constant volume (in units of N_e) versus the temperature T for the single-strand human chromosome Ch22 considering 512 nucleotides. The inset in Figure 4.8a depicts the case for the human chromosome Ch22 with 1,024 nucleotides. (b) Chemical potential μ versus the temperature T. After Albuquerque et al. [36]

when $T \to 0$ when compared to the quasiperiodic structures. This probably is due to the fact that, in contrast to a real fractal, human chromosomes present a common compositional structure with two characteristic scales – the largest one corresponding to long, homogeneous DNA segments (the isochores) and the smallest one corresponding to small- and medium-scale genomic elements. The inset presents the case for the human chromosome Ch22 with 1,024 nucleotides.

The chemical potential depicted in Figure 4.8b strongly resembles the one obtained for RS sequence, which means that the energy distribution per particle is very similar in these two cases. This qualitative resemblance is an indication that a real DNA chain can, at least in principle, be modeled through substitutional sequences, like FB and RS. Also, the lack of an oscillatory behavior around a medium value (the spectral or fractal dimension of the system), a common feature presented in the classical Maxwell-Boltzmann calculation shown in Section 4.2, clearly indicates that the statistic considered (Fermi-Dirac), which forbids more than one particle per state (excluding the spin), plays a decisive role on the collective behavior of electrons propagating in real and modeled DNA chains.

4.5 Quantum Statistics: The Double-Strand DNA Structure

Consider now the double-strand DNA structure discussed in Chapter 2. The thermodynamic behavior can be now directly obtained from its electronic DOS depicted in Figure 2.6, following the lines presented in the previous section.

Figure 4.9 depicts a log–log plot of the normalized specific heat spectra at constant volume (in units of the number of non-interacting Fermi particles N_e)

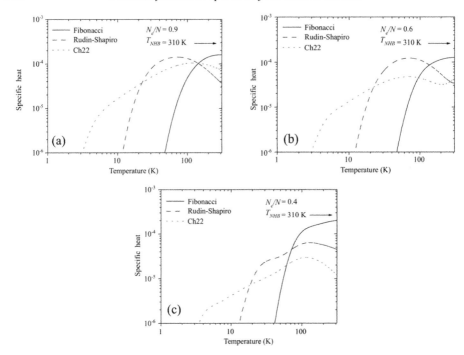

Figure 4.9 The log–log plot of the ESH spectra (in units of N_e) against the temperature T (in Kelvin) for the double-strand DNA chains modeled by a Fibonacci sequence (solid line), Rudin-Shapiro sequence (dashed line), and the double-strand DNA human chromosome Ch22 (dotted line). Three values of the band fillings N_e/N are considered – namely (a) $N_e/N = 0.9$; (b) $N_e/N = 0.6$; (c) $N_e/N = 0.4$. The limit of the temperature scale (right-hand side) represents the normal human body temperature $T_{NHB} = 310$ K. After Albuquerque et al. [36]

versus the temperature T for the double-strand DNA chains modeled by a Fibonacci sequence (solid line), Rudin-Shapiro sequence (dashed line), and the double-strand DNA human chromosome Ch22 (dotted line). The selected band fillings were: $N_e/N = 0.9$ (Figure 4.9a), 0.6 (Figure 4.9b), and 0.4 (Figure 4.9c) – for all sequences studied.

Broadly speaking, Figure 4.9 shows that an increased disorder (Fibonacci $\rightarrow$ Rudin-Shapiro $\rightarrow$ DNA human chromosome 22) gives rise to a structured specific heat C_V, with a different band filling N_e/N and temperature T dependence. Although the existence of a structure in the DNA heat capacity at low temperatures has already been demonstrated experimentally, it was strictly assigned to the difference in hydration and/or structural transitions related to the various DNA conformations. Our theoretical/computational analysis indicates that only the C_V behavior of a more disordered nucleotides arrangement can approach that of the human chromosome 22. This last finding supports the idea of Schrödinger [105], in which he predicted that a gene or perhaps a whole chromosome thread represents an aperiodic solid.

Furthermore, at these band fillings ($N_e/N = n/10, n = 4, 6, 9$), the Fermi energy is located in a dense region of the energy spectrum. Therefore, there are empty states closer to the ground state, and these can be thermally occupied even at very low temperatures. For a periodic infinite crystal, the energy spectrum yields a linear temperature dependence (in the low-temperature regime) of the ESH.

However, although quasiperiodic systems may not being classifiable in the non-linear physics context, they do exhibit a multifractality in their spectra and, instead of the expected linear temperature behavior, the internal energy scales as a power-law $U - U_0 \propto T^{1+\phi}$, and consequently $C_V \propto T^\phi$. In our case, these ϕ exponents are equal to 0.12 (Fibonacci sequence), 0.15 (Rudin-Shapiro sequence), and 0.23 (Ch22 DNA finite segment), no matter the value of the band fillings N_e/N.

This universality class of the specific heat decay exponentially at low temperatures, as far as the band fillings N_e/N are concerned, can be understood on basis of a simple multifractal scale argument. For small thermal excitations, each particle can absorb an energy of the order of T. The number of particles that can be excited corresponds to the number of states in an energy range of the order of T around the Fermi energy.

Hence, the observed specific heat exponents ϕ lie within the range of values of the singularity strength exponent $(\alpha_{min}, \alpha_{max})$ defined by the so-called multifractal $f(\alpha)$ spectrum [106], which in turn strengthens the preceding scaling analysis, unveiling a relationship between the low-temperature power-law decay of the ESH of a molecular system with multifractal spectrum and the underlying energy distribution singularities – disregarding the values of N_e/N and, of course, any finite size effect. This finding may provide a useful tool for the analysis of the low-temperature thermodynamic behavior of more robust protein models modeled by a quasiperiodic system.

There are some other features in the temperature dependence of the specific heat that deserve to be stressed:

(a) At temperatures around the normal human body temperature $T_{NHB} = 310$ K, a striking difference is observed: while the ESH for the Fibonacci sequence shows a peak, regardless the value of the band filling N_e/N, the same does not occur for the RS and Ch22, which have similar behavior.

(b) The RS and Ch22 structures show a peak at the temperature around 100 K with similar profiles.

(c) At low temperatures, the ESH falls linearly to zero, faster for the Fibonacci sequence than for the Rudin-Shapiro sequence, which is faster than the DNA human chromosome 22.

4.6 Nonextensive Thermodynamics

It is by now well established that the powerful standard Boltzmann-Gibbs (BG) statistical mechanics and its associated thermodynamics are valid only when certain conditions are satisfied. The typical situation occurs for microscopic dynamics exhibiting strong chaos – i.e., the positive largest Lyapunov exponent and, consistently, the usual thermodynamic extensivity. This is the scenario that typically occurs for short-range-interacting many-body Hamiltonian systems.

On the other hand, a vast class of natural and artificial systems exists for which the largest Lyapunov exponent vanishes, a situation that is referred to as weak chaos. Weak chaos is typically associated with power-law decay (instead of exponential decay), sensitivity to the initial conditions and relaxations, fractal or multifractal occupation of phase space, and thermodynamic nonextensivity – i.e., phenomena involving long-range interactions (see [107] and the references therein for a recent review).

Taking into account the preceding requirements, a possible generalization of BG statistical mechanics was proposed many years ago by Tsallis [108], on the basis of the following distribution

$$p_q(E) = \left[1 - (1 - q)\beta E\right]^{1/(q-1)}, \qquad (4.17)$$

where $p_q(E)$ is the probability that the system has energy E, $\beta = 1/k_B T$, and q, the entropic index (intimately related to and determined by the microscopic dynamics), characterizes the degree of non-extensivity, a number that is believed to have some relationship to the intrinsic characteristics of the system. When $q \to 1$, that expression recover, the well-known Boltzman-Gibson distribution. The entropy of the system follows [108]:

$$S_q(E) = k_B \sum_{i=1}^{W} \frac{p_i^q}{1 - q}, \qquad (4.18)$$

where W is the total number of possible microstates of the system. Observe that for the $q < 0$ case, care must be taken to exclude all those possibilities whose probability is not strictly positive; otherwise, $S_q(E)$ would diverge. Such care is not necessary for $q > 0$.

This generalization of BG statistical mechanics is usually referred to as nonextensive statistical mechanics. Its denomination nonextensive comes from the following property: if we have two probabilistically independent systems A and B – i.e., $p_{ij}(A + B) = p_i(A)p_j(B)$ – we straightforwardly verify that

$$S_q(A + B) = S_q(A) + S_q(B) + (1 - q)S_q(A)S_q(B). \qquad (4.19)$$

Consequently, $q = 1$, the BG case, corresponds to extensivity, whereas $q < 1$ ($q > 1$) corresponds to superextensivity (subextensivity), where the nonnegativity of q has been taken into account.

Within nonextensive statistical mechanics, many of the previously cited anomalous systems have found an interpretative frame. In particular, several authors have recently reported the investigation of thermodynamical properties associated with systems that exhibit long-range correlated structures with hierarchical or fractal structures.

The first results showed that the specific heat of quasiperiodic spin chains presents logarithmic-periodic oscillations in the low-temperature region. Similar results were found for the specific heat properties associated to hierarchical structures or to the specific heat corresponding to the Heisenberg model with quasiperiodic exchange couplings at some circumstances. All these examples have in common that the corresponding energy spectra exhibit fractal properties.

The energy spectra with fractal structure present an additional interesting feature: quasiperiodic sequences, often used to model quasicrystals, are known to have energy spectra with fractal properties, similar in structure to fractal sets of the Cantor type. This is the reason why the results obtained from studies performed on energy spectra of the Cantor type have been used to explain the properties of the specific heat of Fibonacci sequences, either modeled as one-dimensional (1-D) tight-binding Hamiltonians (as discussed previously in this book) or as superlattices. Also, the properties of the specific heat associated to fractal spectra present similar properties when quantum, fermionic, or bosonic statistics are considered [109, 110].

This section reports the study of the ESH at low temperatures, considering nonextensive distribution of long-range correlated quasiperiodic (Fibonacci and Rudin-Shapiro types) DNA molecules, as well as the real genomic Ch22 DNA sequence.

Defining the internal energy as

$$U \equiv \sum_i^W p_i^q E_i, \tag{4.20}$$

the optimization of the entropy, defined as $S_q = k \ln_q \Omega$ (Ω being the number of accessible states), yields

$$p_i = \frac{e_q(-\beta E_i)}{Z_q}, \tag{4.21}$$

with the partition function written as

$$Z_q \equiv \sum_{j=1}^W e_q(-\beta E_j). \tag{4.22}$$

In the preceding expressions,

$$\ln_q(x) \equiv \frac{x^{1-q} - 1}{1 - q}; \quad \ln_1 x = \ln x, \tag{4.23}$$

$$e_q(x) \equiv [1 + (1 - q)x]^{1/(1-q)}; \quad e_1^x = e^x. \tag{4.24}$$

After a straightforward calculation, and taking the DOS $\rho(E) = 1$, we can write Z_q as

$$Z_q = \frac{1}{\beta(2 - q)} \sum_{i=1,3,\ldots}^{2N-1} \left[e_q(-\beta E_i)^{2-q} - e_q(-\beta E_{i+1})^{2-q} \right], \tag{4.25}$$

where the index i with odd (even) value refers to the bottom (top) of the energy band. The specific heat can then been derived, yielding

$$C_q(T) = 1 + \frac{\beta f_q}{Z_q} - \frac{g_q^2}{Z_q^2}. \tag{4.26}$$

Here,

$$f_q = \sum_{i=1,3,\ldots}^{2N-1} \left[E_i^2 e_q(-\beta E_i)^q - E_{i+1}^2 e_q(-\beta E_{i+1})^q \right], \tag{4.27}$$

$$g_q = \sum_{i=1,3,\ldots}^{2N-1} \left[- E_i e_q(-\beta E_i) + E_{i+1} e_q(-\beta E_{i+1}) \right]. \tag{4.28}$$

We address now the specific heat spectra obtained in Eq. (4.26). In Figure 4.10, we show several profiles of the specific heat $C_q(T)$ for a Fibonacci quasiperiodic

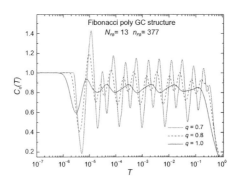

Figure 4.10 Several profiles of the specific heat as a function of the temperature T in logarithmic scale corresponding to a Fibonacci quasiperiodic DNA obtained for different values of the the entropic index q. We have considered energy spectra corresponding to the 13th Fibonacci generation, which means a number of nucleotides $n_{FB} = 377$. After Albuquerque et al. [36]

DNA fragment obtained for different values of the entropic index q. We focus our attention in the low-temperature region, where the specific heat spectra start to present an oscillatory behavior due to a superposition of Schottky anomalies corresponding to the scales of the energy spectrum [111].

Although the number of oscillations observed in the spectra is related to the number of nucleotides n_{FB}, once n_{FB} depends on the hierarchical generation of the Fibonacci sequence (more oscillations appear as n_{FB} increases), we consider the energy spectra corresponding to the 13th Fibonacci generation, which means a number of nucleotides $n_{FB} = 377$. In general, we see that when q decreases, the oscillations become more pronounced, as expected.

As a remark, we may notice that different values of q might correspond to different (multi-) fractal structures. An illustration of such behavior can be seen in nonlinear unimodal maps, where an analytic connection exists between the entropic index q and the multifractality [112]. A discussion of the effect of the fractal properties of the spectrum in the amplitude of the oscillations of the specific heat $C_q(T)$ can be found in Reference [113].

Let us define the separation between consecutive local maxima (or minima) of the specific heat spectra for a given q as

$$\Delta_q = \log_{10} T_{i+1} - \log_{10} T_i, \tag{4.29}$$

where T_i stands for the temperature value for which $C_q(T)$ reaches its ith local maxima or minima. For $q \neq 1$ (the nonextensive case), the period is constant. As q departs from the unity, the distribution of scales presented in the oscillation spectrum becomes wider and better behaved. Therefore, one should expect a higher regularity in the corresponding specific heat. When $q = 1$ (the BG case), the period is not constant, depending on the particular oscillation considered. Furthermore, it leads to a wider distribution of periods due to the presence of a larger diversity of scales.

Similarly, let us define the amplitude of the oscillations as

$$A_{\pm} = \pm C_q(T_{\pm}) \mp <C_q(T)>, \tag{4.30}$$

where $C_q(T_{\pm})$ stands for a local maximum (minima) of $C_q(T)$, and $<C_q(T)>$ is the mean value around which $C_q(T)$ oscillates log-periodically. The mean value $<C_q(T)>$ can be given approximately by the minimum singularity exponent $\alpha_{min} = 0.835$ [102].

Besides, when deterministic fractal spectra are considered (like the Fibonacci one), the oscillations of the specific heat, although perfectly regular and periodic, are nonharmonic, the nonharmonicity being reflected in their amplitudes. For decreasing q, the amplitudes start to be nonconstant and depend on the particular oscillation considered.

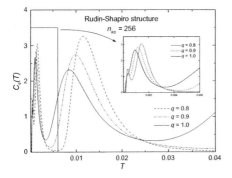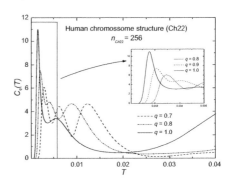

Figure 4.11 Left: Specific heat as a function of the temperature T corresponding to a Rudin-Shapiro quasiperiodic DNA structure obtained for different values of the entropic index q, and considering a number of nucleotides $n_{RS} = 256$. The inset shows a log-plot curve to emphasize the erratic oscillation profile. Right: same as before but for the single-strand Ch22 DNA mode. After Albuquerque et al. [36]

These results suggest that when $q \neq 1$, instead of single values of period and amplitude, we have distributions of these values. On the other hand, for $q = 1$ (the BG case), although the spectrum presents nonperiodic behavior, the amplitudes are almost constant and independent on the particular oscillation considered.

A different scenario appears when one consider the other quasiperiodic structure studied here (i.e., the modeling of the DNA molecule by the RS sequence), which is depicted in Figure 4.11 (left). Similarly to the Fibonacci case, there are oscillations in the region near $T \to 0$ (which are better shown by the inset of the figure, where a log-plot curve is presented). Although these oscillations can be interpreted as Schottky anomalies, as in the Fibonacci case, the log-plot curve $C_q(T)$ versus T in the inset does not show a log-periodic behavior. Instead, it shows an erratic profile, which can be attributed to the more disordered structure of the Rudin-Shapiro sequence. For comparison purposes, we present in Figure 4.11 (right) the specific heat behavior of the human chromosome Ch22 chain. As in the two previous cases, the specific heat spectra show oscillations in the low-temperature region due to Schottky anomalies (better shown by the inset of Figure 4.11), with erratic behavior in their log plots instead of the log-periodic behavior found in the Fibonacci case.

4.7 DNA Denaturation

A remarkable amount of studies published over recent decades (for a review see Reference [114]) shed some light on the dynamics of DNA that affect the thermal properties of its double-helix structure, eventually including its melting at high temperature.

One of the most important theoretical methods assumes a Hamiltonian approach based on the Peyrard-Bishop (PB) model [115], considering that the DNA macromolecule consists of two spring–mass chains in which the transverse stretchings between its complementary base pairs are represented by a 1-D Morse potential simulating the hydrogen bonds. By using a transfer integral technique, they determined the temperature dependence on the interstrand separation that initiates the denaturation.

However, DNA double strands denature quite abruptly with temperature. To take into account the sharp DNA denaturation, an extension to the original PB model was introduced afterward, considering either the intrinsic discreteness of the molecule [116] or the cooperative effects through anharmonic nearest-neighbor stacking interactions (anharmonic PB) [117]. Furthermore, a nonlinear dynamical model for DNA thermal denaturation based on the finite stacking enthalpies used in thermodynamical nearest-neighbor calculations was developed [118]. It was then shown that the finiteness of stacking enthalpies is responsible for the sharpness of calculated melting curves. The specific heat and the entropy were also calculated using three different models – namely (a) the harmonic PB model, (b) the anharmonic PB model, and (c) their own anharmonic model.

In all these works, the authors considered the DNA isolated from any environment. Notwithstanding, the DNA molecule is a highly negatively charged polyelectrolyte both in double-stranded ("native") and in single-stranded ("denatured") forms, displaying considerable sensitivity to ionic surroundings during various structural transitions, as well as when interacting with charged species.

The DNA–solvent interaction was first investigated by using a molecular dynamics simulation within the harmonic PB model. Later on, Weber [119], using a similar approach, also showed that when solvent interactions are included, a sharp denaturation of the DNA double-helix structure is obtained without considering any anharmonic nearest-neighbor stacking interactions. An anharmonic PB model strategy was later developed by modeling the solvation interaction by assuming a Gaussian barrier in the usual on-site Morse potential. For an adenine–thimine (AT) homogeneous chain, he observed that the barrier not only changes the phase transition but also has a great influence in the dynamics of localized excitations. Twisted DNA model was then considered [79] using a path integral formalism to study the thermodynamics of a short fragment of heterogeneous DNA interacting with a stabilizing solvent on the temperature range in which denaturation takes place.

Very recently, a thermal denaturation of the harmonic PB model with an external potential was investigated [120]. The authors used a transfer integral method to obtain the partition function and evaluated the effect of an external potential on the stretching of the hydrogen bonds using a time-independent perturbation method.

They showed that, owing to the external potentials, the denaturation process occurs at lower temperatures.

In this section, we want to investigate the thermal properties of the PB model with a solvent interaction to obtain not only the stretching of the hydrogen bonds between the nucleobases pairs but also some thermodynamical functions, such as the specific heat and the entropy.

The nonlinear model Hamiltonian for a chain of N heterogeneous double-strand nucleobases, assuming a harmonic coupling between the neighboring base pairs and including the effects of the solvent interaction, the so-called Dauxois-Peyrard-Bishop (DPB) model, can be written as [121]

$$H = \sum_{n=1}^{N} \left[H(y_n, y_{n-1}) + V_{Sol}(y_n) \right], \tag{4.31}$$

$$H(y_n, y_{n-1}) = (1/2)m(dy_n/dt)^2 + V_M(y_n) + V_S(y_n, y_{n-1}). \tag{4.32}$$

Here, y_n, the transverse stretching of the nth base pair (AT or GC) in the DNA chain, is the degree of freedom for the 1-D model and measures the relative base-pair separation from the ground-state position. Also, m is the assumed common mass of the nucleobases, and $V_M(y_n)$, the Morse potential modeling the transverse hydrogen bonds interaction linking the two strands, is defined by

$$V_M(y_n) = D_n[\exp(a_n y_n) - 1]^2, \tag{4.33}$$

where D_n (a_n) is the dissociation energy (inverse length) of a base pair at site n. They account for the heterogeneity in the DNA sequence.

The stacking coupling $V_S(y_n, y_{n-1})$ between the intra-neighboring nucleobases along the DNA's strands can be given by [122]

$$V_S(y_n, y_{n-1}) = (kg/2)[(y_n - y_{n-1})^2 + 4y_n y_{n-1} \sin^2(\theta_{n,n-1}/2)], \tag{4.34}$$

$$g(y_n, y_{n-1}) = 1 + \rho \exp[-\alpha(y_n + y_{n-1})]. \tag{4.35}$$

In the preceding equations, k is the backbone harmonic coupling defined by $k = m\omega^2$, with ω being the harmonic phonon frequency. The nonlinear parameters ρ and α are considered to be independent of the type of nucleobases (G, C, A and T) at the sites n and $n \pm 1$. Their widely numerical values used in the calculations are $\rho = 1$ and $\alpha = 0.35\text{Å}^{-1}$.

A relevant feature here is the torsional effects, responsible for the helicoidal DNA structure, which are taking into account the $\sin^2(\theta_{n,n-1}/2)$ term in the stacking potential $V_S(y_n, y_{n-1})$. Here, $\theta_{n,n-1}$, as defined in Chapter 3, is the angle between two adjacent base pairs at the n and $n-1$ sites along the molecule backbone. As this effect was already properly considered in Section 3.3, we consider here only the untwisted term of Eq. (4.34) – i.e., $\theta_{n,n-1} = 0$ (the situation of perfectly parallel neighboring bonds) – in which large fluctuations are possible since the corresponding stacking energy coupling $V_S(y_n, y_{n-1})$ is low on the energy scale set by the Morse potential $V_M(y_n)$ plus the solvent interaction potential $V_{Sol}(y_n)$.

Observe that when $g(y_n, y_{n-1}) = 1$ and $\theta_{n,n-1} = 0$, $V_S(y_n, y_{n-1})$ reduces to the ordinary potential energy, recovering the harmonic PB model. When the molecule is closed, $\alpha y_n \ll 1$ for all site n. On the other hand, when either $\alpha y_n < 1$ or $\alpha y_n > 1$, the corresponding hydrogen bond breaks, and the nucleobases moves out of the stack, reducing the electronic overlap.

Finally, V_{Sol}, the solvent interaction potential, is given by [119]

$$V_{Sol}(y_n) = -f_s D_n[\tanh(y_n/\lambda_s) - 1], \tag{4.36}$$

where $f_s D_n$ is the maximum amplitude of V_{Sol} and λ_s is the solvent interaction factor. This solvent potential combined with the Morse potential results in a single barrier of height (width) of the order of $f_s D_n$ (λ_s).

Figure 4.12 depicts the base-pair interaction potential $U(y_n) = V_M(y_n) + V_{Sol}(y_n)$ (in meV) as a function of the AT base-pair stretching y_n (in Å), considering an AT base-pair whose dissociation energy (inverse length) D_n (a_n) is equal to 30 meV (4.2 Å). The full line represents the Morse potential (with no solvent contribution, i.e., $f_s = 0$), while the dotted line was drawn for a solvent (width) barrier factor $f_s = 0.1$ ($\lambda_s = 6.0$ Å). Observe that the solvent contribution has the effect to enhance (by the term $f_s D_n$) both the energy of the equilibrium configuration and

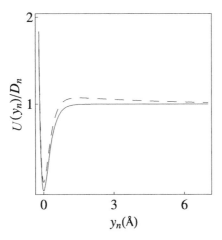

Figure 4.12 The base-pair interaction potential $[U(y_n) = V_M(y_n) + V_{Sol}(y_n)]/D_n$ (in meV) as a function of the AT base-pair stretching y_n (in Å). The full line represents the Morse potential, and the dotted line was drawn for a solvent (width) barrier factor $f_s = 0.1$ ($\lambda_s = 6.0$ Å). Observe that for sufficiently large y_n both potentials will become the same. After Macedo et al. [100]

the height of the barrier below which the base pair (AT) is confined. Besides, the maximum value of the barrier height occurs at the treshold stretching y_n, around which a base pair may open and then become fully dissociate.

In the limit of a large system ($N \to \infty$), the classical thermodynamic properties of the harmonic part of the Hamiltonian, Eq. (4.32), without the solvent interaction term, can be described exactly in terms of a transfer integral as [123]

$$t_{n,n-1}(y_n) = \int \exp\left[- \beta H(y_n, y_{n-1})\right]\psi_n(y_{n-1})dy_{n-1}, \tag{4.37}$$

where $\beta = 1/(k_B T)$, with k_B being the Boltzmann's constant.

Solving Eq. (4.37), we find

$$t_{n,n-1}(y_n) = \exp(-\beta E_n^{(0)})\psi_n(y_n), \tag{4.38}$$

where $E_n^{(0)}$ is given by

$$E_n^{(0)} = \frac{1}{2\beta}\ln\left(\frac{\beta k}{2\pi}\right) + \frac{a_n}{\beta}\left(\frac{D_n}{k}\right)^{1/2} - \frac{a_n^2}{4\beta^2 k}. \tag{4.39}$$

The partition function is then determined by

$$Z_n^{(0)} = (2\pi m/\beta)^{N/2}\exp(-N\beta E_n^{(0)}), \tag{4.40}$$

yielding the following free energy

$$F_n^{(0)} = (-1/\beta)\ln Z_n^{(0)} = N\left[E_n^{(0)} - \frac{1}{2\beta}\ln\left(\frac{2\pi m}{\beta}\right)\right]. \tag{4.41}$$

The mean stretching $< y_n^{(0)} >$ then follows as

$$< y_n^{(0)} > = \int y_n[\psi_n^{(0)}(y_n)]^2 dy_n, \tag{4.42}$$

with $\psi_n^{(0)}$ being the ground-state wave function associated to the ground-state eigenvalue $E_n^{(0)}$.

Other thermodynamic properties of interest are the specific heat at constant volume,

$$C_n^{(0)} = k_B \frac{d^2[T \ln Z_n^{(0)}]}{dT^2} = Nk_B\left(2 + a_n^2 k_B T/2k\right), \tag{4.43}$$

and the entropy,

$$S_n^{(0)} = k_B\frac{d[T \ln Z_n^{(0)}]}{dT} \tag{4.44}$$

$$= Nk_B\left[\ln\left[(m/k)(1/\beta)^2\right] + [a_n^2/2\beta k] + [2 - a_n(D_n/k)^{1/2}]\right].$$

Let us now take into account the influence of the solvent interaction $V_{Sol}(y_n)$, considering it as a perturbation. Making using of a time-independent perturbation theory, one finds that the corrections, up to first order, to the ground state's eigenvalue $E_n^{(0)}$ and normalized eigenfunction $\psi_n^{(0)}$ are given by

$$E_n^{(1)} = E_n^{(0)} + \int_{-\infty}^{+\infty} [\psi_n^{(0)}(y_n)]^2 V_{Sol}(y_n) dy_n, \tag{4.45}$$

$$\psi_n^{(1)} = \psi_n^{(0)} + \sum_i \frac{\langle \psi_i | V_{Sol} | \psi_n^{(0)} \rangle}{(E_n^{(0)} - E_i)} \psi_i. \tag{4.46}$$

The mean stretching $< y_n^{(1)} >$, considering the solvent contributions, is

$$< y_n^{(1)} > = \int y_n [\psi_n^{(1)}(y)]^2 dy_n. \tag{4.47}$$

In a similar way, we can find the corrections to the specific heat and the entropy due to the solvent effects,

$$C_n^{(1)} = C_n^0 - NT \frac{d^2[\Delta E_n^{(1)}]}{dT^2}, \tag{4.48}$$

where $\Delta E_n^{(1)} = E_n^{(1)} - E_n^{(0)}$. As for the entropy, we find

$$S_n^{(1)} = S_n^{(0)} - N \frac{d[\Delta E_n^{(1)}]}{dT}. \tag{4.49}$$

Now let us investigate the profiles of the stretching $< y_n^{(1)} >$, the specific heat $C_n^{(1)}$ and the entropy $S_n^{(1)}$, by considering the effects of the solvent interaction potential V_{Sol}. We shall use four values of the solvent term f_s : $f_s = 0$ (no solvent interaction potential), 0.05, 0.10, and 0.15. In all figures, we considered the physical parameters related with an AT base pair – namely $D_n = 30$ meV, $a_n = 4.2$ Å^{-1}. The backbone harmonic coupling k is taken to be 60 meV/Å^2.

Figure 4.13a shows the variation of the mean stretching $< y_n^{(1)} >$, given in Å, against the temperature. There is a sharp melting transition profile, no matter the value of the solvent term f_s. From there, one can infer that for $f_s = 0$ the melting temperature (i.e., the temperature in which the thermal variation of the mean stretching $< y_n^{(1)} >$ goes to infinity) reaches the value $T_M = 354$ K, an acceptable value for the denaturation temperature for DNA systems, which occurs in the range 353–373 K [114]. Recall that this value for the melting temperature was obtained from the simple harmonic PB model (with no solvent term) through a sharp melting curve without the inclusion of other features such as nonlinearities, helicoidal geometry, bending, etc., which are key factors to understand the DNA thermal behavior. Similar results were obtained in [118] but at higher melting temperatures

($>$400 K), considering the anharmonic PB model with an additional interaction potential. By increasing f_s the melting temperature decreases, going from 354 K for $f_s = 0$ to 336 K for $f_s = 0.15$, indicating that the solvent interaction potential lowers the denaturation temperature. In the calculation of the mean stretching $< y_n^{(1)} >$, we considered terms up to $i = 3$ in the perturbation series defined by Eq. (4.46).

In Figure 4.13b, we show the behavior of the specific heat at constant volume $C_n^{(1)}$ given in units of Nk_B as a function of the temperature for the same values of the solvent term f_s. We have taken into account corrections up to the first order for the eigenvalue $E_n^{(1)}$, as specified by Eq. (4.45). For $f_s = 0$, we observe that the

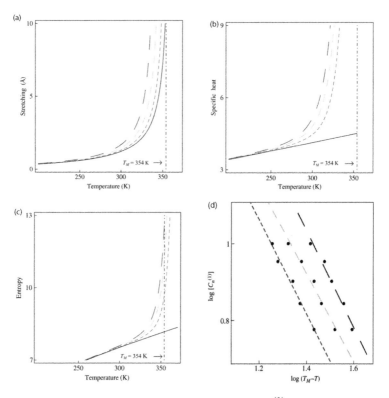

Figure 4.13 (a) The mean hydrogen bond stretching $< y_n^{(1)} >$, (b) the specific heat $C_n^{(1)}$, and (c) the entropy $S_n^{(1)}$, all in units of Nk_B, as a function of the temperature for several values of the solvent term f_s: $f_s = 0$ (straight line), $f_s = 0.05$ (small dashed line), $f_s = 0.10$ (medium dashed line), and $f_s = 0.15$ (large dashed line). In all figures, the vertical dotted line indicates the melting temperature ($T_M = 354$ K) for the harmonic PB model. (d) Log–log plot of the of the specific heat $C_n^{(1)}$, in units of Nk_B, versus ($T_M - T$), in the vicinity of the melting temperature T for the same values of the solvent term f_s – namely $f_s = 0.05$ (small dashed line), $f_s = 0.10$ (medium dashed line), and $f_s = 0.15$ (large dashed line). After Macedo et al. [100]

specific heat increases linearly with the temperature T until it reaches the melting temperature $T_M = 354$ K. Beyond this point, the DNA molecule is denatured. In Reference [116], it was observed that the specific heat increases linearly up to 500 K, decreases abruptly and remains constant as the temperature T increases, while in Reference [118], it exhibits a sharp peak at a melting temperature of around 350 K. By increasing the solvent term f_s, we observe that the specific heat starts to deviate from the linear behavior and tends to diverge for temperatures lower than 350 K.

Similar results are shown in Figure 4.13c for the entropy $S^{(1)}$ in units of Nk_B. From there, one can see that the entropy varies linearly with the temperature T according to the harmonic PB model, presenting asymptotic behavior for temperatures lower than 350 K and different values of f_s. Again, one must remember that the melting temperature T_M imposes an upper limit for the entropy profile. Adding the solvent interaction leads to discontinuity profiles similar to those found in the specific heat case. Anyhow, the first-order or very narrow second-order phase transition of the harmonic PB model with solvent interaction is granted by the divergence of both the specific heat and the entropy. The difference observed in the value of the transition temperature results from the approximate method used.

It is important to notice that, in the regime of weak solvent potential, the mean-field-like behavior of the specific heat (a discontinuous jump) is replaced by a divergent specific heat, whose nature is quite important. Figure 4.13d reports the profile of this divergent dependence through a log–log plot of the specific heat versus $(T_M - T)$ for several values of the solvent term f_s (0.05, 0.10, and 0.15), leading to the power-law behavior profile $C_n^{(1)} \propto (T_M - T)^{-\phi}$, with a power-law exponent $\phi = 1.262$ for $f_s = 0.05$, $\phi = 1.190$ for $f_s = 0.10$, and $\phi = 1.260$ for $f_s = 0.15$, defining a kind of universality class from a statistical physics viewpoint (if we neglect the small variation of the power-law exponents). Soliton physics mechanism can offer a reasonable interpretation of such thermodynamic behavior.

Summing up, in this section we have studied the thermodynamics properties of a finite segment of an heterogeneous DNA molecule around its melting transition temperature. Our model Hamiltonian contains a solvent interaction term that enhances the base pairs' dissociation energy and stabilizes the hydrogen bonds between complementary strands. Assuming an adenine–thymine sequence, we have found a sharp DNA melting transition as a striking common feature for all thermodynamical potentials discussed here. Although the sharpness of the melting transition temperature T_M is far from being established, both theoretically and experimentally (a recent neutron scattering study points to a smooth transition [124]), the fact that the melting profiles show some sudden, even steplike, increments certainly may indicate the overall transition as sharp, in accordance with our theoretical prediction. Recently, the effect of sodium ion concentration on thermal stability was systematically studied for a set of 92 duplex DNA oligomers showing T_M independent of the DNA concentration and length of the oligonucleotide [125], in agreement with previous theoretical findings, in which the normalized transmission coefficient was hardly affected by the polymer length, the type of base pair, or the temperature [126].

4.8 Conclusions

We have discussed in this chapter a theoretical model to study the ESH spectra of single- and double-strand DNA molecules, considering classical and quantum statistics. The nucleotides G, A, C, and T were arranged to form two artificial sequences, namely the Fibonacci and Rudin-Shapiro sequences, both with long-range correlations. We have also considered the sequence of natural DNA as part of the human chromosome Ch 22.

For all structures studied here, the oscillatory profile occurs in the low-temperature region. They also depend on the type and the size of the sequence used to model the DNA molecule. However, the low-temperature behavior strongly depends on the sequence applied in the construction of the system. For the Fibonacci DNA chains, there is an even–odd parity of the specific heat oscillations, while for the Rudin-Shapiro chains, no parity emerges from the specific heat oscillations. Besides, a well-defined log-periodicity was found for the Fibonacci specific heat profile, while erratic log-plot behavior is the main signature of the Rudin-Shapiro case.

Furthermore, a discussion involving nonextensive statistics (beyond Boltzmann-Gibbs), a scenario referred to as weak chaos, was also presented, as well as a description of the DNA denaturation thermodynamic properties.

In the experimental side, heat changes induced by protein unfolding, protein association, ligand binding, and reactions with other biological molecules can now be measured routinely. The two main techniques are differential scanning calorimetry (DSC), which measures sample heat capacity with respect to a reference as a function of temperature, and isothermal titration calorimetry (ITC), which measures the heat uptake/evolution during a titration experiment (for a good description of them, see the review Reference [127]). The third major tool is a thermodynamic calorimetry. Unfortunately, none of these tools are able to probe directly the electronic contribution to the specific heat of biological molecules: they encompass all contributions, including the vibrational one. Nevertheless, the theoretical predictions shown here can be tested experimentally, at least at the important low-temperature regime, considering these tools at the disposal of biophysicists and biochemists.

5

Properties of the DNA/RNA Nucleobases

5.1 Introduction

Only about 10 years after the publication of the landmark work of Watson and Crick on the helical DNA structure, Eley and Spivey [128] proposed a pathway for rapid, 1-D charge separation in double-stranded DNA through $\pi-\pi$ interactions. The dream of using DNA as a nanoscale semiconductor was born, but after all these years, contradictory results were obtained, as highlighted in Chapters 2 and 3. DNA can act as a high conduction wire, a proximity-induced superconductor, a semiconductor, or even an insulator.

DNA and RNA nucleobases are demonstrating to be versatile materials in natural electronics (or bioelectronics) and photonics [129], a very promising field of research in which biological materials are employed to build environment-friendly microelectronic and optoelectronic devices. In comparison to the DNA/RNA molecules, the electron affinities of the nucleobases cover a wider energy interval, which gives additional room for the design of new applications.

In this chapter, we consider the crystalized forms of the four DNA nucleobases (cytosine, guanine, adenine, and thymine) and the RNA nucleobase uracil, employing a dispersion-corrected GGA exchange-correlation functional. Optimized geometries, Kohn-Sham band structures and orbitals, atomic charges, and optical properties (optical absorption and the complex dielectric function) were obtained. Optical absorption measurements were also performed to compare to the theoretical calculations and to estimate the main bandgap of nucleobases in solid state. The Δ-sol scheme of Chan and Ceder [130] is employed to correct the Kohn-Sham bandgaps, sensibly improving the theoretical calculations to compare with the experimental values [131].

As a result, we show that lattice parameters employing the GGA dispersion corrected functional are much closer to the X-ray experimental data available in the literature in comparison to the LDA case and surpass the HOMO-LUMO energy gap estimates of recently published works [132]. As a matter of fact, we have obtained energy gaps for the cytosine, guanine, adenine, and thymine anhydrous crystals, respectively, which agree with the performed optical absorption measured values and are much better than the values previously calculated using DFT-LDA.

The effective masses obtained for the carriers point to guanine as a promising candidate for the development of optoelectronic nanobiodevices in the ultraviolet/visible range, while the complex dielectric function of the nucleobase crystals is strongly anisotropic for distinct polarization states of the incident light. An estimate of the binding energy of the Frenkel exciton in solid-state nucleobases was also accomplished using the time-dependent density functional theory (TDDFT) calculations [133, 134], predicting an exciton binding energy for adenine very close to experiment.

5.2 Experimental Procedure and Computational Details

Powders of all anhydrous nucleobases – namely cytosine 99% (C3506), guanine 99% (G11950), adenine 99% (A8626), thymine 99% (T0376), and uracil 99% (U0750)) – were obtained from sigma-Aldrich without further purification, and X-ray measurements were performed to check their crystalline structure. They were mixed with KBr to form pellets for each nucleobase. The ultraviolet optical absorption spectra of those pellets were measured by transmittance employing a Varian Cary 5000 UV-visible NIR spectrophotometer with solid sample holders. The wavelength range of the measurements varied from 200 to 800 nm (50,000 down to 12,500 cm^{-1}). Background removal was accomplished by comparing the absorption spectrum of a pure KBr pellet, with baseline corrections being made if necessary.

The initial anhydrous crystal structures of the nucleobase crystals used here were extracted from previously published experimental measurements. For anhydrous cytosine, X-ray diffraction [135] reveals an orthorhombic unit cell with number of nucleobases Z = 4 and space group P212121, where the amino nitrogen atom participates in two relatively long hydrogen bonds to carbonyl oxygen atoms from neighbor molecules. It has two symmetrically intercalated stacking planes: (201) and (−201) (see Figure 5.1 C).

The anhydrous guanine crystal (see Figure 5.1 G), on the other hand, was investigated by using synchrotron radiation. It has a monoclinic unit cell with Z = 4 and space group P21/c, and it is held by N–H $\cdots$ N and N–H $\cdots$ O hydrogen bonds, leading to a stacking of molecular sheets along (102) planes interacting through π–π effects.

The crystal structure of anhydrous adenine was recently determined by using single-crystal X-ray diffraction, obtaining a monoclinic unit cell with Z = 8, space

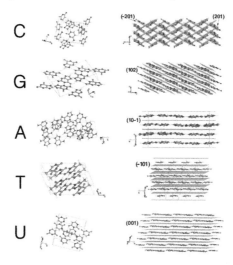

Figure 5.1 Nucleobases cytosine (C), guanine (G), adenine (A), thymine (T), and uracil (U) anhydrous crystal unit cells (left) and stacking planes (right). After da Silva et al. [131].

group P21/c with the molecules connected through N–H $\cdots$ N hydrogen bonds, and stacking along (10-1) planes (see Figure 5.1 A).

Thymine anhydrous crystals are monoclinic with space group P21/c42. Their unit cell has four molecules and exhibits twofold screw axes with molecular units being connected through two N–H $\cdots$ O=C hydrogen bonds creating infinite chains along the b direction (see Figure 5.1 T) and (-101) stacking planes.

Lastly, uracil anhydrous crystals are monoclinic with Z = 4 and space group P21/a as determined X-ray diffraction measurements [136]. This crystal is formed by the (001) plane stacking of molecular layers interacting through N–H $\cdots$ O hydrogen bonds with its pyrimidine ring being similar in size to that of adenine hydrochloride (see Figure 5.1 U).

In order to optimize the geometry of the unit cells, the plane-wave DFT code CASTEP software was used. Two exchange-correlation functionals were adopted in the simulation, as described in Chapter 1: the generalized gradient approximation of Perdew-Burke-Ernzerhof (GGA-PBE) [12] and the local density approximation (LDA) [7]. The semiempirical dispersion correction energy term of Tkatchenko and Scheffler (TS) [21] was added to the first functional to estimate van der Waals noncovalent interactions.

Dispersion-correction schemes can be classified into three main groups, depending on the type of mathematical treatment, namely

(a) Semiclassical description of the dispersion interaction, adding the dispersion energy between atom pairs to the electronic energy, such as in the GGA-TS approach [21], or the D3 proposal of Grimme [19].

(b) Nonlocal density-dependent dispersion corrections, which use the electron density to evaluate the dispersion energy and employ an exchange-correlation functional kernel dependent on two nonseparable electron coordinates (van der Waals density functionals, vdW-DF) [137].

(c) Effective one-electron potentials, which describe London dispersion by local properties using atom-centered external potentials and semilocal density functionals, such as the flexible functional form [18].

Groups (a) and (b), in particular, are very accurate for various noncovalent systems and, when used in combination with the methods of group (c), lead to very high accuracy by taking into account long- and short-range electron correlation.

The core electrons were represented by ultrasoft (LDA functional) and norm-conserving (GGA+TS approach) pseudopotentials while valence electrons with orbital configurations H 1s1, C 2s2 2p2, N 2s2 2p3, and O 2s2 2p4 were taken into account explicitly during the calculations. The Kohn-Sham orbitals were expanded in two plane-wave basis sets with energy cutoffs of 500 eV and 830 eV for the LDA and GGA computations, respectively. These values were selected after performing a convergence study of the per atom forces in a previous set of geometry optimizations.

Self-consistency was assumed to be achieved when the total energy (electron eigenenergy) variation through three successive iterations was smaller than 10^{-6} eV/atom (0.5×10^{-6} eV). Geometry optimization thresholds were set to

(a) maximum force per atom smaller than 0.03 eV/Å,
(b) total energy variation smaller than 10^{-5} eV/atom,
(c) maximum atomic displacement smaller than 0.001 Å,
(d) maximum stress component smaller than 0.05 GPa.

The Broyden-Fletcher-Goldfarb-Shanno (BFGS) iterative method [138] was used to relax both the lattice parameters and internal atomic coordinates. Mulliken's [139] and Hirshfeld's [140] population analysis calculations were also performed on the electron density to evaluate the electric charge partition to each atom in the nucleobases crystals.

As it is well known, DFT bandgap values tend to underestimate the real bandgap of solid-state samples [141]. So, in order to improve our gap predictions, we have also used the Δ-sol scheme [130], which generalizes the delta self-consistent field theory (ΔSCF) for molecules to periodic structures taking into account the dielectric screening properties of electrons, decreasing the mean absolute error of Kohn-Sham gaps by 70% on average. The Δ-sol gaps were used to adjust the calculated optical absorption spectra for comparison to our corresponding experimental data.

The software DMOL3 code74 was employed to find optimal geometries for the individual nucleobases solvated in water to estimate the binding energy of the Frankel exciton in the respective nucleobase solid-state systems. Geometry optimizations for this case were carried out using the GGA+TS approach [21] and a double numerical plus polarization (DNP) basis set obeying the following convergence thresholds: (a) maximum force per atom smaller than 0.002 Ha Å^{-1}; (b) total energy variation smaller than 10^{-5} Ha; (c) maximum displacement per atom less than 0.005 Å. Water solvation was taken into account through the COSMO (COnductor-like Screening MOdel) method [142]. The first optically active excited state of each nucleobase molecule solvated in water was optimized using TDDFT calculations [133, 134] within the GGA+TS approach.

The binding energy of the Frenkel exciton E_b^{exc} can be approximated by considering a single nucleobase molecule solvated in water and taking the difference between its fundamental gap E_g^{mol} and the optical gap E_g^{opt}, i.e.,

$$E_b^{exc} = E_g^{mol} - E_g^{opt}, \tag{5.1}$$

$$E_g^{mol} = E(N+1) + E(N-1) - 2E(N), \tag{5.2}$$

$$E_g^{opt} = E^*(N) - E(N). \tag{5.3}$$

Here, $E(N)$ is the total energy of the relaxed molecule with N electrons, N being the number of electrons in the neutral molecule. Also, $E^*(N)$ is the first optically active excited state total energy, achieved after structural relaxation.

5.3 Crystal Structures

Figure 5.2 depicts the variation of the unit cell total energy as we modify independently each lattice parameter for the nucleobase crystals from their equilibrium values. For cytosine (top left), the curves obtained by varying the lattice parameters a and b are very similar, while the curve for the lattice parameter c is practically flat. This indicates that it is much easier to compress this material along the lattice parameter c in comparison with the lattice parameters a and b, an effect related to the existence of a network of hydrogen bonds (six per molecular unit) approximately aligned to the ab plane of the crystal, while the molecular interactions along the c-axis are mostly noncovalent and, therefore, weaker, leading to a smaller dispersion of the respective total energy curve.

The same observation on the total energy behavior can be made for thymine (middle right) and uracil (bottom), mostly because there is a clear noncovalent stacking of parallel molecular layers along c and four hydrogen bonds per molecule almost aligned to the a and b directions in both phases. In the case of guanine (middle left), the lattice parameters b and c curves are close to each other, and the

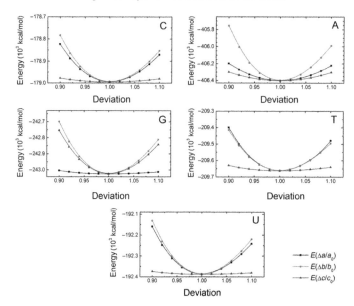

Figure 5.2 Nucleobases cytosine (C), guanine (G), adenine (A), thymine (T), and uracil (U) anhydrous crystals: variation of the total unit cell energy with changes in the lattice parameters. The black (light gray, dark gray) squares (circles, triangles) depict the calculated variation of the energy as a function of the relative lattice parameter variation $\Delta a/a_0$ ($\Delta b/b_0, \Delta c/c_0$). Solid curves interpolate the calculated data as a visual aid. After da Silva et al. [131]

curve for a is almost flat as a consequence of its eight hydrogen bonds per molecule disposed along the bc plane and the action of dispersive forces only between its molecular layers, which are almost perpendicular to the a-axis.

Lastly, for adenine (top right), the three curves – for a, b, and c – are more distinct than for the other nucleobases, the a and c curves being closer to each other. The curve for the lattice parameter c displays the smallest dispersion about the minimum of the energy, but apparently not so small as the smallest dispersion plots found for the other nucleobase crystals, indicating stronger dispersive interactions parallel to the c direction for this system. The curvature difference between the a and b curves is probably due to the existence of more hydrogen bonds (six per molecule) nearly aligned to the b-axis than to the a-axis, leading to a larger curvature for the total energy when the b lattice parameter is changed.

Under the energetic point of view, the binding energy per molecule is a measure of the strength of the interactions between the nucleobases in the solid phase. It can be evaluated as follows: let E_{cell} be the total energy of the unit cell, Z be the number of molecular units contained in it, and E_{mol} be the total energy of an isolated nucleobase molecule. The binding energy per molecule E_{bpm} is given by

$$E_{bpm} = E_{cell}/Z - E_{mol}. \tag{5.4}$$

The GGA+TS calculations predict the following values for E_{bpm} (in eV): -1.855 for cytosine, -2.510 for guanine, -1.837 for adenine, -1.615 for thymine, and -1.598 for uracil, with the following sequence of binding energy strengths: $G > C > A > T > U$. This result is in agreement with the data of our previous work [143] using the LDA exchange-correlation functional for the DNA nucleobases only but contrasts with the simulations of Šponer et al. [144] for guanine, adenine, cytosine, and uracil stacked dimers using the second-order Moller-Plesset approach, MP2 ($G > A > C > U$).

However, as our calculations predict a very small difference between cytosine and adenine binding energies (18 meV or 0.41 kcal/mol), while the MP2 simulations predict a difference of the same order (22.5 meV or 0.50 kcal/mol) one can be certain that the binding interactions between cytosine and adenine units are very similar in strength. Besides, the thermal stability of the nucleobase thin films reveals the sequence, $G > C > A > T > U$, in agreement with the order of binding energies we found in our LDA and GGA+TS simulations, notwithstanding the fact that they are obtained for a bulk and not a small periodic structure. As a matter of fact, even if it is preferable to carry out thin-film DFT simulations for direct comparison to these systems, their thicknesses, typically of the order of 100 nm, would demand a very high computational cost.

5.4 Electronic Band Structure

Figure 5.3 depicts the Kohn-Sham band structures near the main bandgap for the nucleobase crystals. The same Brillouin zone path was used for monoclinic guanine, adenine, thymine, and uracil, following the high symmetry points $\Gamma(0,0,0) \rightarrow$ $Y(0,1/2,0) \rightarrow A(-1/2,1/2,0) \rightarrow \Gamma \rightarrow C(0,1/2,1/2) \rightarrow E(-1/2,1/2,1/2) \rightarrow \Gamma \rightarrow$ $Z(0,0,1/2) \rightarrow D(-1/2,0,1/2) \rightarrow \Gamma \rightarrow B(-1/2,0,0)$.

For orthorhombic cytosine, the reciprocal space path sequence of points was $\Gamma(0,0,0) \rightarrow Z(0,0,1/2) \rightarrow T(-1/2,0,1/2) \rightarrow \Gamma \rightarrow U(0,1/2,1/2) \rightarrow R(-1/2,1/2,1/2)$ $\rightarrow \Gamma \rightarrow X(0,1/2,0) \rightarrow S(-1/2,1/2,0) \rightarrow \Gamma \rightarrow Y(-1/2,0,0)$. Of the five nucleobase crystals, two have a direct gap, namely cytosine ($\Gamma \rightarrow \Gamma$, 3.47 eV) and guanine ($B \rightarrow B$, 2.86 eV), while adenine, thymine, and uracil have indirect gaps, the first being 3.04 eV ($\alpha_A \rightarrow \Gamma$), the second being 3.50 eV ($\alpha_T \rightarrow \beta_T$), and the last being 3.45 eV ($\alpha_U \rightarrow \beta_U$). The α and β points are used as labels that indicate the valence band maximum and conduction band minimum, respectively, of the subscripted nucleobase when these points do not coincide with a high symmetry point in the first Brilllouin zone.

The origin of the different gap types cannot be assigned to the crystal structure only, as, of the four monoclinic nucleobase crystals with same space group

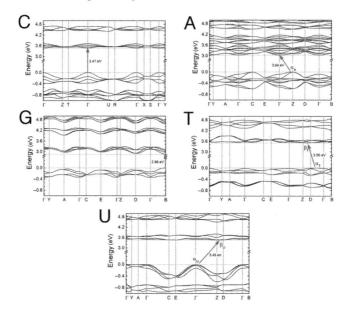

Figure 5.3 Kohn-Sham band structures near the main bandgap of the nucleobase anhydrous crystals: cytosine (C), guanine (G), adenine (A), thymine (T), and uracil (U). The arrows indicate the most important transitions and corresponding energy gap values. After da Silva et al. [131]

symmetry (adenine, guanine, thymine, and uracil), three have indirect gap and one (guanine) has a direct main electronic transition, while the orthorhombic crystal (cytosine) has a direct main bandgap. It also seems that the presence or absence of oxygen atoms in the nucleobase does not play a role to determine the nature of the gap (four nucleobases are oxygenated, and two of their crystals have indirect gap, while the other two have a direct gap). No correlation was found for the gap type due to the presence of pyrimidine and imidazole rings either, which implies that the nature of the main bandgap in these materials must result from a more complex interplay between molecular properties and intermolecular couplings.

To better understand the electronic states at the valence and conduction band extrema, we have calculated the Kohn-Sham molecular orbitals for the isolated molecules, considering an equivalent supercell with spacing larger than 10 Å between molecular images. Our approach is based on the GGA+TS exchange-correlation functional. Atom labels used in our analysis follow the description shown in Figure 5.4.

The HOMO and LUMO orbitals thus obtained are shown in Figure 5.5, together with the respective HOMO-LUMO gaps for each nucleobase. For cytosine, the HOMO-LUMO gap is the smallest among the five nucleobases, namely 3.31 eV, with the HOMO orbital consisting mainly in the superposition of O1, N1, and N2

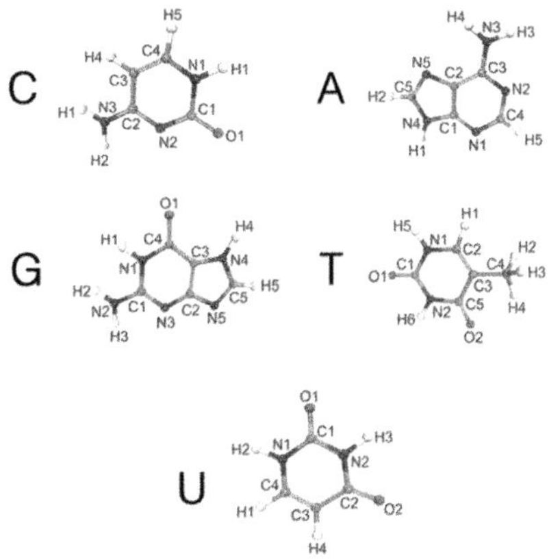

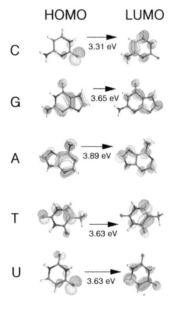

Figure 5.4 Labels used to identify the atoms of cytosine (C), guanine (G), adenine (A), thymine (T), and uracil (U). Carbon (C), hydrogen (H), nitrogen (N), and oxygen (O) atoms are depicted in dark gray, white, black, and light gray, respectively. After da Silva et al. [131]

HOMO   LUMO

C   3.31 eV

G   3.65 eV

A   3.89 eV

T   3.63 eV

U   3.63 eV

Figure 5.5 HOMO and LUMO molecular orbitals of cytosine (C), guanine (G), adenine (A), thymine (T), and uracil (U). The HOMO-LUMO gaps are indicated. Also, the shaded gray areas identify positive (dark gray) and negative (light gray) phases of the wavefunction, respectively. After da Silva et al. [131]

2p states, while the LUMO involves N3 2p and a C2-C3 π orbital. For guanine, we have 3.65 eV for the HOMO-LUMO gap, O1, N2 2p, C2-C3-C4, C1-N3, and C5-N4 π states contributing strongly to the HOMO and O1 2p, and C3-C4, C2-N5 π orbitals forming the LUMO.

In the case of adenine, one can observe the largest HOMO-LUMO gap of all nucleobases, namely 3.89 eV, with the HOMO orbital consisting mainly in the superposition of π orbitals along the hexagonal ring (C1–C2–C3–N2 and C4–N1), and a N3 2p orbital. For the LUMO, the π orbitals at the hexagonal ring are replaced by 2p contributions perpendicular to the ring plane at all atoms, except C1 and C2. Thymine and uracil have the same HOMO-LUMO gap, namely 3.63 eV, and somehow similar contributions to the frontier orbitals: O1, O2, N, and C 2p states contribute to the HOMO, and C–C π orbitals contribute to the LUMO.

Our results compare well, except for the relative magnitudes of guanine and adenine, with those presented by Gomez et al. [129], who predicted HOMO-LUMO gaps (in eV) of 3.6, 3.9, 3.8, 3.7, and 3.7 for cytosine, guanine, adenine, thymine, and uracil, respectively, following the sequence of decreasing values $G > A > T \approx U > C$.

The Kohn-Sham frontier states for the nucleobase crystals at the Γ point, shown in Figure 5.6 (HOVB is highest occupied valence band and LUCB is lowest unoccupied conduction band states at $k = 0$), are very similar to the pure molecular orbitals, pointing to a strong noncovalent character in the interaction between neighbor

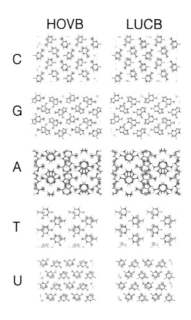

Figure 5.6 Highest occupied valence band (HOVB) and lowest unoccupied conduction band (LUCB) Kohn-Sham states at $k = 0$ (Γ point of the Brillouin zone) of cytosine (C), guanine (G), adenine (A), thymine (T), and uracil (U) anhydrous crystals. The shaded gray areas identify positive (dark gray) and negative (light gray) phases of the wave function, respectively. After da Silva et al. [131]

molecular units, with small wave-function overlap. The solid-state gaps tend to be smaller than the molecular gaps, with the exception of the cytosine crystal, which has a direct gap that is 0.15 eV larger than its isolated molecule.

The electric charge of each atom in a crystal unit cell can be estimated by using charge partition techniques such as Mulliken population analysis (MPA) and Hirshfeld population analysis (HPA). Hirshfeld population analysis, in particular, is capable of better predicting Fukui function indices [145], which are used to estimate the reactivity trends of a molecule better than traditional Mulliken charge partitions. It is also able to minimize the loss of information due to the formation of chemical bonds, notwithstanding its tendency to underestimate charge values. This last limitation, however, can be removed by using an iterative approach successfully implemented for the solid state [146]. Here, we discuss only the GGA+TS results for the Hirshfeld charges of each nucleobase molecule in the crystal shown in Figure 5.7.

In all nucleobase crystals, except for adenine, there are concentrations of negative charge around the oxygen atoms, with cytosine exhibiting the largest absolute value (in electron charge units, −0.25) and uracil exhibiting the smallest (−0.20). The nitrogen atoms also tend to be negatively charged, with the N2 atom of the

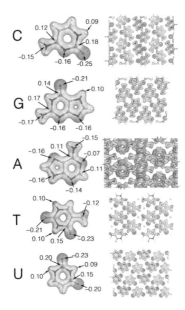

Figure 5.7 Constant electron density isosurfaces for cytosine (C), guanine (G), adenine (A), thymine (T), and uracil (U) molecular units (left) in the solid state (right), with their electrostatic potential projected onto each isosurface increasing from the light gray to the dark gray shadows. Hirshfeld charges for some atoms are indicated. After da Silva et al. [131]

guanine crystal reaching -0.17. Carbon and hydrogen atoms, on the other hand, have a positive charge. For example, C1 has a charge of 0.18 in cytosine and 0.17 in guanine, C4 in adenine has a charge of 0.11, C5 (C2) in thymine (uracil) has a charge of 0.15.

Hydrogen atoms, on the other hand, have charges of about 0.10 in all nucleobase crystals. The charge distribution shown in Figure 5.7 suggests that cytosine, thymine, and uracil have a higher dipole moment than guanine and adenine. As guanine and thymine are the only crystals with a direct gap, the first less polarized than the second, it seems that one cannot establish correlations between the degree of charge separation in the molecular units of the crystalline systems and gap types. However, the DFT bandgaps of cytosine, thymine, and uracil are very close and about 0.5 eV larger than the gaps of adenine and guanine, suggesting that the degree of charge polarization is related to the magnitude of the bandgap (and, consequently, the onset of optical absorption). A possible mechanism by which this could happen is that the polarization increase produced by optical excitation in the crystal demands a larger amount of energy in a system already in a high-polarization state if the excitation involves the creation of a localized electron-hole pair in a spatial configuration energetically disfavored by the polarization geometry of the crystal. Finally, the charge distribution inside a unit cell (right side of Figure 5.7) reveals how electrons are spread in such a way as to maximize the electrostatic binding interaction (minimizing total energy), with the negatively charged groups of one molecule being close to positively charged groups of its neighbors.

Figure 5.8 shows the per-atom electronic partial density of states for the nucleobase crystals between -1.0 and 4.9 eV, spanning the uppermost valence bands and the lowest-energy conduction bands. One can note that, for all nucleobases, the top of the valence band has a significant contribution from the nitrogen atoms, especially adenine, originating from N 2p levels. O 2p states have a similar (smaller) contribution for thymine and uracil (cytosine and guanine), while C 2p states have a strong role in the valence states for cytosine, adenine, thymine, and uracil. The electronic states at the bottom of the conduction band are also dominated by C 2p contributions for cytosine, adenine, thymine, and uracil, but O 2p levels are not relevant in the case of cytosine although they contribute to the conduction band for thymine and uracil in a proportion similar to N 2p orbitals.

5.5 Effective Masses

Band structure plots at the valence and conduction extrema along directions perpendicular ($\perp$) and parallel ($\parallel$) to the molecular stacking planes of the solid-state nucleobases are shown in Figure 5.9. The electron energy dispersion curve at the

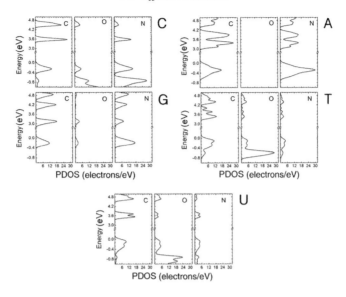

Figure 5.8 Electron per atom densities of states for the nucleobases cytosine (C), guanine (G), adenine (A), thymine (T), and uracil (U) anhydrous crystals near the main bandgap. Contributions from carbon (C), oxygen (O), and nitrogen (N) are also shown. After da Silva et al. [131]

valence band maximum and conduction band minimum can be fitted to a parabola in order to estimate the carrier effective masses, relevant for charge transport. In part, our motivation to evaluate them here is the lack of understanding of the conductivity behavior of DNA strands, which are formed from the stacking of distinct nucleobases.

The perpendicular mass is measured along directions perpendicular to the stacking planes of each nucleobase crystal. The parallel mass, on the other hand, was calculated along some selected in-plane hydrogen bonds for each system. For cytosine, which has a direct gap at the Γ point ($k = 0$), effective masses parallel to its molecular planes are smaller than along the perpendicular direction, while the electron effective masses are larger and more anisotropic than the hole masses ($m_h = 4.8$ and 3.3 for the perpendicular ($\perp$) and parallel ($\parallel$) cases, respectively; the free electron has mass $m = 1.0$). In the case of guanine, with its direct main transition B $\rightarrow$ B, the perpendicular electron and hole effective masses are almost equal to each other ($m_{e,\perp} = 5.2$, $m_{h,\perp} = 5.3$), as occurred in [143]. Along the parallel direction, the same behavior can be observed but with smaller effective mass values.

Adenine, thymine, and uracil have indirect gaps, so their band extrema occur at different points in reciprocal space. For adenine, both electron effective masses (perpendicular and parallel) are very large and anisotropic, with $m_{e,\perp} = 31.4$ and

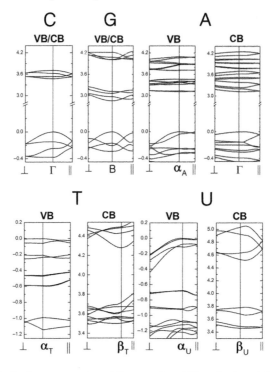

Figure 5.9 Nucleobases cytosine (C), guanine (G), adenine (A), thymine (T), and uracil (U): Kohn-Sham band structures near the smallest bandgaps along directions perpendicular (⊥) and parallel (∥) to the stacking planes. After da Silva et al. [131]

$m_{e,\parallel} = 10.6$, values respectively larger and smaller than the corresponding ones in [143], while its hole is relatively light, with m_h varying from 3.2 (⊥) up to 5.6 (∥); previous LDA calculations predicted a very large parallel hole effective mass (greater than 40).

In the case of thymine, the perpendicular hole effective mass is very large (30, LDA value 15), and the parallel mass is much smaller (2.0), while the electron effective masses are not very large (6.1 and 3.4 for the perpendicular and parallel cases, respectively), very close to the LDA estimate of [143].

Finally, for uracil, the carriers moving parallel to the stacking plane are relatively heavy (with an electron mass of 14.4 and a hole mass of 10.4), while those moving along the perpendicular direction have reduced effective masses in comparison (3.8 for $m_{h,\perp}$ and 6.1 for $m_{e,\perp}$). These results suggest that the guanine crystal, with an experimental direct gap of 3.60 eV (from our optical absorption measurements) and effective masses varying between 2.5 and 5.3, is more promising as a semiconductor material with potential optoelectronic applications, followed by the cytosine crystal (with a direct gap of 4.05 eV).

This contrasts with the conventional picture using molecular affinity trends, which points to guanine as an electron blocker. Adenine, on the other hand, approaches insulator behavior with its large electron effective masses, while thymine and uracil can exhibit semiconducting characteristics due to the presence of not-so-heavy electrons and holes. As a matter of fact, there is a report on the use of adenine and thymine in organic light-emitting diodes, the former as an electron blocking layer and the latter as a hole transport layer [132], but no work has been published on the potential of guanine as an optically active matrix in the ultraviolet(UV)/visible range.

5.6 Absorption Spectra

Figure 5.10 plots our experimental and theoretical optical absorption curves of the five nucleobase anhydrous crystals. To estimate the optical bandgap from the measured data, we have taken into account the electronic gap type (whether direct

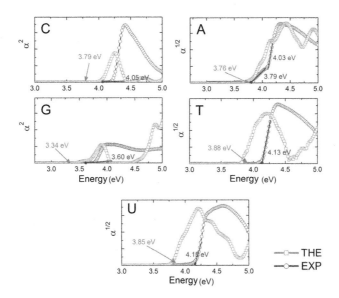

Figure 5.10 Optical absorption of cytosine (C), guanine (G), adenine (A), thymine (T), and uracil (U) anhydrous crystals near the main bandgap. For the direct gap materials, the square of the optical absorption is plotted, while for the indirect ones, the square root of the optical absorption is shown. The squares gray curves (THE) are the absorption calculated theoretically, here using the GGA+TS approach with the Δ-sol correction. The circles gray curves (EXP) were obtained from our experimental measurements. The black small straight lines interpolate the curves near the absorption onsets and reveal the gap estimates for the experimental data. Observe that for the case of adenine, there are two gaps due to the Frenkel exciton. After da Silva et al. [131]

or indirect) predicted by the computational simulations. For a direct (indirect) electronic gap material, the onset of the optical absorption α is proportional to the square root (square) of the photon energy, so in this case, we have plotted the square (square root) of α and performed a linear fit to find the main bandgap (circles gray lines in Figure 5.10).

In the cases of the direct gap characteristics depicted by cytosine and guanine, the absorption onset used is barely visible, so we instead prefer to consider the theoretical curves to estimate its approximate location, obtaining 4.05 eV for the former and 3.60 eV for the latter. For adenine, thymine, and uracil, their optical gaps were estimated to be 4.03, 4.13, and 4.15 eV, respectively. The theoretical curves, on the other hand, were adjusted to the Δ-sol bandgaps by a rigid shift. One can see that they are in reasonable agreement with our experimental data.

The electronic transitions from the valence band to the conduction band contributing to the onset of the optical absorption can be related to molecular orbital transitions: for cytosine and adenine, with density of states near the bandgap originating mostly from N and C 2p orbitals (see Figure 5.8), we can assign a $\pi \rightarrow \pi^*$ character.

In the case of guanine, thymine, and uracil, a strong contribution from non-bonding O 2p electrons in the valence band is present, corresponding to a $\pi \rightarrow \pi^*$ molecular transition to the absorption onset. These results must be contrasted with the absorption maximum mostly due to $\pi \rightarrow \pi^*$ transitions observed in the molecular nucleobases solvated in water.

Our optical absorption measurements compare well with those of [129] for the nucleobase films, with exhibit onsets at 4.0 eV for cytosine, 3.6 eV for guanine, 3.8 eV for adenine, 4.1 eV for thymine, and 4.2 eV for uracil.

5.6.1 Frenkel Exciton Binding Energy

Estimates of the Frenkel excitonic binding energies (E_b^{exc}), as well as fundamental gaps (E_g^{mol}), optical gaps (E_g^{opt}), and relative oscillator strengths (f) for the nucleobase molecules solvated in water are shown in Table 5.1 by using TDDFT. The fundamental gaps range from 4.14 eV (guanine) to 4.62 eV (cytosine), increasing following the sequence G < U < T < A < C, while the optical gaps start at 3.61 eV (guanine) reaching up to 4.39 eV (adenine) in the sequence of increasing energy values, G < U < T < C < A.

The Frenkel exciton binding energy estimate, on the other hand, is the largest for thymine (0.79 eV) and the smallest for adenine (0.22 eV), increasing in the order A < G < U < C < T. Oscillator strengths are in general very small, except for adenine. This is an interesting result, as only the experimental optical absorption of the adenine crystal (Figure 5.10, top right) exhibits two absorption onsets, one starting at about 3.79 eV and another at about 4.05 eV.

Table 5.1 *Molecular gaps, Frenkel exciton energies, and oscillator strengths from the nucleobases solvated in water. After da Silva et al. [131].*

Nucleobase	E_g^{mol} (eV)	E_g^{opt} (eV)	E_b^{exc} (eV)	f
C	4.62	3.87	0.75	10^{-2}
G	4.14	3.61	0.53	10^{-6}
A	4.61	4.39	0.22	1
T	4.55	3.76	0.79	10^{-4}
U	4.35	3.70	0.65	10^{-5}

Considering the first onset as due to the Frenkel exciton, one can estimate an exciton binding energy of about 0.26 eV, and even a bit smaller if we linearly interpolate the second absorption onset, the optical gap is 4.03 eV, leading to an exciton binding energy of 0.24 eV, which is relatively close to the theoretical Frenkel exciton binding energy prediction of 0.22 eV for adenine in Table 5.1.

5.6.2 Dielectric Function

The complex dielectric function $\epsilon = \epsilon_1 + \epsilon_2$, as a function of the incident photon, was calculated for the nucleobase crystals in the cases of incident polarized light and light interacting with a simulated polycrystalline sample (POLY).

The real part of the dielectric function ϵ_1 is obtained through the Kramers-Kronig transform due to causality:

$$\epsilon_1(\omega) = \frac{1}{\pi} P \left(\int_{-\infty}^{+\infty} \frac{\epsilon_2(\omega')}{\omega' - \omega} d\omega' \right), \tag{5.5}$$

where P is the Cauchy principal value of the integral.

The imaginary part ϵ_2 of the dielectric function is calculated by using the Fermi's golden rule for time-dependent perturbations, i.e.,

$$\epsilon_2(\omega) = \frac{2e^2\pi}{\epsilon_0 V} \sum_{k,v,c} |< \psi_k^c \mid \vec{u} \cdot \vec{r} \mid \psi_k^v >|^2 \, \delta(E_k^c - E_k^v - \hbar\omega). \tag{5.6}$$

Here, k, v, c, and $\vec{u}$ represent, in this order, the DFT electronic wavevector in the reciprocal space: the valence band, the conduction band, and the vector defining the polarization of the incident electric field. E_k^c (E_k^v) stands for the energy of the electron with wavevector k at the conduction (valence) band, ω is the photon angular frequency, and V is the unit cell volume.

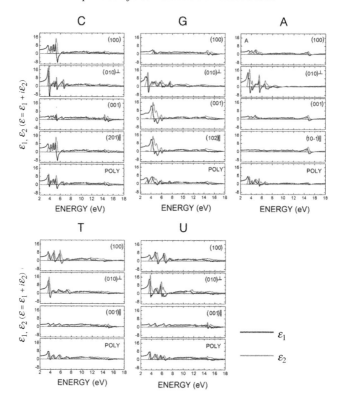

Figure 5.11 Complex dielectric function $\epsilon = \epsilon_1 + \epsilon_2$ of cytosine (C), guanine (G), adenine (A), thymine (T), and uracil (U) anhydrous crystals. The real (imaginary) part ϵ_1 (ϵ_2) is in black (gray). Dielectric function curves for light polarized along some selected crystalline planes and for a polycrystalline sample (POLY) were obtained, as well as curves for light polarized along the crystalline planes perpendicular ($\perp$) and parallel ($\parallel$) to the molecular stackings. After da Silva et al. [131]

Figure 5.11 shows the calculated complex dielectric function considering light polarized along the 100, 010, and 001 directions without the Δ-sol gap correction. For cytosine, guanine, and adenine, the directions 201, 102, and 10-1, respectively, were also taken into account as they define the crystalline molecular stacking planes. For thymine and uracil, the 001 polarization corresponds to the planar stacking.

A quick look shows that the complex dielectric function is very anisotropic, with well-structured and intense maxima and minima for the 010 direction, which is perpendicular to the crystalline planes in all nucleobase systems, but very attenuated along other polarization states (001 for cytosine, adenine, thymine, and uracil and 100 for guanine). For adenine, as a matter of fact, there is a very pronounced attenuation along the 100, 001, and 10-1 directions in comparison with 010.

The values of $\epsilon_1(\omega = 0)$ for the nucleobase polycrystals are 3.03 (cytosine), 3.12 (guanine), 2.65 (adenine), 2.68 (thymine), and 2.87 (uracil). For the DNA nucleobases, these results are in good agreement with the experimental measurements of Silaghi et al. [147]. The maximum values of ϵ_1 along the 010 direction occur at 3.68 eV ($\epsilon_1 = 17.6$), 2.94 eV ($\epsilon_1 = 9.36$), 3.35 eV ($\epsilon_1 = 13.9$), 3.61 eV ($\epsilon_1 = 12.5$), and 3.61 eV ($\epsilon_1 = 14.1$) for cytosine, guanine, adenine, thymine, and uracil, respectively. After these maxima, with increasing energy, ϵ_1 decreases sharply, becoming negative at 3.80 eV (cytosine), 3.40 eV (guanine), 3.67 eV (adenine), 3.78 eV (thymine), and 3.80 eV (uracil). The complex part of the dielectric function ϵ_2, on the other hand, is proportional to the optical absorption already discussed in the beginning of this section.

5.7 Conclusions

In this chapter, we have presented the results for optical absorption measurements and DFT GGA+TS (dispersion-corrected) calculations for the anhydrous crystals of the five nucleobases of the DNA/RNA molecule. The structural, electronic, and optoelectronic properties of orthorhombic cytosine and monoclinic guanine, adenine, thymine, and uracil were consistently evaluated to and their main Kohn-Sham bandgaps were corrected using the Δ-sol scheme. The GGA+TS lattice parameters have shown a large improvement over the LDA data in comparison to the experimental data, the mean error decreasing from about 12% down to less than 2% as we switch from the LDA to the GGA+TS exchange-correlation functional.

The GGA+TS computations have also allowed us to estimate the binding energy per molecule in the nucleobase systems, with the following sequence of stability (in decreasing binding strength): G > C > A > T > U, in agreement with the LDA data and the experimental thermal stability of nucleobase films [129], but differing with respect to MP2 results (A > C) [144]. However, the GGA+TS absolute difference of binding energies for cytosine and adenine binding energies is as small (0.41 kcal/mol) as it is in the MP2 simulations (0.50 kcal/mol).

For the electronic band structure, the inclusion of dispersion effects has not changed the gap type of the nucleobase crystals in comparison to the LDA predictions, with cytosine and guanine exhibiting direct bandgaps (3.79 and 3.34 eV, respectively, with Δ-sol correction) while adenine, thymine, and uracil have indirect bandgaps (Δ-sol values: 3.76, 3.88, and 3.85 eV, in the same order).

A comparison between the bandgaps estimated from optical absorption and the theoretical values shows that Δ-sol figures improve the gap by almost 70%. The agreement between our calculated and experimental optical absorption curves is reasonable, especially for adenine, notwithstanding the fact that DFT is a ground-state theory. The HOMO-LUMO energy gaps estimated using the GGA+TS approach are in good agreement with previous reports, except for the relative ordering of guanine and adenine.

We have also estimated the binding energy of the Frenkel exciton in the nucleobase crystals using TDDFT calculations performed for water-solvated molecules as an approximation, obtaining the largest exciton binding energy for thymine (0.79 eV), but with a very small oscillator strength f. The nucleobase with the largest oscillator strength, in contrast, was adenine, with corresponding exciton binding energy of 0.22 eV, very close to the evaluation found after analysis of its measured optical absorption spectrum (0.24 eV).

Charge population analysis, on the other hand, reveals that electrostatic effects are essential for the stabilization of the nucleobase crystals. Estimated effective masses are, in general, anisotropic, with large (relatively small) electron (hole) effective masses for adenine. The guanine crystal, with lighter carriers and a direct gap of 3.60 eV is a promising semiconductor for optoelectronic devices, followed by cytosine. The other nucleobase crystals may have semiconductor features depending on the carrier type.

Finally, the complex dielectric function for polarized incident light reveals a strong optical anisotropy, with significant features being observed along the 010 direction, normal to the crystalline molecular planes for all nucleobases, and a strong attenuation for parallel polarization. The results described here can be helpful to provide a deeper understanding of potential biodevices built using nucleobase crystals in their design.

6

Molecular Electronics

6.1 Introduction

The theoretical treatment of biologically active molecules forming the proteins is a mature, challenging, and active research area. Driven by important advances of quantum mechanical methods and increasing hardware performance, and mirrored by an astonishing number of computational studies devoted to their structural and conformational behavior, this subject has become an exciting field of research. The main difficulties in this area involve taking into account the topology of the structures, as well as the high level of precision required to characterize them, which often present a high degree of complexity, and must, therefore, be approached with approximate methods. In the protein domain, the helical motif is a ubiquitous conformation of amino acids in their structure (helices constitute 20%–30% of the secondary protein structure), and its formation is a fundamental step of the protein folding process.

To date, many molecules with important electronic properties have been identified, and more with desired properties are being synthesized in chemistry laboratories. In addition to their electronic properties, many molecules possess rich optical, magnetic, thermoelectric, electromechanical, and molecular recognition properties, which may lead to new devices that are not possible by using conventional materials or approaches. Despite many unsolved issues, important and solid advances have been made so far, mainly during the twenty-first century (see [148] and the references therein).

All these advances were achieved due to the ability to integrate human-made solid-state nanostructures with the biomolecular systems, opening up revolutionary means for the electrical and optical characterization of new devices, with great impact on a wide variety of medical and biological applications. Accordingly, nowadays, nanostructure-biomolecule complexes – including quantum dots (QDs), carbon nanotubes (CNTs), oligopeptides and proteins – have been designed, fabricated, modeled, and characterized.

The idea of replacing electronic components with molecules is not new. In 1974, Aviram and Ratner [149] were the first to suggest an organic molecular system showing current rectification, composed by a donor and an acceptor group attached by a carbon bridge with single bonds. Since then, several works have been done in the fields of molecular electronics and nanoelectronics [150].

Although molecular electronics has been proposed as an alternative to silicon in post-CMOS (complementary metal–oxide–semiconductor) devices, molecules with unique functions may have applications that are complementary to the silicon-based microelectronics. The main advantages of such devices are their lower current and power operation, leading to a cheaper and simpler fabrication, together with versatility in usage, mechanical flexibility, and the possibility to be incorporated into flexible plastic structures. The main disadvantages, however, are their short lifespan due to degradation, as well as their reactivity with water and other substances, demanding the design of effective packaging systems.

Encompassing all the properties summarized earlier, the so-called molecular electronics technologies of today offer a viable alternative to overcome the difficulties arising from the biological materials, as well as those associated with the continuing shrinking of electronic devices in the silicon-based technology. Usually, molecular electronics take full advantage of the unique properties of their molecular components, leading to applications that may be complementary to conventional electronics, instead of trying to replace them. This complementarity, indeed, points to applications much more diversified than the simple miniaturization of electronic circuits. The possibilities include mainly the connection of traditional electronics to biological tissues, allowing the creation of neural chips, implants, prostheses, and devices designed to extend the capabilities of the human body, among many other possibilities [151].

On the other hand, amino acids are important organic compounds composed of amine ($-NH_2$) and carboxylic acid ($-COOH$) functional groups, along with a side chain specific to each of them. Their key elements are carbon, hydrogen, oxygen, and nitrogen, though other elements are found in their side chains. Besides, amino acids are the basic units of proteins, performing several important functions such as neurotransmitters and the formation of hormones, drugs, methylation, etc. Short chains of amino acid monomers, whose covalent chemical bonds are formed when the carboxyl group of one amino acid reacts with the amino group of the other, are known as oligopeptides.

Results of several electron-transfer studies through oligopeptides and proteins suggest that the efficiency of electron transport is strongly influenced by the length of the peptide, the nature of the scaffold, and the amino acid sequence [152]. The presence of hydrogen bonding also influences the electron-transfer rates.

Our motivation to present in this chapter the electron transport through different biological molecules is based on the fact that it may be controlled electrically, magnetically, optically, mechanically, chemically, and electrochemically, leading to various potential device applications. To reach the ultimate goal in device applications, experimental techniques to fabricate an *electrode–molecule–electrode* junction, and theoretical methods to describe the electron transport properties are being developed today.

We first consider a numerical study of the electronic transport through oligopeptide chains composed of two amino acid pairs, namely alanine–lysine (Ala–Lys) and threonine–alanine (Thr–Ala), respectively, sandwiched between two platinum electrodes [153]. Our results show that factors such as the oligopeptide chain length and the possible combinations between the amino acid residues are crucial to the diode-like profile of the current–voltage (I×V) characteristics, whose asymmetric curves were analyzed using the inverted rectification ratio (IRR).

Afterward, our focus changes to the de novo-designed α3-polypeptide by gene engineering [154], as well as its variants 5Qα3 and 7Qα3. Their charge transport properties are investigated with a tight-binding model Hamiltonian where *input* parameters (amino acid vertical ionization and dipeptide hopping energy) were obtained by performing ab initio calculations within the DFT [155]. Furthermore, considering the energy spectra of the α_3-helical polypeptide and its 5Qα_3 and 7Qα_3 variant structures as the energy spectrum of a single fermionic system, we will calculate and analyze the temperature-dependent electronic specific heat (ESH) at constant volume $C_V(T)$ of the polypeptides in focus [156].

Finally, charge transport properties of single-stranded microRNAs (miRNAs) chains associated with autism disorder were investigated. Current–voltage (I×V) curves of 12 miRNA chains related to the autism spectrum disorders were calculated and analyzed. We have obtained both semiconductor and insulator behavior, as well as a relationship between the current intensity and the autism-related miRNA bases sequences, suggesting that a kind of electronic biosensor can be developed to distinguish different profiles of autism disorders [157].

6.2 Molecular Diode

A diode or rectifier is an important component in conventional electronics, allowing an electric current to flow in one direction, but blocks it in the opposite direction. Building diodes using single molecules has been pursued by many groups (for a review, see Reference [158]).

The basic structure of early molecular diodes consists of a donor and an acceptor separated by a σ-bridge, with σ being some saturated covalent bond linking the

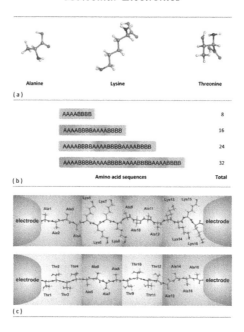

Figure 6.1 (a) The three amino acids used here: alanine (Ala), lysine (Lys), and threonine (Thr), with their corresponding 3-D structures. (b) The oligopeptide chains obtained by juxtaposing two building blocks (A, B), meaning the (Ala, Lys) and (Thr, Ala) amino acids, respectively. The length of these chains varies from eight to 32 amino acids. (c) Schematic illustration of the oligopeptide chains (Ala, Lys) and (Thr, Ala) sandwiched between two electrodes, mimicking the second sequence in (b), each of them with 16 amino acids in the total. After Oliveira et al. [153]

donor and acceptor, providing a tunneling barrier between them. In this *donor-σ-acceptor* structure, the diode behavior was expected to occur as a consequence of different thresholds at positive and negative bias voltages.

To illustrate this picture, let us consider the electronic transport through oligopeptide chains formed by the combination of two by two of three amino acids: alanine (Ala), lysine (Lys), and threonine (Thr). The two oligopeptide chains studied are the Ala–Lys and Thr–Ala, in their neutral and non-solvated conformation (see Figure 6.1a). These amino acids were gathered in groups of four building blocks each, in a linear molecular geometry of non-equilibrium, forming oligopeptide chains with eight, 16, 24, and 32 amino acids, covalently linked to two platinum electrodes (see Figure 6.1b and c).

The electronic transport properties of such molecules are usually described by two approaches: ab initio calculations and model-based Hamiltonians. The former can provide a detailed description but is currently limited to relatively short molecules, while the latter is less detailed but allows for describing systems of

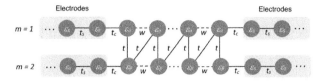

Figure 6.2 Schematic representation of the oligopeptide chains as illustrated in Figure 6.1c. Here, ε is the different on-site energies, and t, w are the hopping terms. After Oliveira et al. [153]

realistic length. Here, as our focus is mainly on the qualitative properties of a protein–nano junction, we choose a mathematical framework based on an effective tight-binding model, as described in Chapter 2, without any environment- or complex contact-related effects. In this way, keeping the formalism as simple as possible, the total Hamiltonian of the structure can be written as (see Figure 6.2)

$$H_{total} = H_{oligopeptide} + H_{electrode} + H_{coupling}. \tag{6.1}$$

The first term of Eq. (6.1) describes the intrachain charge propagation through the oligopeptide chain and is given by

$$H_{oligopeptide} = \sum_{n=1}^{N} \left[\varepsilon_\alpha^n |n,1\rangle\langle n,1| + \varepsilon_\beta^n |n,2\rangle\langle n,2| \right]$$

$$+ \sum_{n=1}^{N} w \left[|n,1\rangle\langle n \pm 1,1| + |n,2\rangle\langle n \pm 1,2| \right]$$

$$+ \sum_{n=1}^{N} t \left[|n,1\rangle\langle n,2| + |n,2\rangle\langle n,1| + |n,2\rangle\langle n \pm 1,1| + |n,1\rangle\langle n \pm 1,2| \right], \tag{6.2}$$

where the variable $\varepsilon_{\alpha,\beta}^n$ is the on-site ionization energy of the respective base α, β = alanine (Ala), lysine (Lys), and threonine (Thr) at the nth site.

The energies $\varepsilon_{\alpha,\beta}^n$ are chosen from the ionization potential of their respective bases: $\varepsilon_{Ala} = 9.85$, $\varepsilon_{Lis} = 9.50$, and $\varepsilon_{Thr} = 10.20$ (all units in eV) [159]. Furthermore, t (w) represent the hopping intra (inter)-chain term between adjacent amino acids residues $n, n \pm 1$ in the oligopeptide chain.

All electronic parameters were derived from molecular orbitals close to the HOMO or upper valence band edge. These results were obtained by first-principles calculations using the software Gaussian 09 within the DFT framework through

$$t_{n,n\pm1} = 0.5\left[\left(E_{n,n\pm1}^{HOMO} - E_{n,n\pm1}^{HOMO-1} \right)^2 - \left(E_n - E_{n\pm1} \right)^2 \right]^{1/2}, \tag{6.3}$$

where the two-state model based on the energetic splitting between the HOMO and HOMO $-$ 1 in a system of two amino acids was employed [160]. The $E_{n,n\pm1}^{HOMO}$ and $E_{n,n\pm1}^{HOMO-1}$ are, respectively, the first- and second-highest occupied molecular orbital energies calculated for each base combination formed by $n, n \pm 1$ amino acids residues of the oligopeptide chain. For these amino acid, the individual site energies E_n and $E_{n\pm1}$ (vertical ionization energies) were determined experimentally and reported in the literature [159].

When the difference in the preceding square brackets is negative, we adopted the expression already used in Chapter 3 (Eq. (3.34)):

$$t_{n,n\pm1} = 0.5 \left(E_{n,n\pm1}^{HOMO} - E_{n,n\pm1}^{HOMO-1} \right). \tag{6.4}$$

Furthermore, recent studies show that the conventional DFT functionals, commonly used to model charge transport in organic materials, underestimate the values of the hopping terms in comparison to long-range corrected density functional [161]. Due to that, a new density functional, a Coulomb-attenuated hybrid exchange-correlation functional (CAM-B3LYP), has recently been developed specifically to properly predict molecular charge-transfer spectra. It is a skilled Coulomb-attenuation scheme able to solve charge transfer excitations in the zinc bacteriochlorin-bacteriochlorin complex and in the dipeptides model. It has also been successfully applied in intermolecular charge transfer transitions, predicting the occurrence of charge transfer bands in the porphyrins and chlorophylls, and the identification of an unexpectedly large capacity of the aliphatic bridge to electronically connect the arylamines. To take into account these remarks, we used here the CAM-B3LYP functionals with Dunning's correlation consistent basis sets (cc-pVDZ) to optimize the structures of the oligopeptides and to calculate their HOMO energies [88]. Lastly, the interchain hopping between the amino acid bases is considered to be $w = 0.1t$, in eV.

The second term of Eq. (6.1) is related to the two semi-infinite metallic electrodes:

$$H_{electrode} = \sum_{n=-\infty}^{0} \sum_{m=1}^{2} \left[\varepsilon_s^n |n,m\rangle\langle n,m| + t_s |n,m\rangle\langle n \pm 1, m| \right]$$

$$+ \sum_{n=N+1}^{\infty} \sum_{m=1}^{2} \left[\varepsilon_s^n |n,m\rangle\langle n,m| + t_s |n,m\rangle\langle n \pm 1, m| \right]. \tag{6.5}$$

Here, ε_s^n (t_s) is the ionization energy (hopping term) of the electrode. We consider a platinum electrode whose ionization energy $\varepsilon_s = 5.36$ eV is related to the metallic work function of this metal, and $t_s = 12.0$ eV.

Finally, the third term of Eq. (6.1) describes the contacts between the oligopeptide and the semi-infinite metallic electrodes, yielding

$$H_{coupling} = \sum_{m=1}^{2} t_c \Big[|0,m\rangle\langle 1,m| + |N,m\rangle\langle N \pm 1,m| \Big], \qquad (6.6)$$

where $t_c = 0.317$ eV represents the hopping amplitude between the AC (DC) electrodes and the ends of the oligopeptide, with N being the number of amino acid residues in the chain considered.

With the tight-binding Hamiltonian given before, one can evaluate the $I \times V$ characteristics by applying the Landauer-Büttiker formulation described in Chapter 3. By properly locating the Fermi-level energy close to some of the characteristic resonances of the electrode–oligopeptide–electrode–nano junction, as the voltage drop is switched on, the transmission coefficient becomes voltage dependent.

The main current–voltage ($I \times V$) characteristics shown in Figure 6.3a and b present an insulator region for -4 V $\leq V \leq 4$ V and nonlinear and asymmetric regions indicating transitions toward saturation currents for $V \leq -4$ V and $V \geq 4$ V. It is easy to see that the general shape of the $I \times V$ curves are clearly asymmetric, depicting diode-like behavior.

Figure 6.3a shows that the current intensity increases with the applied negative (positive) voltages and reaches a maximum of -29.3 μA at $V = -15$ V (19.3 μA at $V = 15$ V), respectively. The same occurs to the $I \times V$ curves depicted in Figure 6.3b, but now with a lower maximum intensity of -5.8 μA at $V = -15$ V (2.3 μA at $V = 15$ V), respectively.

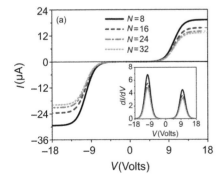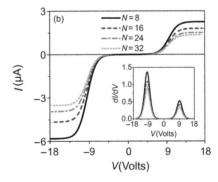

Figure 6.3 Current–voltage ($I \times V$) characteristic curves for the two oligopeptide chains: (a) Thr–Ala and (b) Ala–Lys, with N being the total number of the amino acids in the chain. The inset shows the transconductance $(dI/dV) \times V$. After Oliveira et al. [153]

In both cases, the maximum values for the current intensities were found for the total number of the amino acid residues in the chain $N = 8$. Note that the current intensity is inversely proportional to the length of the oligopeptide chains.

Lastly, the insets in Figure 6.3a and b show the transconductance ($dI/dV \times V$) of the devices, which are highly nonlinear. In contrast with previous work [162], they do not show negative differential resistance – i.e., $dI/dV < 0$. This phenomenon is a tunneling-related effect that was originally observed in silicon-based heterostructures and can be masked while the current is still small.

The non-negligible rectified currents, depicted in Figure 6.3a for the oligopeptide chain Thr–Ala, are 9.9 ($N = 8$), 7.8 ($N = 16$), 7.3 ($N = 24$), and 6.5 ($N = 32$), respectively (all units in μA). For the oligopeptide chain Ala–Lys (Figure 6.3b), the rectified currents are 3.6 ($N = 8$), 2.9 ($N = 16$), 2.3 ($N = 24$), and 2.1 ($N = 32$), respectively (all units also in μA). Here, N is the total number of the amino acids in the oligopeptide chain. Note that the absolute value of these rectified currents may be changed by varying the electronic hopping integrals, as well as by the way the oligopeptide chains are in contact with the electrodes.

Observe that the I×V curves present a particular asymmetry whose shape as a function of the polarity of the applied voltage constitutes the rectification of the current by the molecule.

To quantify the asymmetry of I×V characteristic in Figure 6.3a and b, we defined the *rectification ratio* (RR) by $R(V) = |I(V)/I(-V)|$, where $I(V)$ and $I(-V)$ represent the forward and reverse currents for the same voltage magnitude, respectively. However, because the structures studied in this work display higher current intensity in the region of negative voltage, we use instead an *inverted rectification ratio* $IRR = 1/R(V)$, to measure the degree of asymmetry of the oligopeptide chains.

We start from the I×V curves in Figure 6.3a for each oligopeptide chain, as illustrated in Figure 6.1b and c. It is found that, in all curves, an obvious rectification is observed with the rectifying direction along the negative bias direction – i.e., the electrons prefer to flow from the alanine group to the threonine one at a negative bias.

The $(1/R)$–V curve shows the maximum value of 1.58 at 6.4 V for the shortest chain ($N = 8$), as depicted in Figure 6.4a. We can also observe that, in this case, the $(1/R)$–V curves are inversely proportional to the length of the oligopeptide chains but with very close values. They can be divided into three regions namely: (a) 0 V $\leq$ V < 6 V, (b) 6 V $\leq$ V < 15 V, and (c) 15 V $\leq$ V < 18 V.

In the first region, the IRRs increase continuously from 0 to 6 V (maximum value), with a very small but progressive separation between them. In the second one, the $(1/R)$–V curves remain parallel and decrease softly. The third one is the saturation region.

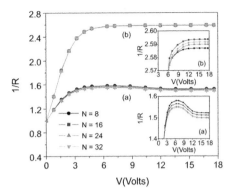

Figure 6.4 The inverted rectification ratios (IRR) corresponding to the (I×V) curves depicted in Figure 6.3a and b, respectively. The insets show magnification of the respective curves. After Oliveira et al. [153]

Figure 6.4b, which is related to Figure 6.3b (Ala–Lys pair), displays the same rectifying direction as those observed in Figure 6.3a. However, while the current intensity values in Figure 6.3b are much smaller than those observed in Figure 6.3a, the $(1/R)$–V curve depicted a maximum value of 2.59 at 12 V for the chain with $N = 16$ (see inset (a) of Figure 6.4). Unlike what was observed in Figure 6.4a, in this case, there is no relationship of inverse proportionality between the IRRs and the length of the chains (see inset (b) of Figure 6.4).

For the Ala–Lys pair case, the $(1/R)$–V curves can be divided into two regions of interest: (1) 0 V $\leq$ V < 12 V and (2) 12 V $\leq$ V < 18 V. In the first region, they increase continuously from zero until reaching the saturation value at 12 V (maximum value). The second one is the saturation region. Differently from the previous case, here, the curves are practically indistinguishable, suggesting the lack of relationship with the chain length.

In summary, by using an effective tight-binding model, we have theoretically investigated the transport properties of a molecular device made up of an oligopeptide chain directly coupled to two platinum electrodes. The I × V characteristics are discussed in terms of their on-site ionization and electrode energies, as well as their different hopping parameters. Our calculations reveal that the asymmetry observed in the I×V curves clearly indicate a diode-like profile in the two structures analyzed in this work. However, the values obtained for the I×V and $(1/R)$–V curves give quite different aspects for the oligopeptide chains formed by the threonine–alanine (Thr–Ala) and alanine–lysine (Ala–Lys) pairs: while the Thr–Ala pair presents a variation in the maximum current intensity six times greater than the Ala–Lys one, the latter has a IRR maximum value nearly two times greater than the former. Besides, we observed also a proportional relationship between the diference of the ionization energies of the oligopeptide chains (Thr–Ala or Ala–Lys) and the values obtained for the $I{\times}V$ and $(1/R)$–V curves. Such characteristics make these two oligopeptide chains good candidates for an artificial prototype of a molecular diode.

6.3 Alpha3-Helical Polypeptide and Its Biochemical Variants

Alpha helix is an usual shape that amino acid chains will form, being a common motif in the secondary structures of proteins [163]. It is characterized by a tight right-handed twist in the amino acid chain leading to a rod shape. Hydrogen bonds between the hydrogen (oxygen) in an amino (carboxil) group on the amino acid cause this structure.

In particular, the biased α3-polypeptide is a 21-residue amino acid structure formed by three repeats of the seven-residue (heptad) amino acid sequence Leu–Glu–Thr–Leu–Ala–Lys–Ala, defining an α-helical bundle structure through hydrophobic interaction between Leu residues. The α3-helical polypeptide has the ability to form fibrous assemblies that are observed by transmission electron microscopy and atomic force microscopy [154]. Although differing only by the Ala $\to$ Gln substitution at the fifth or seventh position of the α_3-peptide amino acid sequence (Leu–Glu–Thr–Leu–Ala–Lys–Ala)$_3$, the $5Q\alpha_3$ variant forms fibrous assemblies more attenuated than those of the α_3-peptide, while the $7Q\alpha_3$ variant does not form fibrils at all (see Figure 6.5).

An important issue of the formation of fibrous assemblies in polypeptides is related to its association with amyloidosis-like diseases, like Alzheimer's, Parkinsons, and Creutzfeldt-Jakob diseases, whose common pathology includes fatal transmissible spongiform degeneration and characteristic formation of plaques

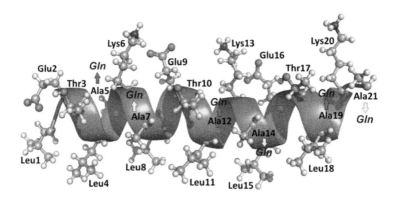

Figure 6.5 The α_3-helical (Leu–Glu–Thr–Leu–Ala–Lys–Ala)$_3$ polypeptide. The gray and white arrows indicate the mutations Ala5Gln and Ala7Gln in the first, second, and third heptad sequence, giving rise to the $5Q\alpha_3$ and $7Q\alpha_3$ variants, respectively. After Mendes et al. [156]

in the brain tissue. The protein-only hypothesis states that the infectious agent is a protein, named prion, which is a pathogenic isoform that is seemingly able to convert the normal isoform in an autocatalytic process. Two conformations of this protein are important for characterizing the disease, namely the normally folded host-encoded cellular protein and an abnormal dangerous pathogenic conformation. The latter form is hydrophobic, has a tendency to form aggregates, and may be found in different strains.

Considering that, it is our aim in this section to investigate how charge transport in peptides can be a useful tool for the development of biosensors to probe the onset of these diseases. Taking into account a tight-binding transport Hamiltonian, with hopping energies of the polypeptides calculated within the DFT framework, we obtain their current–voltage characteristics (Section 6.3.1), as well as their ESH at constant volume spectra (Section 6.3.2), to set up a pattern to distinguish them, and suggest a new biosensor devices as diagnostic tools for amyloidosis-like diseases.

6.3.1 Electronic Structure

The charge transport properties of the biased $\alpha3$-polypeptide and its (5Q, 7Q) $\alpha3$ variants are calculated using Dyson's equation together with a transfer-matrix treatment as described in Chapter 2. Considering the molecular circuit depicted in Figure 6.6, the following electronic tight-binding model Hamiltonian can be expressed as

$$H = \sum_n \epsilon_n |\psi_n\rangle \langle\psi_n| + \sum_{n,m} V_{n,m} |\psi_n\rangle \langle\psi_m|, \tag{6.7}$$

which describes the carrier moving through the primary peptide structure sandwiched by two electrodes – donor (DN) and acceptor (AC) – with a single orbital per site and nearest-neighbor interactions. Here, ϵ_n is the single carrier energy (isolated amino acid vertical ionization energy) at the orbital ψ_n. The term $V_{n,m}$

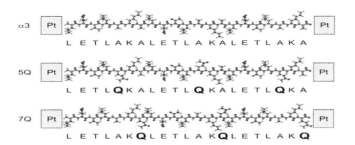

Figure 6.6 Schematic representation of the oligopeptide chains $\alpha3$, $5Q\alpha3$, and $7Q\alpha3$ sandwiched between two platinum (Pt) electrodes.

is the first-neighbor electronic overlaps (hopping amplitude, obtained from the overlap integrals of the isolated polypeptides). For simplicity, we are considering that the charge carriers are traveling from DN to AC platinum electrodes by hopping between the α_3, $5Q\alpha_3$, and $7Q\alpha_3$ peptides, discrete molecular orbitals.

The vertical ionization energies and hopping terms for isolated amino acids and polypeptides, respectively, were obtained by first-principles calculations considering only the primary amino acid sequence of them.

The isolated amino acid and polypeptide conformers of smaller energies were found through a minimization energy process within the DFT framework using the DMOL3 code [134]. The generalized gradient approximation (GGA) Perdew-Burke-Ernzerhof exchange-correlation functional [12] was adopted, and a double numerical plus polarization basis set (DNP) was chosen to expand the Kohn-Sham wave functions. All electrons, valence and core, were taken into account.

The geometry optimization convergence thresholds were 10^{-5} Ha for the total energy variation, 0.002 Ha/Å for the maximum force per atom, and 0.005 Å for the maximum atomic displacement. In the case of the amino acid conformers, they are in good agreement with the lowest-energy amino acid conformers found by other authors [164].

To obtain the vertical ionization energies of the amino acids, the total energy differences between their N-electron ground state and their excited states of smaller energies were computed. The lowest single-electron excitations are the (first) amino acid vertical ionization potential $IE = E(N-1) - E(N)$, where the ionized amino acids with one missing electron are characterized by the total energies $E(N-1)$.

The calculated amino acids vertical IEs of the $\alpha3$-helical peptides and the $(5Q,7Q)\alpha3$ variants are (all units in eV) as follows: alanine, 9.25; glutamine, 8.52; lysine, 8.00; glutamic acid, 8.63; threonine, 9.05; leucine, 8.85. The hopping terms, on the other hand, are estimated through Eq. (6.4); together with the first HOMO energies $E_{n,m}^{HOMO}$ and $E_{n,m}^{HOMO-1}$, they are depicted in Table 6.1. The charge carriers are supposed to be traveling between the biased platinum electrodes by hopping through discrete molecular orbitals of the $\alpha3$-helical peptides and the $(5Q,7Q)\alpha3$ variants. The energy of the platinum electrode is 5.36 eV, which is related to the work function of this metal.

Although the $\alpha3$-helical peptides and its $(5Q,7Q)\alpha3$ variants differ only by the Ala $\rightarrow$ Gln substitutions, their $I \times V_{bias}$ characteristics should change considerably due the degree of fibrils formation. This allows to distinguish the $\alpha3$, $5Q\alpha3$ and $7Q\alpha3$ peptides, and consequently to point the existence (or absence!) of fibrous assemblies, meaning that charge transport can be used as a tool for characterizing them.

Figure 6.7 shows the behavior of the current in the positive and negative biased peptides. For both the positive and negative polarity, there is a characteristic ohmic

Table 6.1 *HOMO energies and the hopping terms for the the n,m amino acid residues of the α3-helical polypeptide. After Bezerril et al. [155].*

n,m	$E_{n,m}^{\text{HOMO-1}}$ (eV)	$E_{n,m}^{\text{HOMO}}$ (eV)	$V_{n,m}(eV)$
Leu–Glu	−5.960	−5.503	0.229
Glu–Thr	−6.147	−5.906	0.121
Thr–Leu	−6.046	−5.638	0.204
Leu–Gln	−5.815	−5.414	0.201
Gln–Lys	−5.643	−5.179	0.232
Lys–Ala	−5.485	−5.120	0.183
Ala–Leu	−6.147	−5.830	0.159
Leu–Ala	−5.959	−5.254	0.353
Ala–Lys	−5.494	−5.052	0.221
Lys–Gln	−5.725	−5.324	0.201
Gln–Leu	−6.151	−5.418	0.367

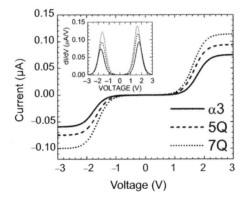

Figure 6.7 The current (in μA)-voltage (in V) characteristics profiles through the biased α_3 (dotted), $5Q\alpha_3$ (dashed), and $7Q\alpha_3$ (solid) polypeptides. The inset shows the transconductance $dI/dV \times V$ of the devices. After Bezerril et al. [155]

region for $-1.0 \leq V_{bias} \leq +1.0\,\text{eV}$ and nonlinear regions indicating transitions toward current saturation for $V_{bias} < -1.0\,\text{eV}$ and $V_{bias} > +1.0\,\text{eV}$. The inset in Figure 6.7 shows the transconductance $dI/dV \times V$ of the devices, which are highly nonlinear. On the basis of the dI/dV characteristics, the polypeptides α_3, $5Q\alpha_3$, and $7Q\alpha_3$ have an average energy gap of 3.25, 3.32, and 3.53 eV, respectively. They have semiconductors characteristics, as in the case of dry proteins [165].

Each peptide has a characteristic current pattern, which was assumed to depend mainly on its primary structure, depicting that the fibrous/nonfibrous α_3, $5Q\alpha_3$, and $7Q\alpha_3$ peptides can be distinguished by charge transport measurements. As a matter of fact, the α_3 peptide, which has the most fibrous assemblies, shows the smaller current saturation; the $5Q\alpha_3$ variant, which forms fibrous assemblies more attenuated than those of the α_3 peptide, has a current saturation higher than α_3 but smaller than $7Q\alpha_3$; finally, the $7Q\alpha_3$ variant does not form fibrils and shows the highest current saturation.

If the secondary structure of the peptides is considered, the number of charge transport channels should increase due to hydrogen bonding related to the secondary structure, further increasing saturation currents, but not specifically enough to change the order $I(\alpha_3) < I(5Q\alpha_3) < I(7Q\alpha_3)$. Further development on this line is hampered by the absence of crystallographic data allowing the characterization of the fibrous/nonfibrous α_3 and $(5Q,7Q)\alpha_3$ peptides' secondary structure.

Since the formation of fibrous assemblies are characteristic of Alzheimer's, Parkinson's, and Creutzfeldt-Jakob (prion) diseases, for example, our result allows a suggestion that charge transport in polypeptides can be a useful tool for the development of biosensors to probe the onset of amyloidosis-like diseases.

6.3.2 Electronic Specific Heat

Specifically, the quest for a deeper understanding of the protein and polypeptide temperature-dependent heat capacity at constant pressure (volume) $C_{P(V)}(T)$ remains a fundamental challenge. Besides being very promising for the development of applications linking its temperature behavior to biological data, it may pave the road for the development of many devices, envisaging $C_{P(V)}(T)$-based diagnostic tools, since it relates changes on the protein's molecular kinetic (vibrational, rotational, translational, etc.) and electronic (interatomic potential energies, bond stretching, bending, etc.) properties.

As a matter of fact, the thermodynamics of protein and polypeptide conformational changes depends not only on the specific sequence of the amino acids (their primary structure) but also on the polarity and hydrophobic effects, which are related to the protein secondary structure and water effects. Protein unfolding in aqueous solution is usually accompanied by an increase in heat capacity, and this has long been regarded as a somewhat anomalous behavior. In the case of polypeptides, there is, in general, an almost linear increase of their heat capacity with temperature, but several measurements indicate the existence of a boson-type peak in plots $C_P/T^3 \times T$ at helium liquid temperature [166]. Although there are some theoretical models describing the thermodynamic properties of an infinite polyalanine α-helix (few are based in the DFT calculations of the phonon spectra, by the way), no theoretical model was developed to calculate the ESH of proteins and polypeptides so far.

In view of that, and by taking into account the energy spectra of the α_3-helical polypeptide and its $5Q\alpha_3$ and $7Q\alpha_3$ variant structures developed in the previous section, we intend here to contribute further to this subject by considering them as a single fermionic system to obtain their temperature-dependent ESH at constant volume $C_V(T)$. The main features of their ESH and the possibility of using the ESH characteristics also as a tool for early diagnosis of amyloidosis-like diseases will be discussed.

The electronic band structure of biopolymers, like the α-helical peptide studied here, is composed of two main bands of allowed states, separated by an energy gap similar to those found in solid-state semiconductors. The introduction of "defects" may generate states within the gap, substantially changing their electronic band structure. These "defects" in polypeptides originate from a residue replacement in the primary amino acid sequence due to a mutation and can be investigated, for example, by using the renormalization scheme described in Chapter 2. Defects can profoundly modify the electronic states, mimicking complex structures.

In order to avoid unnecessary numerical and mathematical treatment, we will use here the same electronic Hamiltonian (Eq. (6.7)) as in the previous section. The calculation of the ionization energies and the polypeptide frontier molecular orbitals follows the same procedure of the previous section, summarized in Table 6.1.

We first determine the electronic DOS depicted in Figure 6.8. Rather than traces of bands, the DOS profile for each peptide is fragmented, showing a number of discrete strongly localized bunches of states that are believed to reflect their 1-D band structure. Observe that the number of van Hove singularities is much larger for the fibrous (α_3-helical and its $5Q\alpha_3$ variant) peptides than for the nonfibrous $7Q\alpha_3$ variant one. Surely, this fact will be reflected into the ESH spectra discussed later.

The chemical potential $\mu = \mu(N_e/N, T)$ can be computed as a function of the temperature and the band filling N_e/N from Eq. (4.14) of Chapter 4 and can then be

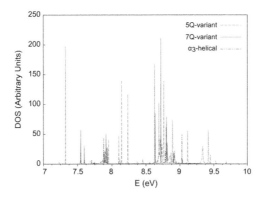

Figure 6.8 The electronic DOS in arbitrary units plotted against the energy E (in eV) for the α_3 polypeptide (chain-dotted line) and its $5Q\alpha_3$ (solid line) and $7Q\alpha_3$ (dashed line) variants. After Mendes et al. [156]

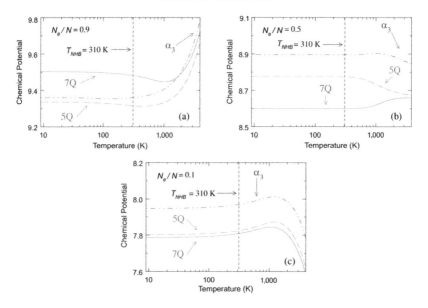

Figure 6.9 Semi-log scale of the chemical potential $\mu = \mu(N_e/N, T)$ versus the temperature T (in K) for the α_3 polypeptide (chain-dotted line) and its 5Qα_3 (dashed line) and 7Qα_3 (solid line) variants. The thin dashed vertical line represents the normal human body temperature $T_{NHB} = 310$ K. Three values of the band fillings N_e/N are considered, namely (a) $N_e/N = 0.9$; (b) $N_e/N = 0.5$; (c) $N_e/N = 0.1$. After Mendes et al. [156]

extracted by numerical methods. Here, N_e is the number of non-interacting Fermi particles (electrons), while N is the total number of one-particle accessible states (electrons and holes). The average internal energy can be found from Eq. (4.15), where the temperature dependence of the chemical potential $\mu(N_e/N, T)$ is explicitly taken into account.

In Figure 6.9, the curves of the chemical potential $\mu(N_e/N, T)$ (the measurement of the energy per particle for a given entropy) against the temperature T (in Kelvin) are depicted for the α_3-helical polypeptide (chain-dotted line) and its 5Qα_3 (dashed line) and 7Qα_3 (solid line) variants. Three representative values of the band fillings N_e/N are considered – namely $N_e/N = 0.9$, 0.5, and 0.1, which are shown in Figure 6.9a–c, respectively. The thin dashed vertical line represents the normal human body temperature $T_{NHB} = 310$ K.

For lower values of T (below the normal human body temperature), there is a transient period for all polypeptides and band fillings, and the chemical potential profile is characterized by a quite flat curve. This situation changes for temperatures higher then T_{NHB} for each band filling considered. In the extreme cases – i.e., for the band filling $N_e/N = 0.9$ and 0.1 – the chemical potential reaches a minimum (maximum) value and then starts to increase (decrease) linearly, following a power-law $\mu(N_e/N,T) \propto T^{\phi}$ ($\propto T^{-\phi}$), with the ϕ exponents equal to 3.26 (2.81) for the α_3 peptide, 3.21 (2.81) for its $5Q\alpha_3$ variant, and 3.14 (2.78) for its $7Q\alpha_3$ variant.

The ESH at constant volume is evaluated by differentiating the average internal energy $U(N_e/N,T)$ with respect to the temperature T, keeping the volume of the system V constant by maintaining a fixed total number of one-particle accessible states N, as expressed by Eq. (4.16). In the limit of high temperatures and/or very low electron densities, the ESH tends to the value obtained through the determination of the partition function using classical Boltzmann-Gibbs statistics. It is worth mentioning that our expression for C_V could be simplified by using a handy general formula for the ESH as a function of the absolute temperature, as discussed by Starikov [167] following a theory developed by Linhart [168]. However, although analytically more demanding, our expression better resembles the ESH profiles of the three peptides considered in this work, depicting the expected oscillations not found in the Linhart theory.

Similarly as it was done in Figure 6.9 for the chemical potential, Figure 6.10 shows the normalized ESH spectra at constant volume (in units of the number of non-interacting Fermi particles N_e times the Boltzmann constant k_B) versus the temperature T (in Kelvin) for the α_3-helical polypeptide (chain-dotted) and its $5Q\alpha_3$ (dashed) and $7Q\alpha_3$ (solid) variants, considering the same band fillings N_e/N $N_e/N = 0.9$ (Figure 6.10a), 0.5 (Figure 6.10b), and 0.1 (Figure 6.10c).

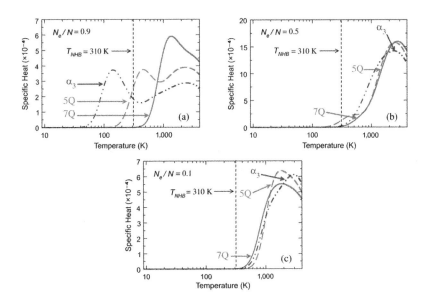

Figure 6.10 Same as in Figure 6.9, but for ESH at constant volume (in units of $N_e \times k_B$). After Mendes et al. [156]

There are some features in the temperature dependence of the ESH that deserve to be stressed. First, for the high temperature limit ($T \to \infty$), the ESH for all polypeptides and band fillings converges and decays according to the power-law T^{-2}; it is an expected result since the systems are bounded.

For high band fillings ($N_e/N = 0.9$) depicted in Figure 6.10a, the Fermi energy falls in a region where the degenerated energy levels are separated by mini-bands, with a predominant electron charge carrier contribution, when compared to its hole charge carrier counterpart. In view of that, as we approach the full-band regime ($N_e/N = 1$), the low-temperature ESH of holes becomes much smaller than that for electrons, and the degeneracies are smaller at the top than at the bottom of the energy band. More importantly, on the other hand, when the temperature decreases, each polypeptide, although differing only by the Ala $\to$ Gln substitutions, presents a unique characteristic ESH profile, allowing us to distinguish one from another and consequently to point the existence (or absence!) of fibrous assemblies. As we can see from Figure 6.10a, the ESH increases up to a maximum value; the corresponding temperature for this maximum value depends not only on the number of band filling N_e/N but also on the molecule type (observe that we are using a log scale for the absolute temperature). After the maximum value, the ESH falls into a low-temperature region and starts, for the fibrous α_3-helical polypeptide (chain-dotted line) and its $5Q\alpha_3$ variant (dashed line), a pattern of log-periodic oscillations, which signals the discrete scale invariance of the spectrum at the vicinity of the Fermi energy, with no counterpart for the nonfibrous $7Q\alpha_3$ variant.

These oscillations persist until the ESH profile starts a linear regime at a temperature below (above) the normal human body temperature $T_{NHB} = 310$ K, reaching zero at a temperature equal to 40 K (100 K) for the α_3-helical polypeptide ($5Q\alpha_3$ variant) case. As a matter of fact, the α_3 polypeptide, which has the most fibrous assemblies, shows the widest ESH profile; the $5Q\alpha_3$ variant, which forms fibrous assemblies more attenuated than those of the α_3 polypeptide, has a ESH profile wider than that of the $7Q\alpha_3$ variant but narrower than that of the α_3 polypeptide; finally, the 7Q variant does not form fibrils and shows the narrowest ESH profile, whose intensity goes to zero at a temperature near to the normal human body temperature T_{NHB}.

After a critical value ($N_e/N \approx 0.85$), the ESH pattern just described disappears. We believe that this occurs because the Fermi energy falls in a dense region of the energy spectrum, and, as a consequence, the ESH displays a power-law decay at low temperatures modulated by complex log-periodic oscillations. The pattern of these oscillations is strongly dependent on the band filling as distinct regions of the energy spectrum are probed in each case. Further, the power-law decay exponent is non-universal. This situation persists until we reach the half band-filling case

$N_e/N = 0.5$, depicted in Figure 6.10b. Now, the Fermi energy is located exactly at the band center with this degenerated level partially filled. As a consequence, the ESH has the typical aspect of non-interacting Fermi systems exhibiting a large gap between the fundamental state and the first excited state. After a maximum value, the ESH goes to zero with a profile showing nearly linear behavior, as is usual for free-electron gases. There is a slight deviation from this behavior only for the fibril $5Q\alpha_3$ variant, which shows a shoulder in its ESH profile at a temperature $T \approx 400\,$K, whose intensity goes to zero at $T = 100\,$K. For the sake of comparison, the ESH strength of the α_3 polypeptide and its 7Q variant, whose profiles do not show oscillations, reaches zero at temperatures equal to $200\,$K and $300\,$K, respectively.

As the band filling continues to decrease, the Fermi energy falls deeper into the energy spectrum composed by degenerated levels separated by energy gaps. The oscillations disappear completely, as one can see from Figure 6.10c ($N_e/N = 0.1$). Quite interestingly, the intensity of the ESH for all polypeptides goes to zero at a temperature close to the normal human body temperature T_{NHB}.

It is important to mention that the logarithmic oscillation behavior in the temperature dependence found in the ESH profiles, no matter the peptide considered, reveals the existence of several energy scales in their spectra. These oscillations are more pronounced at high band fillings because the higher half of the energy spectrum has a more sharply defined sequence of mini-bands. At the lower half of the spectrum, the presence of degenerated levels suppresses the specific heat oscillations, giving clear thermodynamic evidence of the absence of an electron-hole symmetry. It would be interesting to investigate if other Hamiltonian systems would have similar energy spectrum as well as thermodynamic signatures. In particular, the study of collective excitations, such as phonons and magnons, may contribute to a more complete scenario.

In conclusion, aiming to further contribute to the present understanding of the important differences on the fiber formation of the α-helix variants, we have presented in this work a theoretical model to study their ESH spectra, according to the Fermi-Dirac statistics and considering several band filling N_e/N values. For high band fillings ($N_e/N = 0.9$), the differences between the fibrous (α_3-helical and its $5Q\alpha_3$ variant) peptide and nonfibrous $7Q\alpha_3$ variant peptide are amazing. The understanding of fibrous formation is therefore mandatory to the development of effective drugs that could reverse the fibrous assembling presented in the pathogenic protein conformation responsible for the major aspects of the diseases. The present ESH profiles, as well the current–voltage characteristics described earlier, may distinguish the fibrous/nonfibrous α-helical peptide studied here, reinforcing our suggestion to use them as biosensors to probe related diseases.

6.4 Single Micro-RNAs Chains and the Autism Spectrum Disorder

The discovery of miRNAs [169] and their target messenger RNAs (mRNAs) has uncovered novel mechanisms regulating gene expression beyond its central dogma. They are a class of naturally occurring endogenous, noncoding single-stranded RNA molecules, typically 21 to 23 nucleotides in length, which control transcriptional and posttranscriptional regulation of gene expression.

Much emerging evidence has revealed that miRNAs play an instrumental role in a wide range of biological processes, like the cellular processes, including cell growth, differentiation, proliferation, and cell death. In addition, disrupted miRNA function has been proposed to be associated with a number of diseases – namely pathogenesis of cancer, cardiac diseases, various types of infections, and autoimmune/inflammatory diseases, to cite just a few. Besides, it is also linked to psychiatric/neurodegenerative/neurodevelopmental disorders, leading to difficulties in social interactions.

The challenges of studying miRNAs are twofold. First, profiling miRNAs is technically difficult due to their intrinsic characteristics such as short sequence lengths and low abundance. This means that traditional DNA-based methods are not sensitive enough to detect these sequences with any reliability. Second, closely related miRNA family members differ by as little as one nucleotide, emphasizing the need for high specificity and the ability to discriminate between single nucleotide mismatches.

Nevertheless, miRNAs are nowadays intensely studied as useful candidates for diagnostic and prognostic biosensors, and are rapidly emerging as attractive targets for therapeutic intervention and basic biomedical research, such as the molecular control of development and aging of the brain that can be associated with neurodevelopmental disorders.

Among the several possibilities for the study of miRNA chains related to many forms of diseases and disorders, those associated to the autism spectrum disorders are now considered to be very important, since it is believed that genetics is 90% responsible for the risk of a child developing autism [170]. Rare cases may be caused by chemical exposure and other agents that can cause birth defects.

Autism spectrum disorders (ASDs) are a collective term used to describe neurodevelopmental disorders with a pattern of qualitative abnormalities in three functional domains: reciprocal social interactions, communication, and restrictive interests and/or repetitive behaviors. Autism appears to have its roots in the very early brain development, being usually associated with intellectual disability, difficulties in motor coordination and attention, and physical health issues, like sleep and gastrointestinal disturbances. Nevertheless, some persons with ASDs excel in visual accuracy, music, mathematics, and arts, among other skills.

Figure 6.11 Schematic representation of the miRNA chain sandwiched between two electrodes. Here, ε and t are the different on-site energies and hopping terms of the system, respectively. After Oliveira et al. [157]

Although there is no one cause of autism, just as there is no one type of autism, over the last decade, scientists have identified a number of miRNA chains associated with ASDs. Such biomedical improvement opened up the possibilities to seek nanoelectronic devices to unveil how miRNA chains may be associated with the underlying pathophysiology of ASDs, providing a better understanding of the role played by these miRNAs as a biological sensor mechanism.

In this section, we intend to suggest a nanoelectronic device, namely the *electrode–miRNA chain–electrode* junctions depicted in Figure 6.11, by means of its electron-transport properties, taking into account the contributions of the RNA-base system in the miRNA chain through its chain length, the relationship of proportionality between the different nucleotides, and their relative positioning in the chain.

6.4.1 Theory

Traditionally, the RNA molecule is more difficult to model than the DNA one. It has a much more flexible, although manageable, molecular structure, containing all the basic ingredients concerning its stability. It consists of four types of nucleotides: adenine (A), uracil (U), guanine (G), and cytosine (C). Selectively, A (G) forms a base pairing with U (C). Charge migration in RNAs depends on various mechanisms, and characterization and understanding of these mechanisms is a complicated task, leading to a significant reduction of the charge transfer efficiency if its large structural fluctuation is taken into account.

Nevertheless, in a simplified view, one can consider that the main mechanism of charge transfer in RNA molecules lies in the coupling between orbitals belonging to neighboring nucleotides. Analogously to quantum studies on DNA oligomers, as a first approximation, one can describe this charge transfer mechanism by treating each nucleotide as a single entity with a characteristic on-site energy, whose origin comes from an effective $\pi-\pi$ overlap integral describing the aromatic base stacking between adjacent nucleotides, explicitly included in the model by means of the hopping integral.

In this way, the most basic features of RNA electronic structure can be incorporated into an effective tight-binding model Hamiltonian describing the RNA as a linear chain with a tight-binding orbital per site along with a suitably parameterized hopping onto neighboring sites. A more accurate description may require a system–electrode coupling, the phonon contribution and the strength of the charge–phonon coupling, among other refinements, deserving a full study in itself.

To address the influence of the previously mentioned factors on the charge transport efficiency, we used in this work the theoretical framework based on an effective tight-binding model presented previously, which can be given by (see Figure 6.11)

$$H_{total} = H_{RNA} + H_{electrode} + H_{coupling}. \tag{6.8}$$

The first term of Eq. (6.8) describes the charge propagation through the miRNA chain and can be written as

$$H_{RNA} = \sum_{n=1}^{N} [\varepsilon_\alpha^n |n,1\rangle\langle n,1| + t_{n,n\pm1}|n,1\rangle\langle n\pm1,1|], \tag{6.9}$$

where N is the number of nucleotides in the miRNA chain, and ε_α^n is the ionization energy of the respective base α, with α = adenine (A), cytosine (C), guanina (G), and uracil (U), at the nth site. Also, $t_{n,n\pm1}$ are the hopping parameters between the adjacent nucleotides $n, n \pm 1$ in the miRNA chain.

The second term, related to the two semi-infinite metallic electrodes, reads

$$H_{electrode} = \sum_{n=-\infty}^{0} [\varepsilon_s^n |n,1\rangle\langle n,1| + t_s|n,1\rangle\langle n\pm1,1|]$$

$$+ \sum_{n=N+1}^{\infty} [\varepsilon_s^n |n,1\rangle\langle n,1| + t_s|n,1\rangle\langle n\pm1,1|]. \tag{6.10}$$

Here, ε_s^n (t_s) is the ionization energy (hopping parameter) of the electrode.

Finally, the third term describes the coupling between the miRNA chain and the semi-infinite metallic electrodes, yielding

$$H_{coupling} = t_c[|0,1\rangle\langle1,1| + |N,1\rangle\langle N+1,1|], \tag{6.11}$$

where t_c is the boundary-hopping parameter between the miRNA chain and the semi-infinite metallic electrodes. For simplicity, it is considered to have the same value regardless of the boundary nucleotide (see Figure 6.11).

The values of the ionization energy for each nucleotide in the miRNA chains are $\varepsilon_A = 8.24, \varepsilon_C = 8.87, \varepsilon_G = 7.75$ (as used in Chapter 2), and $\varepsilon_U = 9.32$ (all units in eV). Again, as in Chapter 2, we consider a platinum electrode whose ionization energy is $\varepsilon_s = 5.36$ eV. Also, the hopping parameters are $t_c = 0.63$ eV and $t_s = 12.0$ eV [60, 61].

The hopping parameters $t_{n,n\pm1}$ between the adjacent nucleotides $n, n \pm 1$ in the miRNA chain are listed in Table 6.2. They were obtained by first-principles

Table 6.2 *Values of the hopping parameters (in eV)*
between adjacent nucleotides in the miRNA chain.
After Oliveira et al. [157].

$X|Y$	A	C	G	U
A	0.167	0.189	0.118	0.355
C	0.188	0.308	0.266	0.138
G	0.269	0.279	0.240	0.630
U	0.466	0.295	0.477	0.059

calculations using the software Gaussian 09 within the DFT framework, together with the CAM-B3LYp functional with Dunning's correlation, as in Section 6.2.

It is important to mention that the sugar–phosphate units were replaced by methyl groups in all base pairs, aiming to model the electronic effect of the sugar–phosphate backbone. Inclusion of the sugar–phosphate backbone did not significantly affect the values of the hopping term and site energies, since the HOMOs were found to be almost entirely (>96%) localized on the nucleobases, while the density on the methyl group was very small, in agreement with earlier findings [171].

6.4.2 Results

Consider now that the miRNA chains are further assumed to be connected to two semi-infinite electrodes. Access to their electronic conductance can be performed, as before, by measuring their I×V characteristics.

Let us now examine closely the 27 miRNA chains related to ASDs used in the present work. The database of all of them is depicted in Table 6.3, where the first column lists the identity (ID) of the miRNA (hsa-miRNA-number), the second depicts the nucleotide sequence, and the third shows the order of conductivity classified into five groups: (I) high, (II) medium, (III) low, (IV) very low, and (V) zero conductivity. It is easy to see that they differ only by the amount of A (adenine), C (cytosine), G (guanine), and U (uracile) nucleotides presented on their chains.

Current–voltage characteristics of the first 12 sequences are plotted in Figure 6.12a–d. Generally speaking, they present a staircase profile since the potential barrier between the metallic contacts and the miRNA chain was set to zero. Furthermore, there is a characteristic insulator region for $-4V \leq V \leq 4V$, and nonlinear regions indicating transitions toward saturation currents for $V \leq -4V$ and $V \geq 4V$ in all frames. All of them have semiconductor characteristics, as in the case of the peptides α_3 previously studied in this chapter.

Table 6.3 *The database of ASD miRNAs chains used in this book. After Oliveira et al. [157].*

ID	Sequence	Group
559	UAAAGUAAAUAUGCACCAAAA	I
455-3p	GCAGUCCAUGGGCAUAUACAC	II
484	UCAGGCUCAGUCCCCUCCCGAU	II
486-3p	CGGGGCAGCUCAGUACAGGAU	III
193b	AACUGGCCCUCAAAGUCCCGCU	III
486-5p	UCCUGUACUGAGCUGCCCCGAG	III
212	UAACAGUCUCCAGUCACGGCC	III
106b	UAAAGUGCUGACAGUGCAGAU	III
132	UAACAGUCUACAGCCAUGGUCG	III
652	AAUGGCGCCACUAGGGUUGUG	IV
106a	AAAAGUGCUUACAGUGCAGGUAG	IV
320a	AAAAGCUGGGUUGAGAGGGCGA	IV
128	UCACAGUGAACCGGUCUCUUU	V
148b	UCAGUGCAUCACAGAACUUUGU	V
381	UAUACAAGGGCAAGCUCUCUGU	V
431	UGUCUUGCAGGCCGUCAUGCA	V
15b	UAGCAGCACAUCAUGGUUUACA	V
539	GGAGAAAUUAUCCUUGGUGUGU	V
21	UAGCUUAUCAGACUGAUGUUGA	V
15a	UAGCAGCACAUAAUGGUUUGUG	V
27a	UUCACAGUGGCUAAGUUCCGC	V
23a	AUCACAUUGCCAGGGAUUUCC	V
95	UUCAACGGGUAUUUAUUGAGCA	V
93	CAAAGUGCUGUUCGUGCAGGUAG	V
432	UCUUGGAGUAGGUCAUUGGGUGG	V
181d	AACAUUCAUUGUUGUCGGUGGGU	V
7	UGGAAGACUAGUGAUUUUGUUGU	V

Regarding their conductivity, however, they can be divided into four groups, namely high (group I, depicted in Figure 6.12a), medium (group II, depicted in Figure 6.12b), low (group III, depicted in Figure 6.12c), and very low (group IV, depicted in Figure 6.12d) conductivity, respectively. The other 15 sequences, which form group V in Table 6.3, were not considered here because they present zero conductivity.

To further identify the role played by the conductivity on the miRNA chains, we have plotted as an inset in Figure 6.12a–d the transconductance $dI/dV \times V$ of the devices, which are highly nonlinear. Again, the same fingerprint found in the $I \times V$ curves is preserved. Surely, these conductivity curve profiles distinguish the ASD miRNA chains, signaling a charge transport device tool to characterize them.

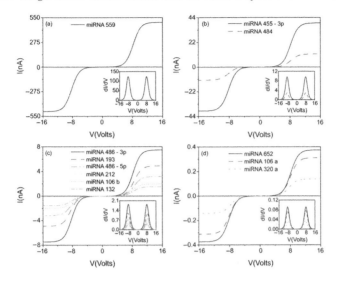

Figure 6.12 Current–voltage (I×V) curves for the 12 single-stranded miRNA chains related to ASDs. The insets show the transconductance $(dI/dV) \times V$ of the devices. After Oliveira et al. [157]

One aspect that deserves attention that was not fully explored so far is the electrodes. Not only the intrabase electronic coupling but also the electrodes considerably affect the charge transport. The preferred material for electrodes is gold or platinum, to avoid oxidation and degradation. For gold electrodes, sulphur, amine, or a number of lone-pair species are typically used to attach the molecule; for platinum electrodes, lone-pair species are most often used. In the present study, we use platinum electrodes, but nothing prevents different electrodes being tested in miRNA chains, such as the gold ones, in future studies.

Summing up, using an effective tight-binding model, we have investigated the qualitative features of the coherent charge transport through ASD miRNA chains connected between two electrodes. The transmission coefficient and I×V curves are discussed in terms of their on-site ionization and electrode energies, as well as their different hopping parameters. Several meaningful conclusions can been derived:

(a) Among the 27 ASD miRNA chains investigated here and described in the scientific literature, with their respective lengths in the range from 21 to 23 nucleotides, only 12 chains present current–voltage characteristics. Their profiles are depicted in Figure 6.12a–d.

(b) Considering that the four nucleotides have different ionization energies, the local chemical reactivity is sequence and scale dependent. For instance, accumulation of U (uracile) nucleotides, as those chains with six or more Us (see Table 6.3), independent of their length or whether they have Us repeated in the same chain, present a zero conductivity behavior. The ASD hsa-miRNA-7 sequence (23 nucleotides), with four Us repeated in a total of 10, appears to be the limit, below which the device becomes an insulator.

(c) Taking into account short- and long-range correlations between base pairs provides more information to further distinguish the random distributions to further complex sequences, like those described by a quasiperiodic approach, whose long-range correlations might be also associated with some biological properties.

(d) The ratio of As (adenines), Cs (cytosines), Gs (guanines), and Us (uraciles) in a particular ASD miRNA chain of nucleotides, its relative position in the chain, and its length define a limit from which a given nucleotide sequence can be considered either a semiconductor or an insulator.

Such studies clearly show that correlations are present at different scales and that they are strongly sequence dependent, playing an important role in the charge transfer mechanism. To corroborate even more this assertion, we have considered I×V curves for a large number of randomized sequences with the same amount of nucleotide composition as their corresponding miRNA chains. The profiles of some representative examples are shown in Figure 6.13a–d and point to a critical dependence on the exact structure of the sequence. Accordingly, we believe that the

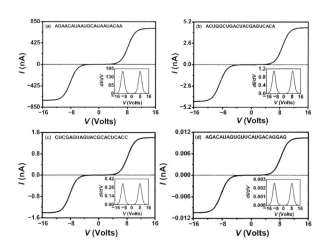

Figure 6.13 Current–voltage (I×V) curves for uncorrelated random RNA sequences with the same amount of nucleotide composition of some miRNA chains as the representative examples of the randomized sequences associated with the original: (a) miRNA 559 (group I); (b) miRNA 455-3p chain (group II); (c) miRNA 106a chain (group III); and (d) miRNA 106a chain (group IV). The insets show the transconductance $(dI/dV) \times V$ of the devices. After Oliveira et al. [157]

current saturation profiles discussed here are a step forward in the understanding of the autism neurobehavioral disorders and a useful tool for the development of biosensors to probe them.

6.5 Conclusions

Although direct measurement of charge transport in nucleic acids (DNA and RNA) and proteins (like the α_3 polypeptide considered here) is still a highly controversial topic due to the very challenging level of required manipulation at the nanoscale, it is believed to play a crucial role in many phenomena in living organisms. Among them, besides those discussed in this chapter, we can name the location of DNA lesions by base excision repair glycosylases and the regulation of tumor-suppressor genes, such as p53, by detection of oxidative damage.

No doubt, changes in the electronic structure of nucleic acids and proteins induced by the vast majority of mutations, as measured by a change in their charge transport, plays a role in fundamental biological and biochemical processes, leading to early diagnosis (like for ASD) and detection of mutation hotspots (like for prion-like diseases). Damaged bases in DNA are known to lead to errors in replication and transcription, compromising the integrity of the genome. In addition, biological molecules can facilitate charge transport over long molecular distances, being sensitive to a wide variety of lesions that perturb the DNA base stacking. We hope our biosensor prototype may help future experimental and engineering findings in this sense.

7

Amino Acid Anhydrous Crystals

7.1 Introduction

The last two decades have witnessed the development of many diverse applications demanding materials with specific electronic properties and advances for the integration of nanotechnology to biological systems. Organic materials and molecular crystals formed either from the nucleotide bases (guanine, adenine, cytosine, thymine, and uracil) or from the 20 amino acids that appear in the genetic code – stabilized by hydrogen bonds (H-bond), van der Waals (vdW), and dipolar electrostatic interactions (known as salt bridges) – are very promising due to their simpler fabrication processes, lower cost, flexibility, and ability to be easily incorporated into other materials. Unfortunately, they also tend to degrade with time and react with water or other substances in undesirable ways, which must be prevented with the use of adequate containers. Besides, in a crystallized structure, they form localized electronic states in which electrons move from one molecule to the next through quantum tunneling, which leads to a more complex charge transport than in usual semiconductor devices.

It is well known that a large proportion of our cells, muscles, and tissues are composed of amino acids, which are classified according to the locations of the central structural functional groups. They also play a key role in the transport and storage of nutrients, with great influence on the function of organs, glands, tendons, and arteries. In addition, they are essential for wound healing, tissue repair, and the removal of not only all types of nutrients but also the waste deposits produced in connection with the metabolism.

Recently, there have been efforts to employ amino acid molecules and crystals in biosensors and optoelectronic devices, although the interest remains focused mainly on the characterization of their polymorphs by, for instance, nuclear magnetic resonance and electron paramagnetic resonance spectroscopies. For example, a study of the adhesion of amino acids on a series of inorganic surfaces including

insulators and semiconductors, as well as the investigation of the stability of pro-
teogenic amino acids employing an array of semiempirical and ab initio methods,
was successfully achieved [172]. The electrical control of the amino acid ionization and the conformation of proteins arranged on semiconductor surfaces, which
might produce new types of biodevices, follow afterward. In additions, Stroscio
and Dutta have described advances in human-made synthetic nanostructures integrated with biological molecules and systems, their properties, characteristics, and
functions [173].

Despite its relevance, however, the electronic, optoelectronic, and vibrational
properties of few amino acid crystals have been measured so far, with results suggesting that some of them are wide-bandgap semicondutors and others could be
small-bandgap insulators. It was also demonstrated, by using a DFT framework,
that anhydrous crystals of the DNA bases present a wide-bandgap semiconductor
aspect, as discussed in Chapter 5.

In the solid state, there is a growing interest in the study of amino acid crystals
that has been stimulated by the perspective of understanding a system where the
hydrogen bond plays a fundamental role. As a consequence of this understanding, a better knowledge of some important biological molecules – for example,
proteins – can be obtained. Thus, it is of paramount importance to understand
these fundamental physical aspects of the amino acids for the development of
sustainable bio-organic electronic and optoelectronic devices, which could advance
the development of biodegradable, biocompatible, low-cost, and mass-produced
electronic components.

In particular, the vibrational spectra of amino acids is a rather useful tool in a
great number of biochemical studies involving proteins, enzymes, and their reactions that can be investigated by IR (infrared) and Raman techniques. There are
some reasons for a theoretical and experimental interest of these vibrational properties, such as:

(a) A detailed characterization of the interactions between polar and nonpolar
    molecules, which is fundamental for our understanding of biological functions.
(b) The knowledge of the Raman polarizability tensor, which can give informations
    about the charge transfer mechanism and intermolecular interactions.
(c) The Raman and infrared vibrational spectroscopy, to help the understanding of
    the molecular conformation and the nature of hydrogen bonding in proteins.
(d) The estimation of some thermodynamics parameters based on the phonon-
    related properties – such as heat capacity, entropy, enthalpy, and free energy –
    associated with the hydrogen bond in the crystals.

IR and Raman spectroscopy have also contributed to characterize the vibrational modes of the amino acid crystals, searching for polymorphism, due to their

importance in crystal engineering and the pharmacological industry. Besides, they can be quite important to the understanding of the structures of amino acids under high-pressure conditions, as pointed out by several works on different amino acids.

Motivated by this scenario, in this chapter, we present a quantum chemistry approach within the DFT computations by using the local density approximation (LDA) and the generalized gradient approximation with scatter correction (GGA+TS) to investigate the structural, electronic, optical, and vibrational properties of some important amino acid anhydrous crystals [174]. The electronic (band structure and density of states) and optical absorption properties are discussed to interpret the calculated light absorption performed at room temperature [175]. The real and imaginary parts of the dielectric function are also presented, as well as the absorption spectrum and the refractive index [176]. The infrared and Raman spectra are obtained for some of them, and their normal modes are assigned [177, 178]. Finally, a comparison between the vibrational properties of anhydrous and monohydrated L-aspartic acid crystals is done to unveil the complex role of water on the carrier transport properties in the latter [179].

7.2 Structural, Electronic, and Optical Properties

In this section, we present the structural, electronic, and optical properties of some amino acid crystals. The computational simulations were carried out by using the CASTEP code based on the DFT approach. When appropriate, two distinct exchange-correlation functionals were employed: the LDA and the generalized gradient approximation plus dispersion correction (GGA+TS). The LDA parameterization proposed by Cerpeley, Alder, Perdew, and Zunger [7, 8] was adopted, whereas the GGA functional is the one proposed by Perdew, Burke, and Ernzerhof (PBE) [12]. The PBE functional is quantitatively close in its results to the predictions obtained using the GGA–PW91 functional [10]. The dispersion correction scheme of Tkatchenko and Scheffler [21] was taken into account in the GGA computations to include the effect of van der Waals forces. Norm-conserving pseudopotentials were used to describe the core electronic states of each atomic species in both the LDA and GGA+TS simulations, and the Kohn-Sham orbitals were evaluated using a plane-wave basis set with a converged energy cutoff of 980 eV, its quality being adjusted for consistency as the unit cell volume changes during optimization. Each unit cell was relaxed to attain a total energy minimum, allowing for lattice parameter and atomic position adjustments.

Geometry optimization was performed by varying the unit cell size, angles, and internal atomic coordinates in order to obtain a total energy global minimum. The structure was taken to be converged after the following criteria were satisfied after two successive self-consistent energy iterations: total energy variation smaller than

0.5×10^{-5} eV/atom, maximum force per atom smaller than 0.01 eV/Å, maximum displacement smaller than 0.5×10^{-3} Å, and maximum stress component smaller than 0.02 GPa. The basis set quality was kept fixed despite the changes in the unit cell volume that occurred during the geometry optimization process. The Broyden-Fletcher-Goldfarb-Shanno (BFGS) iterative method or solving unconstrained nonlinear optimization problems was employed to carry out the unit cell optimization [138].

The self-consistent field steps have taken into account tolerances of 5.0×10^{-7} eV/atom for total energy and 0.5×10^{-6} eV for the electronic eigenenergies. From the valence band (VB) and conduction band (CB) curves at their critical points (maxima for VB and minima for CB), the effective masses for electrons and holes can be estimated through parabolic curve fittings, which are directly related to the flatness degree of the bands at critical points.

To evaluate integrals in reciprocal space, a Monkhorst-Pack $2 \times 2 \times 3$ sampling grid was used to achieve a well-converged electronic structure [180]. We note that increasing the sampling to $4 \times 4 \times 6$ using the $2 \times 2 \times 3$ optimized geometry does not lead to significant changes in the unit cell total energy, which varies by less than 0.04%, while the Kohn-Sham band structures remain essentially the same.

In the next subsections, we will perform an analysis of some amino acids, taking into account these comments.

7.2.1 L-Cysteine

Cysteine (symbol Cys or C), the 2-amino-3-thiol propane carboxylic acid, chemical formula $HO_2CCH(NH_2)CH_2SH$, is one of the most important amino acids. As demonstrated by X-ray diffraction and neutron scattering [181], L-cysteine has monoclinic and orthorhombic polymorphs, the latter being the dominant phase in high-purity powder samples, with small amounts of the DL-cysteine and the monoclinic L-cysteine present as impurities [182].

The crystal lattice parameters and atomic positions used as inputs for the calculations were taken from X-ray diffraction and neutron scattering data already published for the monoclinic and orthorhombic cysteine crystals [181]. The asymmetric unit of monoclinic cysteine has two crystallographically independent molecules, L-Cys (A) and L-Cys (B). On the other hand, the orthorhombic form has the thiol H atom disordered over two sites, forming alternate interactions with either a carboxylate O atom or the S atom of another thiol group.

The converged unit cell parameters for monoclinic (orthorhombic) cysteine are $a = 9.453$ Å, $b = 5.192$ Å, $c = 11.429$ Å ($a = 7.993$ Å, $b = 12.246$ Å, $c = 5.432$ Å), being in good agreement with the X-ray diffraction data [181] $a = 9.441$ Å, $b = 5.222$ Å, $c = 11.337$ Å, ($a = 8.116$ Å, $b = 12.185$ Å, $c = 5.426$ Å).

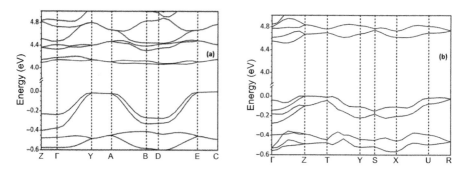

Figure 7.1 The Kohn-Sham electronic band structures of the cysteine poly-morphs: (a) monoclinic and (b) orthorhombic crystal structures. After Cândido-Júnior et al. [174]

The Kohn-Sham electronic band structures of the monoclinic and orthorhom-bic cysteine polymorphs are shown in Figure 7.1a and b, respectively. One can observe that both are indirect bandgap materials, the former with transition energies D → E of 4.06 eV and A → B of 4.08 eV and the latter with transition energies Z → T, T → Y and G → Z of 4.52 eV. The calculated band dispersion flatness at critical points suggests that both cysteine polymorphs have insulator characteristics, which severely limit their use in charge transport applications.

We also have performed optical absorption measurements to estimate the energy bandgap of the orthorhombic cysteine polymorph (the most stable one). Cysteine powder samples (sample C_1) with at least 99% purity were purchased from Sigma-Aldrich and used with no further purification to grow cysteine crystals by the standard slow evaporation method of recrystallization using deionized water as the solvent. The grown crystals were determined to be orthorhombic through X-ray diffraction (data not presented here). The orthorhombic cysteine recrystal-lized samples were then used to obtain a more purified orthorhombic cysteine powder (C2).

The C_1 and C_2 powders were mixed separately with KBr to form pellets. Light absorption measurements were carried out in the C_1-KBr and C_2-KBr pellets using a Varian Cary 5000 UV-visible NIR spectrophotometer equipped with solid sample holders. The absorption spectra of the samples were recorded in the 200–800 nm wavelength range (6.21–1.55 eV). The optical absorption measurements were per-formed by transmittance, with background removal and baseline corrections being made when necessary.

The onset of the absorption coefficient α as a function of the energy in an indirect gap crystal is related to the incident photon energy by [183]

$$\alpha = C[\hbar(\mp\Omega + \omega) - E_g]^{1/2}, \tag{7.1}$$

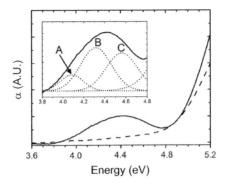

Figure 7.2 Measured optical absorption α (in arbitrary units) of cysteine-KBr pellets made up from the original Sigma-Aldrich cysteine powder C_1 (dashed line), and the recrystallized cysteine powder C_2 (solid line) in the main (stronger) absorption band. The inset shows the deconvolution of the defects related to the weak C_2 band around 4.4 eV, whose maxima peaks suggest inside gap defect energy levels roughly at 4.08 eV (A), 4.32 eV (B), and 4.57 eV (C), respectively. After Cândido-Júnior et al. [174]

where C is a constant, E_g is the indirect bandgap, and $\mp\hbar\Omega$ ($\hbar\omega$) is the energy of the absorbed or emitted (incident) phonon.

The spectra for the optical absorption of the C_1-KBr (dashed line) and C_2-KBr (solid line) pellets are depicted in Figure 7.2. A weak absorption structure centered around 4.4 eV is observed only in the case of the C_2-KBr samples, indicating they are due to the recrystallization process. One can observe a pronounced absorbance increase beginning around 4.9 eV. In the region where the light absorption of the cysteine-KBr pellets increases strongly, we have estimated the indirect bandgap by carrying out a linear fit of the square root of the absorbance, finding a transition energy of 4.62 eV in the case of C_1-KBr pellets.

On the other hand, the energy gap measured in C_2-KBr pellets was estimated at 4.68 eV. The pronounced absorption increase is explained as due to phonon-mediated Z → T, T → Y, and G → Z valence to conduction band electron transfer through the energy gap. The measured energy gap of 4.68 eV for C_2-KBr pellets is larger than the DFT-GGA calculated energy gap 4.52 eV, as expected, due to the well-known trend of DFT methods to underestimate the bandgap of crystals.

Interestingly, the measured energy gap of 4.62 eV for C_1-KBr pellets is closer to the DFT-GGA calculated energy gap. The weaker optical absorption structure beginning around 3.8 eV, which occurs only for the C_2-KBr pellets, can be associated with the existence of defects in the recrystallized cysteine samples. The existence of small monoclinic cysteine crystallites in the recrystallization powders and/or impurities is ruled out by its absence in the OC-KBr pellets. After several curve fitting trials, we have found that at least three deconvolution peaks are required to adequately explain the weaker optical absorption structure, these being assigned to the existence of defect related levels inside the bandgap at 4.08 eV (A), 4.32 eV (B), and 4.57 eV (C), as shown in the inset of Figure 7.2.

7.2.2 L-serine

Serine (symbol Ser or S), pKa $= 2.21$ (9.15) for the α-carboxylic acid group (α-ammonium ion), chemical formula $C_3H_7NO_3$, is one of the 20 natural amino acids. It is a nonessential one, since it can be synthesized from metabolites, including glycine, whose polymerization gives rise to the proteins, biochemical compounds that rule the life-related biological functions. Among the 10 amino acids that can be formed in Miller's atmospheric discharge experiments, it is ranked the sixth in order of decreasing abundance in prebiotic contexts, as predicted by thermodynamic arguments, which was likely reflected in the composition of the first proteins at the time the genetic code originated [184].

In respect to its abiotic synthesis, serine is found in meteorites, spark discharge experiments, and cyanide polymerization experiments. It has two enantiomeric modifications: L-serine and D-serine. In the solid state, serine can be crystallized as L- and D-enantiomorphs under ambient temperature, known today as the following crystal forms: L-Ser anhydrous monoclinic, P212121; L-Ser monohydrated orthorhombic, P212121; and DL-Ser anhydrous monoclinic, P21/a, all of them with atomic number $Z = 4$.

Here, our focus is on the serine anhydrous monoclinic crystal. By taking full advantage of its X-ray diffraction data [185], a monoclinic (P21) unit cell is constructed, and its geometry is optimized with respect to its total energy within the DFT scope. Figure 7.3 shows (i) the L-serine zwitterion molecular structure, (ii) the monoclinic unit cell for the anhydrous crystal, (iii) the pattern of hydrogen bonds around a single molecule inside the crystal, and (iv) a view of molecular stackings along the b-axis.

The Kohn-Sham electronic band structure was evaluated, taking into account a specific path inside the Brillouin zone of the anhydrous L-serine crystals. The LDA full-band structure is shown in the top part of Figure 7.4, together with the partial (per type of orbital) density of states (PDOS). From it, one can see a set of five deep s-like bands between -21.5 and -12.0 eV. Above them, we have the top valence bands, which are mainly p-like. The bottom of the conduction band is originated from the p-orbitals as well (see the sharp peak of the PDOS near 5 eV in Figure 7.4), but s-like contributions are relevant for energies above 6 eV. In the bottom part of Figure 7.4 there is a close-up of the band structure near the Kohn-Sham bandgap, where one can see that both the valence band maximum and the conduction band minimum occur at the Γ ($k = 0$) point. Thus, L-serine anhydrous crystals are direct

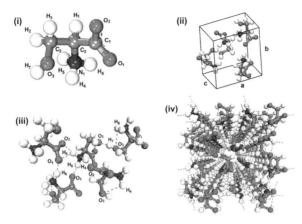

Figure 7.3 (i) The L-serine zwitterionic molecule; (ii) the anhydrous L-serine monoclinic unit cell; (iii) a view of the independent hydrogen bonds of the crystal; (iv) a view of the molecular stackings along the *b*-axis. The atom bonded to the hydrogens H_4, H_5, and H_6 is the nitrogen atom N_1. After Costa et al. [177]

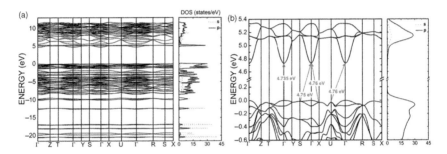

Figure 7.4 The Kohn-Sham electronic band structure of serine anhydrous crystals: (a) in the -21.5 to 13 eV range; (b) in the region around the bandgap energies (gray arrows). The partial density of states for the orbitals s (dotted) and p (solid) are shown on the right for each case. After Costa et al. [177]

bandgap materials, according to the LDA computations, with the direct gap being 4.74 eV. The secondary indirect gaps display slightly larger energies: 4.75 eV between the S point at the valence band (VB) and the Γ point at the conduction band (CB), and 4.76 eV between X (VB) and Γ (CB), and between U (VB) and Γ (CB). The four lowest conduction bands are dominated by contributions from p-orbitals, but with a small contribution from s-levels. The eight highest valence bands, on the other hand, are strongly p-like orbitals, with a very small amount of s-orbital character.

To analyze the most relevant contributions to the electronic states from each atomic species and each functional group of the serine molecule, we have plotted the per-atom PDOS in Figure 7.5 (LDA results). Looking at the top of the valence

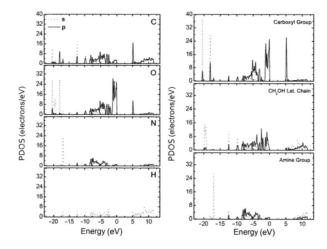

Figure 7.5 Left panel: atoms C, O, N, and H contributions to the PDOS of serine anhydrous crystals calculated by using the LDA approach: s (dotted, gray) and p (solid, black) orbitals. Right panel: contributions per molecular region. After Costa et al. [177]

band, one can see that the electronic energy levels have a strong O-2p character, with a much smaller contribution from the C-2p levels. The bottom of the conduction band, in contrast, is dominated by C-2p levels, but with a significant amount of contribution from O-2p orbitals. For energies higher than 5 eV, H-1s states, together with C-2p states, form the conduction bands. These s-like contributions come mainly from the hydrogen atoms at the lateral chain for energies below 7.50 eV and from the amine group for energies between 7.50 and 12.0 eV. The sharp PDOS peaks of the carboxyl group at the top of the valence band and the bottom of the conduction band are related to the localization of the wave functions corresponding to the HOMO and LUMO of the zwitterionic serine molecule.

The optical absorption curves for distinct incident light polarizations are depicted in Figure 7.6 for both LDA and GGA+TS functionals. Analyzing them, we can see that they present a strong degree of optical anisotropy. For example, in the case of (001) polarized light (the number stands for the crystal plane with which the polarization plane aligns), there is a sharp peak at 8.50 eV, which is much smaller than the (010) and (100) polarized cases. Also, the optical absorption peak structure at the position 6.00 eV observed for the (100) direction is much smaller for the (001) case.

The real and imaginary parts of the dielectric function $\epsilon(\omega)$ are shown in Figure 7.7 for the LDA functional and light polarized along the (100), (010), and (001) directions, as well as for a polycrystalline (POLY) sample. As the imaginary

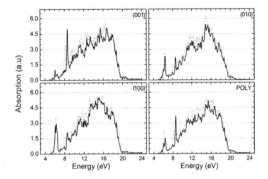

Figure 7.6 The functionals LDA (dotted gray line) and GGA+TS (solid black line) calculated optical absorption of the anhydrous serine crystal along the directions (001), (010), and (100) for the former and (001), (010), and (100) for the latter. The optical absorption for the polycrystal is also shown. After Costa et al. [177]

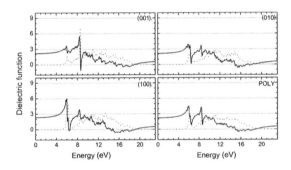

Figure 7.7 Real (solid black line) and imaginary (dotted gray line) components of the LDA calculated dielectric function of the anhydrous serine crystal along the directions (001), (010), (100), and for the polycrystal (POLY). After Costa et al. [177]

part is closely related to the optical absorption, which we already discussed, we will focus now only in the real part, $\epsilon_1(\omega) = \text{Re}(\epsilon)$. If we compare the behavior of $\epsilon_1(\omega)$ along different polarization directions, we can depict again some degree of anisotropy. While the (010) and (100) directions share many similarities, the (001) direction has some remarkable differences in comparison to the other two. For example, the minimum (maximum) near 6.30 eV (5.90 eV) observed for the (010) and (100) cases seems to be shifted upward to 8.30 eV (8.10 eV) for the (001) polarized light, being also just a bit above 2 at $\omega = 0$ for the (001) and (010) polarizations and about 2.20 for the (100) polarization. For most of the 0–23.0 eV energy range, $\epsilon_1(\omega)$ is positive, except for the 14.0–19.0 eV region, where it becomes slightly negative but always larger than -1.

7.2.3 L-Proline

Proline (symbol Pro, P), pKa = 2.0 (10.6) for the α-carboxylic acid group (α-ammonium ion), chemical formula $C_5H_9NO_2$, is one of the 20 amino acids whose polymerization gives rise to proteins. It has a side chain that is twice connected to the protein backbone, forming a five-member nitrogen-containing ring that gives rise to a conformational structure of exceptional rigidity. As a consequence of this inflexibility, proline is unable to occupy many of the main chain conformations easily adopted by all other amino acids, being often found in very tight turns of protein structures – i.e., where the polypeptide chain must change direction.

The formation of the supercoiled triple-helix basis of the 28 types of collagen, the major structural protein of vertebrates formed by connective fibrils that binds and supports all other cellular tissues (see Chapter 11 for details), requires the presence of a repeated glycine–X–Y sequence, the most common being glycine-proline-hydroxyproline, meaning that proline is a major component of collagen structures [186]. Consequently, its importance points to the necessity of a deeper understanding of its individual structural features in vacuum, aqueous media, and the solid state.

As a member of the class of molecular crystals, proline in the solid state is stabilized by hydrogen bonds and van der Waals interactions (there are no electrostatic interactions, known as salt bridges, in this case), which determine their structural, electronic, optical, NMR, and vibrational properties. As a matter of fact, X-ray diffraction data of anhydrous orthorhombic Pro (space group P212121, Z = 4), monohydrated monoclinic Pro (space group C2, Z = 4), anhydrous monoclinic DL-Pro (space group P21/c, Z = 4), and monohydrated orthorhombic DL-Pro (space group Pbca, Z = 8) [53,56] are already available in the literature.

Our focus here is on the orthorhombic anhydrous L-proline crystals, whose lattice parameters are a = 11.550 Å, b = 9.020 Å, c = 5.200 Å, V = 541.741 Å^3. Its structural features are shown in Figure 7.8. The atom labels adopted here are depicted in Figure 7.8a. The pyrrolidine region is nearly parallel to the *ab*-plane (Figure 7.8b), with the C_1–C_2 perpendicular to it pointing nearly along the *c*-axis (Figure 7.8c). A set of hydrogen bonds can be clearly seen along the *c*-direction (Figure 7.8d).

To plot the Kohn-Sham electronic energy levels as a function of the reciprocal space vector **k**, we have chosen a path along the first Brillouin zone passing through a set of high-symmetry points, as depicted in Figure 7.9. For each exchange-correlation functional adopted (LDA, GGA, GGA+TS), a distinct electronic band structure was obtained. However, we present just the GGA+TS band structure in Figure 7.10, as this functional has exhibited the best structural match

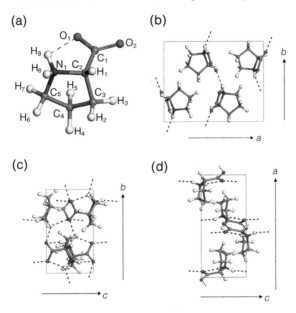

Figure 7.8 (a) L-proline zwitterionic molecule with atom numbers; (b), (c), and (d): different views of the anhydrous L-proline orthorhombic unit cell. Dashed lines represent the hydrogen bonds. After Caetano et al. [176]

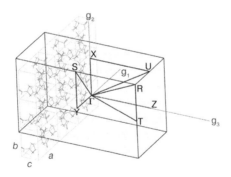

Figure 7.9 First Brillouin zone of anhydrous orthorhombic L-proline crystals. High symmetry points of the reciprocal lattice are indicated. After Caetano et al. [176]

with the experimental data. Looking more closely at the GGA+TS band structure shown in Figure 7.10b, the valence band has four energy maxima at points along the ΓZ, ΓU, ΓR, and ΓT lines, with approximately the same energy, while the conduction band has two energy minima, at T and Z, which also have almost the same energy value. In addition to the main indirect bandgap $\Gamma Z \rightarrow T$ (number 1, in Figure 7.10b) of 4.870 eV, this gives rise to seven more possible electron state transitions with very close indirect bandgaps – namely, in increasing order, 4.872

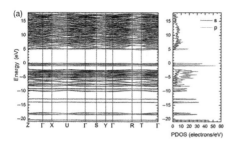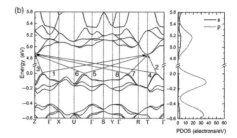

Figure 7.10 GGA+TS calculated band structure of anhydrous L-proline crystals: (a) in the −21 to 18 eV energy range; (b) in the region around the electronic main band gap. Right side panels depict the contributions from s and p orbitals to the partial electronic density of states (PDOS). Eight very close indirect gaps (black arrows) are indicated by numbers in increasing order of energy. After Caetano et al. [176]

(ΓT → T, 2), 4.874 (ΓZ → Z, 3), 4.876 (ΓT → Z, 4), 4.878 (ΓU → T, 5), 4.882 (ΓU → Z, 6, and ΓR → T, 7), and 4.886 eV (ΓR → Z, 8), respectively – covering an energy range of about 12 meV.

The proximity of these bandgaps points to many pathways for the excitation of charge carriers to the conduction band. The valence and conduction band extrema at Γ, on the other hand, are saddle points of the electronic eigenenergy. The top of the valence band has a strong p-like character (see the PDOS inset of Figure 7.10b), while the bottom of the valence band originates mostly from p orbitals with a small contribution from the s states.

If one considers the per-atom and per-group contributions to the electronic states, depicted in Figure 7.11, we have that the top of the valence band has a strong contribution from O 2p levels localized at the COO− group, while the carbon atoms, especially the C_1 atom at COO−, originate the lowest conduction bands in the proline crystal. The amino acid side chain, on the other hand, contributes significantly to the valence bands between −10 and −2 eV, and to the conduction bands above 5 eV. In contrast, the NH2+ group has a negligible contribution to the valence and conduction band extrema.

A comparison of the molecular HOMO and LUMO orbitals with the corresponding states for the crystal (see Figure 7.12), shows that the crystal HOMO at $k = 0$ resembles closely the HOMO state of a single molecule solvated in water, with the wave function concentrating around the COO− group, the C_1–C_2 bond in a σ-like configuration, and a small contribution from the amine group. The HOO orbital for the molecule in vacuum, on the other hand, has a more pronounced contribution from the side chain.

For the LUMO state, both the water solvated and in vacuum configurations are similar, with lobes at the carboxyl, C_1–C_2 π-like bonding, a small contribution from

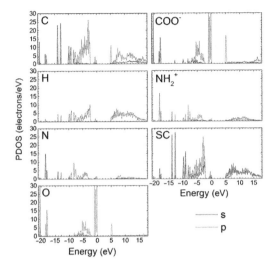

Figure 7.11 The GGA+TS calculated PDOS for the C, H, N, and O atoms and the COOH−, NH₂+, and side-chain groups of proline in the anhydrous crystal form. After Caetano et al. [176]

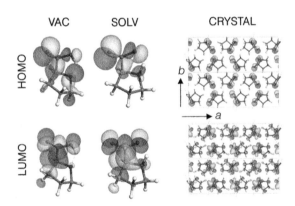

Figure 7.12 GGA+TS calculated HOMO (top) and LUMO (bottom) orbitals for a proline molecule in vacuum (VAC, left), solvated in water (SOLV, middle), and for the orthorhombic anhydrous crystal at $k = 0$ (CRYSTAL, right). After Caetano et al. [176]

the amine group, and s-like contributions at the C_3 and H_2 atoms. In the crystal, the LUMO also includes some s-like structures involving C_4, C_5, and its hydrogen atoms.

For the sake of comparison with the crystal main bandgaps, we have obtained GGA+TS HOMO-LUMO transition energies of 4.91 eV for the isolated molecule without solvation and 5.34 eV with solvation.

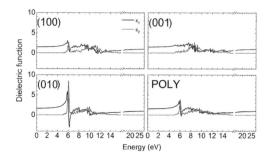

Figure 7.13 Real (black line) and imaginary (gray line) components of the GGA+TS calculated dielectric function of the anhydrous monoclinic L-proline crystal for incident polarized light along the planes (100), (010), (001), and for a polycrystalline sample. After Caetano et al. [176]

The electronic complex dielectric function $\epsilon(\omega) = \epsilon_1 + i\epsilon_2$ of the anhydrous proline crystal depending on the incident photon energy $E = \hbar\omega$ was calculated for polarized incident light with polarization planes (100), (010), and (001) and for a polycrystalline sample, using the same methodology as in the previous section. It is depicted in Figure 7.13. A remarkable degree of optical anisotropy can be seen from it, with the real and imaginary parts of the dielectric function exhibiting a very pronounced oscillation near 6.0 eV along the (010) direction, which is absent from the curves along the polarizations (100) and (001). The maximum of $\epsilon_1(\omega)$ ($\epsilon_2(\omega)$) for the (010) polarization occurs at 5.86 (6.06) eV. The minimum of $\epsilon_1(\omega)$ is at 6.25 eV, and $\epsilon_1(\omega = 0) = 1.63$.

For the (100) polarized radiation, there is an oscillation of both the real and imaginary part of the dielectric function in the 5.4–6.6 eV range of energy, but with a smaller amplitude than the observed for the (010) polarization. Also, $\epsilon_1(\omega = 0) = 1.48$. If the incident light is polarized parallel to the (001) crystal plane, the real part of the dielectric function shows a hump between 4.4 and 8.0 eV, decreasing between 8.0 and 10 eV without becoming negative. The imaginary part of the dielectric function has a set of low-intensity maxima between 5.0 and 7.4 eV and then increases, reaching a maximum at nearly 8.7 eV.

The polycrystalline complex dielectric function resembles more closely the (100) curve but with a wider range of variation for both the real and the imaginary part of the dielectric function.

Looking now to the optical absorption curves shown in Figure 7.14, the same anisotropic behavior is observed for the dielectric function, which is expected, because the optical absorption is proportional to the imaginary part of the dielectric function. Both the (100) and (010) polarizations have a slow onset between 4.9 and 5.8 eV and a pronounced peak between 5.8 and 6.5 eV, with a maximum at about

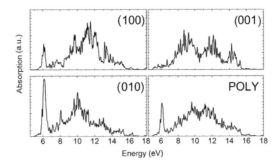

Figure 7.14 GGA+TS calculated optical absorption of the anhydrous monoclinic L-proline crystal for incident polarized light along the planes (100), (010), (001), and for a polycrystalline sample. After Caetano et al. [176]

6.1 eV. These structures are related to the electronic transitions from valence O 2p states at the carboxyl region to H 1s, and C 2s conduction states at the proline side chain. Secondary absorption maxima can be observed for the (010) polarization at 10.0 eV and for the (100) polarization at 11.5 eV. For the (001) direction, we have the most intense absorption peak at 8.7 eV, with a secondary maximum at 9.2 eV. For incident light on a polycrystalline sample, the most important features follow approximately the pattern of the (100) polarization case.

7.3 Infrared and Raman Spectra of the L-Aspartic Acid

The aspartic acid (symbol Asp or D), pKa = 1.99 (9.90) for the α-carboxyl (amino) mode, chemical formula $C_4H_7NO_4$ is also one of the 20 amino acids encoded by the genetic code, being a component of proteins in living beings. It was one of the most abundant amino acids in the primitive earth, having been found in the Murchison meteorite [187]. Aspartic acid was first discovered in 1827 by Plisson, and was synthesized by boiling asparagine, which was isolated from asparagus juice in 1806, with a base [188]. It is a natural nonessential amino acid that provides the amino group in the urea's cycle as well as in the biosynthesis of purine, being also involved in the synthesis of other important biological compounds, such as pyrimidine. As a specific cleavage site for caspases, the left-hand enantiomer L-aspartic acid is a key residue in apoptosis (e.g., programmed cell death) processes, and after more than 30 years, there is strong evidence supporting aspartate (the carboxylate anion, salt, or ester of aspartic acid) as an excitatory neurotransmitter in the central nervous system.

In the solid state, aspartic acid can be crystallized as D- and L- enantiomorphs at room temperature, having the following crystal forms: L-Asp anhydrous monoclinic, P21 and Z = 2; L-Asp monohydrated orthorhombic, P212121 and Z = 4;

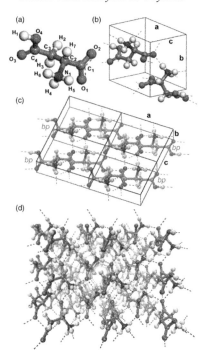

Figure 7.15 (a) The L-aspartic acid zwitterionic molecule; (b) the anhydrous L-aspartic acid monoclinic unit cell; (c) a top view of two superimposed *ab* planes of L-aspartic acid zwitterionic molecules, where one can be transformed into the other by a screw axis. Note that the molecules in the back plane are named *bp*; (d) perspective view of tunnels in anhydrous L-aspartic acid monoclinic crystals. After Silva et al. [175]

and DL-Asp anhydrous monoclinic, $C2/c$ and $Z = 8$. The carbon backbone of the L-aspartic acid molecules in the crystalline phase is practically planar, the average deviation from planarity being less than 0.01 Å.

The L-aspartic acid zwitterionic molecule is depicted in Figure 7.15a, while the monoclinic unit cell of anhydrous L-aspartic acid crystals is shown in Figure 7.15b. Parallel layers of L-aspartic acid molecules can be distinguished, one being converted into another by a screw displacement, as can be inferred from Figure 7.15c. Two adjacent layers are connected through hydrogen bonds occurring between charged groups (COO−, NH3+, and COOH). Within a single layer, the molecules are linked togetherm forming zig-zag chains, which are connected via hydrogen bonds involving the NH3+ group [189]. Finally, a perspective view of the existence of molecular tunnels in L-aspartic acid crystals is shown in Figure 7.15d. There are no intramolecular hydrogen bonds in the crystal because the lattice structure prevents the formation of the internal N1–H4 $\cdots$ O2 bond exhibited by the lowest-energy molecular conformers of the L-aspartic acid. We have considered

Table 7.1 *The optimized parameters of the anhydrous L-aspartic acid crystal unit cell. After Silva et al., [175].*

Approximation	a (Å)	Δa (Å)	b (Å)	Δb (Å)	c (Å)	Δc (Å)	V (Å³)	ΔV (Å³)	β (deg)	$\Delta\beta$ (deg)
LDA	7.448	−0.169	6.712	−0.270	4.971	−0.171	243.606	−25.832	101.370	1.530
GGA	7.722	0.105	7.329	0.347	5.262	0.120	296.092	26.654	96.152	−3.688
GGA + D	7.646	0.029	6.959	−0.023	5.118	−0.024	268.880	−0.558	99.091	−0.749
Expt.	7.617		6.982		5.142		269.438		99.840	

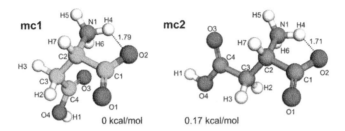

Figure 7.16 L-Aspartic molecular zwitterionic configurations mc1 and mc2. The lowest-formation energy configuration, mc1, is used as a reference to measure the relative formation energy of mc2. Hydrogen bond lengths are also shown. After Silva et al. [178]

729 possible configurations for the L-aspartic acid zwitterionic molecule using the water polarizable continuum model (PCM) to reduce the computational costs. There are only two conformers with formation energies smaller than k_BT (k_B being the Boltzmann's constant) at room temperature, namely the configurations mc1 and mc2 depicted in Figure 7.16.

The optimized parameters of the anhydrous L-aspartic acid crystal unit cell are reported in Table 7.1, which depicts the lattice parameters (in Å), unit cell volume (in Å³), and angle β (in deg) as calculated at the LDA, GGA, and GGA+TS levels. Experimental values for the parameters as measured by Derissen et al. [189], as well as their deviations from the experimental values, are also shown, for the sake of comparison. One can observe the LDA calculated lattice parameters a, b, and c are, respectively, about −0.169 Å, −0.270 Å, and −0.171 Å smaller than their X-ray measured values. This volume shrinking occurs because the LDA functional tends to overestimate interatomic forces. On the other hand, the calculated lattice parameters are always larger than the experimental data (pure GGA, contrary to LDA, tends to underestimate the strength of atomic interactions), with the largest difference being observed for b, 0.347 Å, followed by c, with 0.120 Å. Indeed, of all three approaches (LDA, GGA, and GGA+TS), the pure GGA method is the least accurate to predict the structural parameters of the anhydrous aspartic acid unit cell, notwithstanding the fact that the GGA functional provides a better description of hydrogen bonding in molecular crystals. The GGA+TS approach, in comparison, has the best outcomes, with a, b, and c, respectively, being about 0.029 Å larger, 0.023 Å smaller, and 0.024 Å smaller than the measurements. We must then conclude that the dispersion correction is very important to compensate for the GGA underestimation of interatomic forces in L-aspartic acid anhydrous crystals. The calculated unit cell volumes are 25.832 Å^3 and 0.558 Å^3 smaller than the experimental data, within the LDA and GGA+TS approaches, respectively, and 26.654 Å^3 larger in the GGA case. Besides, the β angle is larger by 1.530° (smaller by -3.688°, -0.749°) in the LDA (GGA, GGA+TS) case in contrast to the experimental value.

The optimized structure through the LDA and GGA approximations were used to perform the density functional perturbation theory (DFPT) calculation, or linear response formalism, to obtain the infrared and Raman spectra as well as its vibrational properties. Linear response provides an analytical way of computing the second derivative of the total energy with respect to a given perturbation. Depending on the nature of this perturbation, a number of properties can be calculated (for example, a perturbation in ionic positions gives the dynamical matrix describing the phonons state; an external electrical field has as a linear response the dielectric functions; etc,).

The infrared absorption intensities are described in terms of a dynamical (Hessian) matrix and Born effective charges (also known as atomic polarizability tensors, ATP) and can be obtained by calculating the phonons at the Γ point ($k = 0$). Raman spectroscopy is used to study the vibrational, rotational, and other low-frequency modes in a system, based on the Raman effect of inelastic scattering of monochromatic light. This interaction with vibrations results in the energy of incident photons being shifted up or down. The energy shift is defined by the vibrational frequency, and the proportion of the inelastically scattered light is defined by the spatial derivatives of the macroscopic polarization. The infrared and Raman spectra of the anhydrous L-aspartic acid crystal are shown in Figure 7.17a and b (0–1,000 cm^{-1}), Figure 7.18a and b (1,000–2,000 cm^{-1}), and Figure 7.19a and b (2,000–4,000 cm^{-1}), respectively, for the crystal experimental (EXP), mc1 (MC1), mc2 (MC2), and crystal theoretical (CRYS) data for comparison.

In Figure 7.17a, one can see the 0–1,000 cm^{-1} wave-number range of the infrared spectra (IR). For wave number below 400 cm^{-1} (dashed line), the IR experimental curve was taken from Matei et al. [190]. This region corresponds to normal modes characterized by torsions, rocking, and deformations across the entire molecular structure. There is good agreement between the experimental curve and the theoretical curve for the crystal. The spectral curves obtained for the single molecules, however, do not agree very well with the measured crystalline spectrum. As a matter of fact, the difference is not only remarkable in the case of lattice modes ($\omega < 200$ cm^{-1}) but also very important for modes around 400, 600,

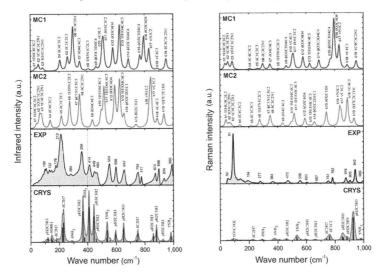

Figure 7.17 (a) Infrared spectra (in arbitrary units) of the anhydrous monoclinic L-aspartic acid crystal in the 0–$1,000$ cm^{-1} wavevector range. Panels MC1 and MC2: DFT-calculated spectra for the conformers mc1 and mc2 depicted in Figure 7.16; Panel EXP: Experimental curve with its most intense absorption peaks indicated (in cm^{-1}); Panel CRYS: DFT-GGA+TS calculated spectrum for the crystal. The normal mode assignments for the theoretical spectra are shown. (b) Same as before but for the Raman spectra. After Silva et al. [178]

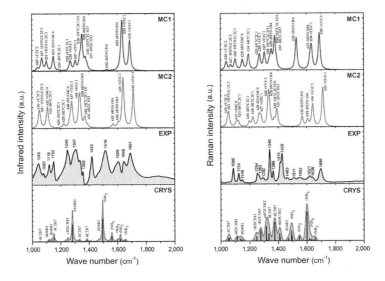

Figure 7.18 Same as in Figure 7.17 but for $1,000$–$2,000$ cm^{-1} wavevector range. After Silva et al. [178]

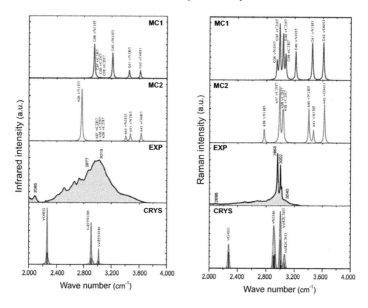

Figure 7.19 Same as in Figure 7.17 but for 2,000–4,000 cm^{-1} wavevector range.
After Silva et al. [178]

and even 800–1,000 cm^{-1}. This result is consistent with the work of Lopes et al.
[191], who performed a study of solid-state anhydrous adenine using vibrational
spectroscopy and DFT computations and concluded that it is necessary to use DFT
periodic approaches to describe the low-energy region of the vibrational spectrum
(<900 cm^{-1}). Consequently, the isolated molecule picture is unable to provide a
good description of the L-aspartic acid anhydrous monoclinic crystal vibrational
properties for wavelengths very much larger than those of the lattice modes –
i.e., in the 200–1,000 cm^{-1} range. The most pronounced IR-active modes in the
experiment appear at 212, 359, 554, and 990 cm^{-1}, which are assigned to the
GGA+TS vibrations at 232 (δC_2NH_3), 365 ($\rho H_3C_3H_2$), 536 (τNH_3), and 982
(τNH_3) cm^{-1}.

Figure 7.17b shows the 0–1,000 cm^{-1} Raman spectra curves following the same
scheme as in Figure 7.17a. As one can see, the relative intensities of the peaks are
in good agreement between the EXP and CRYS curves except for the very intense
experimental peak at 91 cm^{-1}, which in the theoretical curve occurs at 93 cm^{-1}
and is assigned to a $\tau O_1C_1O_2$ motion. Another intense Raman line can be seen at
942 cm^{-1}, being related to the theoretical normal mode at 927 cm^{-1} ($\rho H_2C_3H_3$),
which is shifted down to 934 cm^{-1} in the infrared spectra. The molecular Raman
spectra for the mc1 and mc2 conformers, when contrasted with the experimental
measurements, display fewer similarities than the theoretical Raman curve obtained

for the crystal. Many relative intensities are clearly distinct from the measured spectrum, as occurred for the infrared case.

The infrared spectra (IR) in the 1,000–2,000 cm^{-1} range can be seen in Figure 7.18a. The experimental plot exhibits bands with large broadening and a continuous absorption band above 1,800 cm^{-1} that could not be matched to any feature of the calculated curves. Four maxima – at 1,249, 1,307, 1,516, and 1,691 – can be highlighted because of their intensities. In the theoretical spectra their matches are 1,249, 1,282, 1,496, and 1,656 cm^{-1}, with the last two being assigned to the torsion of the amine group. Interestingly, the peaks at 1,422 and 1,516 cm^{-1} are absent in both the mc1 and mc2 conformers. At the same time, the bands around 1,600 cm^{-1} are due to the amine torsions in the crystal but are related to O_1C_1 stretchings in the molecular spectra (those for the crystal are shifted to higher wave numbers, above 2,000 cm^{-1}).

In the Raman spectra between 1,000 and 2,000 cm^{-1} (Figure 7.18b), three peaks – at 1,340, 1,426, and 1,696 cm^{-1} – are shown to be the most important in the EXP spectrum. The band at 1,340 cm^{-1} corresponds to the IR peak at 1,359 cm^{-1} and originates from the deformation of the C_2H_7 bond relative to the aspartic acid molecule. At 1,426 cm^{-1}, we have an O_4H_1 deformation (1,466 cm^{-1} in the GGA+TS data), and for 1,696 cm^{-1} we have a match with the IR peak at 1,691 cm^{-1}, being assigned to the GGA+TS normal mode at 1,656 cm^{-1} (amine torsion). There is a gap between 1,150 and 1,250 cm^{-1}.

Finally, Figure 7.19a depicts the infrared spectra (IR) for wave numbers between 2,000 and 4,000 cm^{-1}. These are localized modes with a strong molecular component (lattice effects tend to be small). In the experimental curve, there is a lot of thermal broadening. Three maxima – located at 2,085, 2,877, and 3,018 cm^{-1} – seem to be related to the theoretical spectrum for the crystal, the first being assigned to O_4H_1 stretching and the others two involving the asymmetric stretching of NH bonds. The molecular infrared spectra for both mc1 and mc2 conformers do not exhibit any peak to be correlated with the EXP band at 2,085 cm^{-1}. On the other hand, the experimental curve exhibits a tail after 3,300 cm^{-1} that does not appear in the DFT-computed spectrum for the crystal but seems to be connected to a set of NH and OH stretching bands for the mc1 and mc2 conformers.

The Raman spectrum in the 2,000–4,000 cm^{-1} range is shown in Figure 7.19b. It has much less broadening than the infrared absorption shown in Figure 7.19a. Three intense lines are clearly visible at wave numbers of 2,958, 3,002, and 3,040 cm^{-1}, and a barely perceptible band occurs at 2,090 cm^{-1}. The peak at 2,958 cm^{-1} is assigned to NH bond stretching, and the maxima at 3,002 and 3,040 cm^{-1} correspond to $H_2C_3H_3$ symmetric bond stretchings. The small feature at 2,090 cm^{-1}, by the way, can be related to the theoretical peak at 2,301 cm^{-1} that corresponds to the stretching of the O_4H_1. This Raman band is quenched probably due to thermal and anharmonic effects (see the next subsection). Molecular peaks (mc1 and mc2) for wave numbers larger than 3,200 cm^{-1} have no counterparts in the experimental Raman spectrum.

7.3.1 Phonon-Related Properties

The anhydrous L-aspartic crystal has a total of 96 phonon bands that can be decomposed into irreducible representations of factor group C2 as $\Gamma = 48A + 48B$, with two As and one B acoustic branches. In Figure 7.20a, one can see 44 phonon bands in the wave number range of 0–900 cm^{-1} whereas in Figure 7.20b, an enlarged portion of the acoustic bands is depicted, showing 14 dispersion curves. One can see from this that the acoustic modes along $\Gamma \rightarrow D$ and $\Gamma \rightarrow E$ behave in a similar fashion, whereas the phonons along the $\Gamma \rightarrow Z$, $\Gamma \rightarrow Y$, and $\Gamma \rightarrow C$ directions display some anisotropy. The transverse acoustic (TA1 and TA2) modes at Z reach wave numbers 48 and 71 cm^{-1} (maximum of TA2), whereas the longitudinal acoustic (LA) phonon wave number at the same point is 76 cm^{-1}. As we approach Y, wave numbers for the TA branches increase up to 50 cm^{-1}, whereas those for the LA branch reach 59 cm^{-1}. A near-crossing between the LA and TA curves occurs at 83% of the $\Gamma \rightarrow Y$ segment. The maximum of the LA branch is close to the A point, being approximately 77 cm^{-1}, while the maximum of the TA1 curve is at the C point (67 cm^{-1}).

On the other hand, for optical modes, there is a longitudinal optical–transverse optical (LO–TO) splitting at Γ of 27.9 cm^{-1}, with the LO phonon at 79.3 cm^{-1} and the TO phonon at 51.5 cm^{-1}. The TO phonon band maximum occurs nearly in the middle of the $\Gamma \rightarrow D$ segment, corresponding to 84 cm^{-1}, whereas the LO band reaches its peak value in the middle of the $E \rightarrow C$ segment (94 cm^{-1}). In the 0–900 cm^{-1} range, nine phonon gaps can be counted, namely 310–340, 425–440, 480–527, 542–551, 565–595, 604–650, 685–730, 768–850, and 863–882 cm^{-1}.

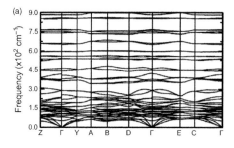 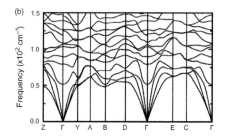

Figure 7.20 Phonon dispersion curves for the anhydrous monoclinic L-aspartic acid crystal in the wavevector ranges of (a) $0 - 9 \times 10^2$ cm^{-1} and (b) $0 - 1.5 \times 10^2$ cm^{-1}. After Silva et al. [178]

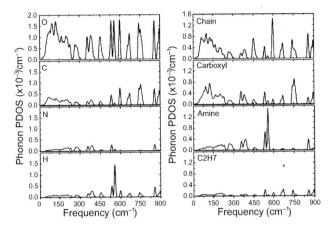

Figure 7.21 Phonon partial densities of states for the anhydrous monoclinic L-aspartic acid crystal: contributions per atom type (left panel) and per molecular region (right panel). After Silva et al. [178]

Figure 7.21 shows the phonon DOS per atom (left) and per atom group (right). In the per-atom analysis, one can see that the main contribution to the vibrations in the anhydrous L-aspartic acid crystal comes from the oxygen atoms. In the wavenumber interval between 0 and 240 cm^{-1}, we have a set of structured broad peaks, with a gap between 240 and 255 cm^{-1}, followed by another broad structure with a maximum at 268 cm^{-1}. Other maxima in the oxygen contribution occur at 358, 384, 452, 534, 556, 598 (highest value), 669, 740, 753, and 855 cm^{-1}. The carbon and nitrogen atoms have smaller phonon DOS values than oxygen. The carbon atoms vibrate more for the phonons at 598, 740, 753, and 856 cm^{-1}, and the nitrogen atom contribution is very small between 570 and 845 cm^{-1}. The hydrogen atoms, on the other hand, produce a very strong peak in the phonon DOS at 556 cm^{-1} and secondary peaks at 535 and 856 cm^{-1}.

Looking now to the contributions from each atomic group, the side chain of the aspartic acid is more active for the phonon bands near 600, 100, 450, 740, and 890 cm^{-1}, in decreasing order, whereas the carboxyl group is more active at about 750, 120, 350, 530, and 900 cm^{-1}. For the amine group, on the other hand, we have two strong contributions at nearly 560 and 530 cm^{-1}, with secondary maxima at 850, 390, 360, and 270 cm^{-1}. Finally, the C_2H_7 (C_2 is the α carbon) group has a more modest role in comparison to that of the other groups, with the most prominent features at 530, 670, 740, 860, and 890 cm^{-1}.

7.3.2 Heat Capacity

The constant volume specific heat C_V of the anhydrous L-aspartic crystal as a function of temperature T can be obtained from the phonon density of states. Its

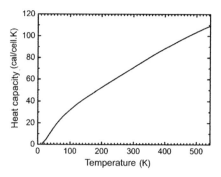

Figure 7.22 Heat capacity curve for the anhydrous monoclinic L-aspartic acid crystal for temperatures up to the melting point of the L-aspartic acid (543 K). After Silva et al. [178]

curve is displayed in Figure 7.22 for a temperature of up to the melting point of the aspartic acid (543 K). One can discern four regimes: the first between 15 and 35 K, where C_V increases more sharply from 1.10 to 13.8 (cal/cell)K; the second between 35 and 185 K, where the rate of increase of C_V becomes smaller, $C_V = 50$ (cal/cell)K at 185 K; the third regime, between 185 and 300 K, C_V grows almost linearly, reaching 72 (cal/cell)K for $T = 300$ K; and the fourth regime occurring above 300 K, where the rate of increase of C_V decreases, with a maximum value of about 150 (cal/cell)K at 1,000 K.

If we perform a comparison with the experimental data for the constant-pressure specific heat C_p from [192] for solid aspartic acid, we see that our C_V curve closely follows its results (for solids, C_p and C_V have almost the same values). For example, for a temperature of 40 K, $C_p = 25.18$ (experimental) and $C_V = 23.9$ (calculated in this work; all units in J/(K mol)). At $T = 100$ K, $C_p = 69.45$, and the DFT estimate for C_V was 67.4. Finally, for $T = 300$ K, $C_p = 155.98$ (experimental) and $C_V = 149$ (calculated; all units in J/(K mol)). These estimates are also within the same order of magnitude as the values found in [193] who studied the phonon dispersion and heat capacity of the α-helical form of poly (L-aspartic acid) using a semiempirical procedure.

7.4 Role of Water on the Vibrational Spectra of L-Aspartic Acid

In this section, we present a comparison between the vibrational (IR and Raman) properties of anhydrous and monohydrated aspartic acid crystals obtained from first-principles calculations carried out within the density functional theory formalism. The unit cell data for the inputs were taken from X-ray diffraction data published by Derissen et al. for the anhydrous phase [189] and Umadevi et al. for the monohydrated crystal [194] and are shown in Figure 7.23a and b.

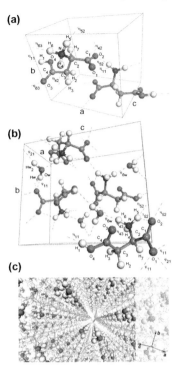

Figure 7.23 (a) Unit cell of the anhydrous aspartic acid crystal, (b) unit cell of the monohydrated L-aspartic acid crystal, and (c) a representation of the water molecule stacking (water columns) along the a-axis of the monohydrated L-aspartic acid crystal. After Silva et al. [179]

Structurally, the L-aspartic acid molecules in the anhydrous phase have their four carbon atoms contained in the same plane. The same planarity, however, is absent in the monohydrated crystal. A network of hydrogen bonds is essential to stabilize both crystals. Following the atomic labeling presented in Figure 7.23a and b, one can see that the anhydrous crystal has four distinct hydrogen bonds, denoted as η_{11} (H_1–O_1), η_{42} (H_4–O_2), η_{52} (H_5–O_2), and η_{63} (H_6–O_3). For the monohydrated L-aspartic acid unit cell, there are six distinct hydrogen bonds, three between L-aspartic acid molecules (η_{11}, η_{43}, η_{52}, the first and the third being reminiscent of the anhydrous structure) and three between a water molecule and an aspartic acid molecule (φ_{11}, φ_{21}, and φ_{6w}). One can note from Figure 7.23c that the water molecules form stacks along the direction of the a-axis, which can enhance hole transport along the $\Gamma \rightarrow Z$ direction, as we will see later.

Some selected regions of the calculated IR and Raman spectra of anhydrous and monohydrated L-aspartic acid crystals are shown in Figure 7.24. The infrared absorption intensities are evaluated from the dynamical matrix and Born effective

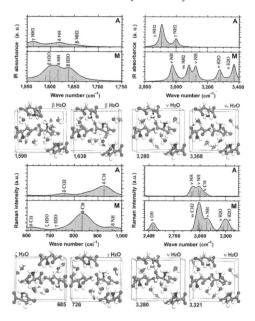

Figure 7.24 Main water vibrational signatures in the infrared (arbitrary unit – two top panels) and Raman (arbitrary unit – two bottom panels) spectra of anhydrous A and monohydrated M L-aspartic acid crystals. Vertical lines indicate the vibrational normal modes and respective intensities. Depictions of the most important normal modes involving water are also presented, with their assignments and frequencies (in cm^{-1}) in the top and bottom of the unit cells, respectively. Here, γ, δ, β, ν, and ζ stand for wagging, deformation, scissoring, stretching, and rocking motions, in this order. Also, ν_s means a symmetric stretching of bonds. After Silva et al. [179]

charges, the latter approximated within the linear response formalism, while the Raman cross section, on the other hand, is numerically calculated from the spatial derivatives of the macroscopic polarization along eigenvectors of each Raman active normal mode, also employing a linear response formalism.

The anhydrous (monohydrated) phase has 93 (225) normal modes with A and B (A, B_1, B_2, and B_3) symmetry. Comparing the IR spectra of anhydrous and monohydrated L-aspartic acid crystals, a series of water vibrational signatures can be pointed out. For example, in the wave-number range between 1,550 and $1,750\,cm^{-1}$, the anhydrous phase has three pronounced peaks at 1,563, 1,620, and $1,656\,cm^{-1}$ – which are assigned, respectively, to NH_3 wagging ($\gamma\,NH_3$), NH deformation ($\delta\,NH$), and NH_2 scissoring ($\beta\,NH_2$).

On the other hand, the monohydrated crystal has the δNH band slightly shifted down to $1,616\,cm^{-1}$ and two peaks related to the scissors movement of H_2O at $1,599\,cm^{-1}$ and $1,638\,cm^{-1}$, also depicted in the unit cell picture just below the infrared spectra curves on the left side of Figure 7.24. In the wave-number range between 2,800 and 3,400 cm^{-1}, two infrared peaks due to NH$_2$ wagging modes are prominent at 2,909 and 3,000 cm^{-1} for the anhydrous crystal spectrum, while for the monohydrated case, NH, NH$_2$, and NH$_3$ stretching bands occur at 2,976, 3,082, and 3,125 cm^{-1}, respectively, together with water stretching normal modes at 3,280 and 3,368 cm^{-1} (see the middle unit cell picture on the right side of Figure 7.24).

The Raman spectra of the monohydrated L-aspartic acid crystal has water-related intensity peaks in the wave-number range between 600 and 1,000 cm^{-1}, as well as between 2,350 and 3,450 cm^{-1}. For the anhydrous crystal, the Raman spectra in those regions exhibit the following features: a COO deformation band near 760 cm^{-1} and a CH deformation band near 930 cm^{-1}; two NH stretching bands near 2,910 and 2,990 cm^{-1}, respectively; and one CH stretching band near 3,060 cm^{-1}. In the lowest wave-number range for the monohydrated crystal, we have two bands related to CH deformations at 620 and 830 cm^{-1}, with the band at 830 cm^{-1} corresponding to the 930 cm^{-1} band of the anhydrous case.

On the other hand, the COO band of the anhydrous crystal is practically absent in the monohydrated structure. Besides, for the latter, a NH deformation band appears at about 960 cm^{-1}, and two small Raman peaks related to water rocking and wagging occur at 682 and 726 cm^{-1}, respectively. In the largest wave-number range, an OH stretching band occurs in the Raman curve for the monohydrated crystal near 2,440 cm^{-1}, and a CH$_2$ symmetric stretch band appears near 3,000 cm^{-1}. NH stretching modes produce high-intensity Raman lines between 3,120 and 3,130 cm^{-1}, about 100 cm^{-1} above the corresponding lines for the anhydrous crystal. Finally, a set of bond stretching normal modes related to water produce a band centered at approximately 3,300 cm^{-1} with most intense Raman peaks at 3,280 and 3,321 cm^{-1}.

7.5 Conclusions

In this chapter, we have presented the structural, electronic, optical, and vibrational (infrared and Raman spectrum) properties of several amino acids anhydrous crystals, an important issue to characterized them. We started depicting the DFT calculations for the monoclinic and orthorhombic cysteine polymorphs within the GGA-PBE framework. Unfortunately, their electronic band structure results suggested that both cysteine polymorphs are small-gap insulators due to the estimated energy gaps and the flatness of the top (bottom) valence (conduction) band curve, thus severely limiting their use for charge transport applications. Their indirect bandgap also prevents their use in the development of optoelectronic devices.

Then we moved to the theoretical/computational investigation of the electronic band structure and density of states, optical absorption, and dielectric function, as well as experimental measurements of optical absorption for L-serine anhydrous crystals by adopting two distinct exchange-correlation functionals: the LDA and the dispersion corrected GGA+TS. The calculated structural properties were in better agreement with the experimental data for the GGA+TS functional, which predicts a unit cell volume only slightly larger (0.32%) than the X-ray diffraction estimates, whereas the LDA functional predicts a unit cell with a volume about 10% smaller. The Kohn-Sham electronic band structure exhibited a 4.73 eV direct bandgap in the case of the LDA computation and an indirect bandgap of about 4.75 eV between the S point at the valence band and the Γ point in the conduction band (the GGA+TS direct gap is 4.80 eV). The optical absorption measurement, on the other hand, allows one to estimate the bandgap of L-serine crystals to be around 4.90 eV; thus, notwithstanding the well-known inaccuracy of DFT in estimating excitation energies, our theoretical estimates are close to the experiment. The top of the valence band originates mainly from O-2p states with a smaller contribution from C-2p levels, whereas the bottom of the conduction band has its strongest contribution from C-2p orbitals, with an almost equivalent contribution from C-2p orbitals. Above 5 eV, the H-1s states have a significant contribution, and O-2p $\rightarrow$ H-1s interband transitions should be responsible for the sharp increase of optical absorption at about 6.00 eV obtained in our measurements. Our main conclusion was that the electron transport in L-serine anhydrous crystals should be more favorable than the hole transport, the crystal behaving as an n-type bandgap semiconductor for the former and an insulator for the latter, except for a few electric field directions. The calculated polarization-dependent dielectric function, on the other hand, revealed a certain degree of optical anisotropy, with the (100) polarization plane exhibiting distinctive features in comparison to the (010) and (001) planes.

Optical absorption measurements and DFT calculations were afterward discussed, aiming a better understanding of the anhydrous orthorhombic crystal of proline, one of the most common amino acids involved in the formation of the collagen triple helix. We have characterized its Kohn-Sham electronic band structure, proved to be very sensitive to the exchange-correlation approximation employed, with all energy gaps predicted to be indirect. A Δ-sol correction was applied to improve the gap estimation, reaching a value of 5.50 eV, which is very close to the estimated gap from the optical absorption measurements, 5.54 eV. The molecular fundamental gap, on the other hand, was evaluated at 5.81 eV using a proline molecule relaxed in water and the GGA+TS approximation. The HOMO and LUMO orbitals in the crystal, on the other hand, closely resemble the characteristics of their water-solvated molecular counterparts in the zwitterionic configuration. Concerning charge transport, electron and hole effective masses at the main band extrema near the gap were, in general, large and anisotropic, with representative conductivity masses of about four free electron masses, indicating that the proline crystal – given its large bandgap – must be an insulating material.

Lastly, the simulated optical properties (complex dielectric function and optical absorption) indicate that anhydrous proline is very anisotropic to polarized incident light, with very prominent structures for (010) polarization, which corresponds to the alignment of molecular dipoles connected through hydrogen bonds along the *b*-direction. Along the *c*-axis, these structures are very attenuated, probably due to the competition between two opposite molecular dipoles – one intramolecular and other intermolecular.

The investigation of the infrared and Raman spectra of the L-aspartic acid anhydrous crystal was next. Before the spectra calculations, the crystal unit cell was optimized using the LDA and GGA-TS exchange-correlation functionals. The structural optimization showed a nice agreement with the X-ray diffraction data when the functional GGA-TS was used, whereas the LDA-optimized structure exhibited lattice parameters much smaller than those measured for the crystal. The infrared and Raman spectra were also evaluated for the two lowest-energy conformations (mc1 and mc2) of zwitterionic aspartic acid solvated in water using the hybrid B3LYP functional for the sake of comparison with the solid-phase simulations. For the wave-number range between 0 and 1,000 cm^{-1}, there is a very good agreement between the crystal theoretical and experimental vibrational spectra (especially for the Raman spectra), whereas the calculated molecular zwitterionic normal modes did not match the experimental curves well. Phonon dispersion curves were calculated as well, and it has been shown that the acoustic phonon branches have some degree of anisotropy, especially along the $\Gamma \to Z$, $\Gamma \to Y$, and $\Gamma \to C$ directions. The phonon density of states reveals a strong contribution from the oxygen atoms to the vibrational modes, followed by carbon and nitrogen. The hydrogen atoms, by the way, produce a very strong peak at 536 cm^{-1} related to NH_3 vibrations. Calculated specific heats were in good agreement with the experimental data available in the literature.

Finally, we have performed a comparison between the electronic band structures, infrared and Raman spectra of anhydrous and monohydrated L-aspartic acid crystals, aiming to investigate the impact of the water on the electronic transport and vibrational spectra of these materials. We conclude that the wet (monohydrated) L-aspartic acid crystals seem to be an n-type insulator and a p-type semiconductor for hole transport along the direction of water stacking, a result not yet achieved experimentally. It also became very clear that molecular water stackings can contribute significantly to improve directional hole transport in hydrated amino acid crystals and similarly ordered bioorganic structures while, at the same time, preventing electron transport due to a very large electron effective masses. Our results suggest that water can have a complex role in the carrier transport properties of biomolecular crystals, in contrast to the general belief that the inclusion of water in materials simply increases their electrical conductance.

8

Protein–Protein Systems

8.1 Introduction

Proteins are basic components of the living cell and consist basically in the joining of amino acid residues through peptide bonds forming one or more polymeric chains (polypeptides). Among their many roles in living organisms, one can mention their action as enzymes that regulate the rate of chemical reactions, as structural blocks that give stiffness and stabilize the cell, as carriers of information in cell signaling, and in the cellular transport of small molecules necessary for metabolic processes. In particular, proteins can interact in many biological functions, such as the regulation of metabolic pathways, immune response, cellular division, replication of nucleic acids, and synthesis of other proteins [195]. The way they interact is dependent on their 3-D geometry that determines which amino acid residues are accessible and the electrostatic patterns that will lead to the stabilization of their complexes. This geometry, on the other hand, seems to be uniquely specified by the sequence of amino acid residues. Proteins formed by multiple subunits can be found in many protein classes, including classical examples such as hemoglobin, core RNA polymerase, pyruvate dehydrogenase, ribossomes, and the tail assembly of the bacteriophage T4.

The proteome is the set of all proteins that an organism or some relevant part of the organism can express at a given moment. It is much more complex than the respective genome, as it involves not only a static structure but a plethora of dynamical factors, such as the development of the organism and its responses to external agents. Networks of interacting proteins (protein interactome) are behind most, if not all, biological activities. Protein–protein interaction (PPI) maps can reveal the details about how the cell works, what proteins bind to one another, and what their functional association is. Methods such as yeast two-hybrid assays, mass spectrometry, correlated mRNA expression, synthetic lethal interactions, and in silico prediction through genome analysis have allowed the research and discovery of many protein interactions in the proteome of yeast [196] and bacteria such as *Helicobacter pylori* [197].

Nuclear magnetic resonance (NMR) measurements can be employed to investigate weak PPIs without the need for crystallization for systems with up to 1,000 kDa, while docking methods can predict PPIs from the atomic coordinates obtained from X-ray diffraction data. The use of a computational scanning using molecular mechanics and an alanine molecular probe, in particular, has allowed the successful description of the contribution of each amino acid residue involved in the interactions between the oncoprotein Mdm2 to the N-terminal of the tumor supressor protein p53. Kinetic aspects of the protein–protein complexation process seem to be strongly affected by long-range electrostactic forces that stabilize their transition states.

One can divide PPIs into those which occur between identical and non-identical polypeptide chains. In the first case, one can have an interface between the protein units which is the same for both proteins (isologous configuration) or distinct (heterologous configuration). In the latter case, different interfaces allow for the possibility of infinite aggregation. It is also possible to classify multiprotein complexes with respect to the stability of their protein units: obligate protomers are those that cannot be found isolated in vivo, while the non-obligate protomers exist independently in living cells. Another way of distinguishing the protein complexes is to consider their lifetime, leading to the classification into transient and permanent complexes. Obligate complexes are usually permanent, while non-obligate complexes may be permanent or transient.

The systematic mapping of protein–protein interactions in the human organism (human interactome map) can produce invaluable information about the topological and dynamical features of the interacting proteome related to biological characteristics of interest. The theoretical modeling of interacting proteins, on the other hand, which employs molecular dynamics and prediction and design tools to obtain optimized protein geometries, is a formidable approach to investigate macromolecular structures and their binding mechanisms, with great potential for streamlining proteomic research [198]. The investigation of potential energy functions adequate to obtain an accurate model of proteins and efficient search algorithms to probe the huge space of configurations, however, remain a challenge to the field. In particular, the transition from low-resolution to high-resolution modeling must overcome difficulties such as the treatment of charge polarization and the flexibility of the protein backbone [199]. Experimental data can provide templates to modeling interacting proteins by homology, leading to hybrid methods for the prediction of protein structure. The existence of structural disorder and polymorphism in the formation of protein complexes, however, implies that a single optimized geometry is unable to account for the properties of these systems. This disorder can be static, due to well-defined conformations, or dynamic, with random geometry fluctuations.

The modulation of PPIs using small molecules can contribute to improve the knowledge of the protein networks, especially for the understanding of pathogenic mechanisms and the creation of new diagnostic tools and therapies. Domain–domain and peptide–domain interaction models usually involve large, flat, and featureless surface areas, so the use of peptides to target their interfaces is interesting, as peptides share many characteristics with the proteins and are easy to synthesize and modify. The mapping of hot spots for peptides in the protein surface at the interaction interface (sites for which mutations cause a significant increase in the binding energy) can be achieved using computational and fragment strategies. The successful development of inhibitors for protein–protein interaction will allow the development of selective drugs able to attack several diseases, but many difficulties remain that will demand significant efforts in the development of computational tools to validate PPI targets.

In this chapter, we intend to highlight some interestig properties of the protein–protein systems, including those related to their interaction as well as the role played by them in the quantum chemistry simulation approaches.

8.2 The Protein Data Bank

Established by the Brookhaven National Laboratory in 1971, the Protein Data Bank (PDB) is a repository of 3-D structures of proteins, nucleic acids, and other relevant biological macromolecules [200] – the most important global macromolecular structure archive. The data files contain information about the atomic coordinates, primary and secondary protein structures, citations, crystallographic details, and links to similar databases. They are available to be used by researchers, teachers, and students of any scientific discipline; they provide invaluable information for many fields of study such as biochemistry, biophysics, and medicinal chemistry, and are the starting point in any structural bioinformatics or structural genomics investigation.

The PDB involves an international effort, with research centers affiliated worldwide. Initially, PDB files were accessed only through the sharing of magnetic media, but with the advent of the World Wide Web (WWW), it is now very easy for any person in any part of the world (with an internet connection) to obtain data sets of interest. Since October 1998, the PDB has been managed by the Research Collaboratory for Structural Bioinformatics (RCSB). PDB deposits can be made through email or by using a specific online application following specific format standards. Validation tools are employed to check for covalent bond distances and angles, consistency of chiral centers, atom and ligand nomenclature, close contacts, sequence comparison, and placement of water molecules [201]. The improvement of the querying capabilities of the PDB demands serious efforts to warrant data uniformity, especially with respect to formatting, nomenclature, and consistency of structural data.

In 2003, the RCSB joined to the Macromolecular Structure Database (MSD) at the European Bioinformatics Institute (EBI) and the Protein Data Bank Japan (PDBj) at the Institute for Protein Research in Osaka University to create the world-wide Protein Data Bank (wwPDB), but the RCSB remains the archive keeper [202]. In 2006, wwPDB was joined by the BioMagResBank, a repository that collects protein data gathered from NMR spectroscopy.

More recently, advances in structural characterization have lead to a larger and more complex data available in the PDB, turning it into an essential resource for drug design. Specific tools for drug discovery were incorporated in the PDB system, such as ligand searches, ligand and structure summaries, binding site visualization, and drug target mapping. A recent study has shown that the PDB resources is used by more than 100 disciplines from molecular biology to the social sciences [203]. More than 200 new drugs approved by the US Food and Drug Administration (USFDA) between 2010 and 2016 were developed with the help of 3-D structural information obtained from the PDB archives. At present (May 2019), the PDB has about 160,000 depositions, with about 6,700 structures containing more than 19,000 atoms. The most part of the data was obtained from X-ray diffraction measurements (about 127,000 proteins, 2,000 nucleic acids), followed by NMR (about 11,000 proteins, 1,200 nucleic acids), and electron microscopy (about 2,300 proteins, 30 nucleic acids).

8.3 Improving PDB through Molecular Dynamics

Classical molecular dynamics (MD) simulations can be employed to improve the quality of PDB data for low-resolution coordinates, such as those obtained for hydrogen atoms in X-ray diffraction measurements, through geometry optimization. Besides, dynamical simulations can be employed to partially replicate the in vitro and in vivo conditions of solvated biological macromolecules. As seen in Chapter 1, classical molecular dynamics consists in the numerical resolution of the Newton's equations of motion considering finite time steps and using a force field to represent the potential energy of the interacting atoms in a given spatial configuration.

There are many codes available to perform this kind of task in the case of biomolecules – for instance, GROMACS [204], AMBER [205], and NAMD [206], to cite just a few. On the other hand, molecular dynamics simulations of proteins including ligands are essential in drug docking and drug design, which often demand data on the ligand topology that can be done automatically using a computer. Initial configurations for the interacting atoms and molecules within certain boundaries can also be obtained by employing specific software such as PACKMOL [207].

Molecular dynamics can be employed to sample the conformational space of proteins and obtain average structures from multiple initial conditions that closely resemble X-ray and NMR data. Protein flexibility can be assessed from several structures of the same system archived in the PDB or from the end points of conformational change available there. In particular, the folding of polypeptides into proteins can be investigated by using classical molecular dynamics calculations, with simulation times of a few microseconds, giving a way to map the energy landscape and find a global energy minimum in good agreement with experimental data from the PDB. In the opposite direction, one can also refine the coordinates of PDB structures obtained from NMR and check the accuracy of structural restraints through MD.

Coarse-grained molecular dynamics of membrane proteins and peptides can improve the data on their structure, especially taking into account how it is difficult to obtain high-resolution measurements for them. Advances in the treatment of solvation and computational power now allow MD to achieve high-resolution refinement of protein data, notwithstanding the fact that accurate folding prediction is not yet possible. Discrete molecular dynamics (DMD) algorithms, which consider discontinuous interaction potentials where the atoms can collide into each other, allow the rapid sampling of the energy landscape to simulate the protein folding process and investigate the details of PPIs. Lastly, accelerated molecular dynamics (aMD) improves the sampling of the conformational space by decreasing the energy barriers between distinct low-energy structures and allowing one to assess rare barrier-crossing events that are hard to find by using conventional MD to investigate protein folding.

Massive molecular dynamics simulations can mimic the behaviour of thousands of protein–ligand geometries to accelerate drug screening from PDB structures, usually performed through docking. Even systems as complicated as enzyme-inhibitor complexes can be reconstructed through MD simulations running in high-performance GPUs [208]. Structural experimental data of high-resolution protein fragments from the PDB repository can be used to guide the molecular dynamics process to estimate native geometries, overcoming the difficulty of finding the global energy minimum. Despite all these advances, some limitations are inherent to the physics employed in all MD methods available: the force fields adopted are crude approximations when compared to the quantum mechanical rules that govern the behavior of molecules.

In order to improve their description, it is necessary to incorporate the quantum effects in some measure. Accurate descriptions of phenomena such as bond break, bond formation, electronic polarization, and charge transfer dynamics are beyond the reach of classical methods. As pure quantum methods are computationally expensive, however, hybrid approaches become more practical. In a hybrid framework, the system under study is partitioned into two regions: a small domain, where quantum mechanical simulations are performed to improve accuracy, and a larger domain, where classical mechanics is applied as in MD.

Improvements on ligand geometries in PDB structures can also be achieved through MD and quantum molecular dynamics (QMD). PDB data with the structure of a protein can be used to test the attachment of novel ligands, which can be manually placed at the binding site. The atomic positions of the ligand and amino acid residues near it at the binding site can be submitted to a MD simulation with harmonic restraints on the C_α atoms, allowing for adjustment of the side chains. The MD snapshots can then be employed to provide initial configurations for high-throughput docking. This and other techniques recently developed advance a new strategy for the investigation of ligand–receptor interactions [209].

8.4 The Dielectric Function of Proteins

In electrostatics, the dielectric function ϵ measures the local polarizability of a medium when an external electric field **E** is present. It is described in its simplest approximation by a local linear relationship between the polarization vector **P** and the electric field **E** (see Figure 8.1).

Within proteins, the electrostatic environment is relevant for processes such as protein folding, interaction with substrates and ligands, and the packing of secondary structures. It is known that the ordering of dipoles related to the peptide bonds is not random but ordered along the α-helix and β-sheet structural units. In contrast, the side chains of the amino acid residues within the protein usually have

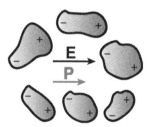

Figure 8.1 An external electric field induces charge polarization in a heterogeneous medium. The dielectric constant quantifies this relationship.

low polarizability, decreasing the dielectric response of the protein. In the case of the residues at the protein surface, ionizable side chains are unevenly distributed and dependent on the pH of the solvent, forming networks of charged atoms. Water molecules contribute to increase the local dielectric constant (ϵ value for water $\approx$ 80), but the shielding is overestimated if one uses the water dielectric constant value in all solvent accessible regions [210].

8.4.1 State of the Art

One can describe the electrostatic interactions of the set of amino acid residues of a protein inside an aqueous solvent through the following Hamiltonian:

$$H = H_{solv} + H_{d-solv} + H_d, \tag{8.1}$$

where H_{solv} is the water solvent Hamiltonian and H_{d-solv} is the term corresponding to the interaction between the amino acid dipoles and the solvent. H_d describes the interacting amino acid dipoles and must include implicitly or explicitly the dipole moment vectors and polarizabilities of each residue, as well as the dipole–dipole interaction. The mean dipole moment of each amino acid residue in the protein is the intrinsic dipole moment of each residue plus corrections including the interaction with the other residues and the solvent. Molecular dynamics simulations allow the determination of intrinsic polarizabilities for each amino acid [211]. Accurate description of each term in Eq. (8.1) is essential to achieve a realistic description of electrostatic effects in protein complexes. In the following subsections, we will present recent strategies and approaches concerning their implementation.

8.4.2 Role of water

Water molecules surround proteins form an hydration layer crucial for protein dynamics. The anomalous properties of water are well known and are a consequence of many singular features: ability to form hydrogen bond networks, small molecular moment of inertia, high dipole moment, and large dielectric constant. The unique combination of these characteristics gives water diverse time scales to operate for its viscosity, rotational dynamics, and solvation of polar molecules. The presence of an interface with another substance modifies locally the hydrogen bond pattern of water, leading to dynamic characteristics distinct from the bulk. In the absence of water interactions, protein structures are unstable and not functional, and DNA cannot be transcribed or replicated [212].

The hydration layer of biomacromolecules, with a thickness between two and three water molecules, must be clearly distinguished from the bulk water around it (see Figure 8.2). For proteins, it can behave as a somewhat rigid scaffold structure,

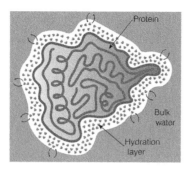

Figure 8.2 The hydration layer of a protein. Water molecules are continuously exchanged with the water bulk around it.

which effectively increases protein size. This picture, however, is only a first approximation, as many experiments and theoretical simulations have shown that the hydration layer is very active, with slow and fast molecular rearrangements coexisting due to strongly and weakly bound water molecules. Simulations of hydration layers in proteins have shown that significant translational and rotational motions occur within the layer, with most molecules exhibiting residence times of tens of picoseconds (ps) and binding energies below 10 kcal/mol. Hydrogen bond dynamics can be slowed down by about 20% near hydrophobic regions of protein surfaces.

Water solvation can be simulated by considering an implicit solvent model where the water is replaced by a dielectric continuum, with lower computational cost, or as an explicit solvent where water molecules are treated individually. In the latter case, a box containing the biomolecules of interest is filled with water molecules, usually in a proportion of 1:10, in terms of number of atoms. The simplest continuum model usually considers a spherical ion in water with hydration free energy given by the Born formula, plus a nonpolar correction:

$$\Delta G = -\left(\frac{\epsilon_w - 1}{\epsilon_w}\right)\frac{kQ^2}{2R} + \Delta G_{\text{np}}, \tag{8.2}$$

where ε_w is the dielectric constant of water, R is the ion radius, and Q is the ion charge. ΔG_{np} is a nonpolar contribution that takes into account the free energy related to the formation of a hole in the water continuum, proportional, in some models, to the solvent accessible area of the molecule. For a more complex structure, the value of ΔG depends on the molecular geometry of the solute. This hydration energy can be added to the Hamiltonian of a classical molecular mechanics force field.

An improvement over the electrostatic part of Eq. (8.2) is the generalized Born (GB) formula, which is the most used in classical molecular dynamics today.

It takes into account the dielectric constants of both the solvent and the solute and is parameterized for the electrostatic size of the latter, considering the atomic charges of each of its atoms as well as their relative distances and radii. Conformational searches within the GB approach are much faster (10–100 times) than those performed in an explicit solvation model.

Another formalism, within the dielectric continuum framework, is to solve the Poisson equation for a given charge density distribution $\rho(\mathbf{r})$ in the solute – i.e.,

$$\nabla \cdot \left[\epsilon(\mathbf{r}) \nabla \phi(\mathbf{r}) \right] = -\rho(\mathbf{r})/\epsilon_0, \tag{8.3}$$

where ∇ is the nabla operator and $\phi(\mathbf{r})$ the electrostatic potential determined by the position-dependent dielectric function $\epsilon(\mathbf{r})$. This dielectric function is often modeled by considering two values: ϵ_1 inside the dielectric boundary and ϵ_2 otherwise. The electrostatic free energy is then given by

$$\Delta G_e = \frac{1}{2} \sum_i Q_i \left[\phi(\mathbf{r})|_{\epsilon(\mathbf{r})} - \phi(\mathbf{r})|_{\epsilon=1} \right], \tag{8.4}$$

where $\phi(\mathbf{r})|_{\epsilon=1}$ is the electrostatic potential obtained in vacuum, without the dielectric.

Considering classic explicit water solvation models, one of the most simple approximations replaces each molecule by rigid point charges at the hydrogen and oxygen atoms kept fixed at the gas phase geometry and subjected to pairwise additive interactions (TIP3P).

Notwithstanding the preceding comments, there is no single classic model of explicit water that is able to reproduce all key properties of water in liquid state. Further advances must include quantum effects at some approximate level, such as density functional theory (DFT. The high computational cost involved, however, prevents its use for systems of more than 100 water molecules interacting with relatively small solutes. At the moment, the best available option is to employ a partitioning of the solvent–solute system in classical and quantum domains (QM/MM approach). Mixing explicit and implicit solvation can also help to significantly increase the speed of the simulations without relevant loss of accuracy.

8.4.3 Homogeneous Case Description

One can model the electrostatic interactions in proteins using three basic dielectric models:

(a) uniform dielectric constant, neglecting effects from distant atoms
(b) distance-dependent dielectric constant, varying between $\epsilon \approx 1$ for very small distances and the bulk value for large distances;
(c) protein in a cavity embedded in a high dielectric constant solvent

However, in the case of a protein–ligand complex, there is no simple relationship between the dielectric constant value and the distance from a ligand or the protein centroid in protein folding simulations. Moreover, these models are very sensitive to the choice of ϵ and suffer from the lack of clear understanding on which factors determine its value. As a matter of fact, the protein dielectric constant is dependent on the model used to describe it and not an universal parameter [213]. Poisson-Boltzmann and GB methods often adopt $\epsilon = 1$ or 2, but other works employ $\epsilon = 1$ to 40 to investigate protein stability. In protein PKa calculations, the value $\epsilon = 4$ is frequently chosen.

8.4.4 Inhomogeneous Case Description

More sophisticated models use the finite differences approach to obtain an electrostatic potential ϕ (**r**) inside the protein, taking into account solvation effects and atomic polarizabilities, dividing the protein into cubic regions with fixed dielectric constant and allowing for the prediction of PKa values [214]. It has be shown, from molecular dynamics simulations, that the dielectric constant in the inner parts of a protein is usually low ($\epsilon \approx 2$–3), while the whole molecule has an average dielectric constant $\epsilon \approx 11$–21, probably due to the polarized side chains at the protein surface (see Figure 8.3). In protein–protein interactions, electrostatic interactions play a very important role, with clusters of charged and polarized residues contributing to improve stability.

As the protein–solvent system is highly inhomogeneous, computing the dielectric function profile ϵ (**r**) is somewhat complicated. One approach is to insert a small probe charge δQ at distinct positions in the system and observe how its free energy, obtained from classical molecular dynamics calculations, is modified as δQ increases. Afterward, a trial dielectric profile ϵ (**r**) is put into the Poisson equation (Eq. (8.3)) and is then numerically solved in order to minimize the difference between the free energy obtained from the ϵ (**r**) profile and the MD data [215].

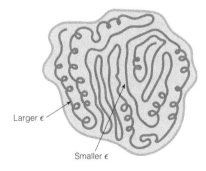

Larger ϵ

Smaller ϵ

Figure 8.3 Inhomogeneous dielectric constant ϵ in a typical protein. The value of ϵ is larger near the protein surface and smaller at the protein core.

In order to describe the variations of the dielectric constant in a protein–ligand system within a fragmentation strategy using DFT (see Section 1.8), one can describe the ligand interactions assuming distinct average values of ϵ as we switch from one residue to another [216] within an implicit solvation model. This means an improvement over previous molecular dynamics studies and protein DTF calculations that commonly consider a fixed value for the protein dielectric constant for all residues.

8.5 The Importance of Protein–Protein Interactions

The biochemical workings inside a single cell are dominated by the complex interaction of proteins with other proteins, polypeptides, lipids, nucleic acids, carbohydrates, and other molecules. Molecular nanomachines perform the most basic tasks for metabolism and reproduction: enzymatic activities, transport, replication of genetic information, DNA transcription and translation, control of molecular and ionic fluxes through the cell membrane, cytoskeleton regulation, signaling and regulatory pathways, etc. For the most part, these nanomachines are formed from multiple modular proteins, composed of distinct domains with enzymatic activity that operate in independent fashion and able to change shape in the presence of specific molecules (folding interactions). These folded interaction domains typically contain less than 150 amino acid residues in a configuration that allows their insertion in a protein loop, leaving their interaction surfaces exposed.

The computational analysis of the interacting proteome (interactome) is a very challenging subject, as the experimental techniques used to investigate it are not very reliable. The topology of the interactome is complicated even further by two facts: the same protein can perform different functions and there are apparently random connections between proteins with distinct activities. Besides, protein–protein interactions cannot be qualitatively predicted from structural features at atomic precision, and the PDB data available does not contain many of the large cell protein assemblies. Lastly, under the dynamical viewpoint, protein–protein complexes are often transient, adding a temporal dimension to the interactome network that must be taken into account to understand how it operates in the living cell.

The starting point for the understanding of the protein–protein complexes, however, is the description of how their contact surfaces interact. As we have seen in Section 8.3, classical molecular dynamics can provide very useful information at the atomistic scale, but its accuracy fails due to the lack of consideration of quantum effects, which demand a much higher computational cost than classical computations. One alternative to address this issue is to consider a fragmentation strategy to estimate the binding energies per amino acid residue at the protein–protein interface, which will be considered next.

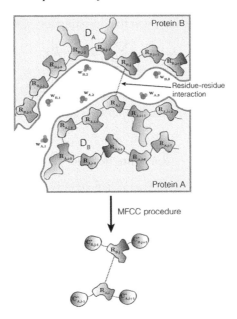

Figure 8.4 MFCC strategy to describe the protein–protein interactions.

A fragmentation strategy for the quantum treatment of protein–protein interfaces using DFT (see Chapter 1) begins by defining the interacting domains of the proteins. One can consider two protein units, A and B, with two interacting domains,

$$D_A = \left\{ R_{A,1}, \ldots, R_{A,N_A}, w_{A,1}, \ldots, w_{A,N_{w,A}} \right\}, \tag{8.5}$$

$$D_B = \left\{ R_{B,1}, \ldots, R_{B,N_A}, w_{B,1}, \ldots, w_{B,N_{w,B}} \right\}. \tag{8.6}$$

Here, $R_{A,i}$ ($R_{B,i}$) is the ith amino acid residue in the A (B) domain, which contains N_A (N_B) residues. Explicit water molecules w_A^j (w_B^j), with $j = 1, \ldots, N_{w,A}$ ($j = 1, \ldots, N_{w,B}$) can be present at the interface and can be associated by close proximity to the A (B) protein (see Figure 8.4).

For each pair of amino acid residues $R_{A,i}$–$R_{B,k}$, one can find their interaction energy as follows:

(a) From PDB data, one constructs an imaginary sphere S that contains both $R_{A,i}$ and $R_{B,k}$.
(b) Considering all residues and water molecules within S, one estimates the average dielectric constant $\epsilon(R_{A,i} - R_{B,k})$ between $R_{A,i}$ and $R_{B,k}$.
(c) Using the conjugated caps approach described in Section 1.8, both residues must be capped and the interaction energy between them evaluated within an implicit solvation model.

The total interaction energy between the domains follows:

$$E_{\mathrm{INT}}\left(D_A - D_B\right) = \sum_{R_{A,i}, R_{B,k}} E_{\mathrm{INT}}\left(R_{A,i} - R_{B,k}\right). \tag{8.7}$$

It is also straightforward to obtain the interaction energy between two peptidic subdomains by limiting the residues $R_{A,i}$ and $R_{B,k}$ to certain subsets of D_A and D_B.

8.6 Conclusions

Many infirmities, such as cancer and neurodegenerative diseases, seem to be related to dysfunctional PPIs. The human protein interactome is not well known and probably involves hundreds of thousands to one million PPIs. Due to its dynamic character, the study of the protein interactome may be more challenging than the deciphering of the human genetic code concluded in 2003. Experimental techniques are not yet mature and do not achieve complete reliability, so it is necessary to appeal to predictive methods using computer simulations to obtain a complete picture or map of PPIs in the human organism [217]. Among them, the most used at present are protein docking and molecular dynamics (classic or hybrid quantum/classic).

Proteins can bond with or without significant conformational change of their contact surfaces in comparison to their isolated monomers. One protein can also interact with a peptide by preserving its shape or by suffering some deformation [218]. All these situations must be taken into account in a consistent methodology to deal with the interactome using computer simulations. In view of this, the PDB is an invaluable resource that provides many structures to be used as starting points of investigation. Through classical molecular dynamics, PDB geometries can be improved and modified, and their conformational space can be probed to find alternative configurations.

Classical methods, however, are unable to provide the accuracy necessary to describe phenomena such as chemical reactions and electrostatic interactions related to electron density changes. These can be only taken into account by including quantum effects in the simulations. Besides, the electrostatic polarization of water-solvated proteins must be considered, so an adequate modeling of their dielectric constant is essential. This modeling can be achieved through an explicit treatment of the solvent (which is very demanding computationally) or through the modeling of the dielectric constant of the system in a continuum model (homogeneous or inhomogeneous).

9

Ascorbic Acid and Ibuprofen Drugs

9.1 Introduction

Ascorbic acid (symbol: AsA; chemical formula: $C_6H_8O_6$) is an organic acid with antioxidant properties to protect human cells from the damage caused by the free radicals, whose L-enantiomer is popularly known as vitamin C. It helps the human body boost the immune system, reducing the risk of chronic diseases like heart diseases, the leading cause of death globally. Additionally, vitamin C, plays an essential role as an enzymatic cofactor for the synthesis of biologically important molecules for the human body – such as collagen, carnitine, catecholamine, myelin, and neuroendocrine peptides – which reduce blood uric acid levels, improve the absorption of iron, and fight oxidative stress and inflammation processes of the central nervous system (see Chapter 13). Although not synthesized by human cells, it is essential for human existence. In fact, a lack of vitamin C can even make you more prone to getting sick.

On the other hand, following the buckminsterfullerene C_{60} discovery [219, 220], it was soon acknowledged that it has unique properties, making it very attractive for biological investigations, with several medicinal applications being proposed. Unfortunately, its poor solubility in polar solvents, which has implications for the formation of aggregates aqueous solutions, is detrimental for biological applications. To circumvent this difficulty, several routes to attach chemical groups to C_{60} were proposed, leading to a wide variety of derivatives with distinct physical and chemical properties. Nevertheless, numerous experiments have established prospective applications for C_{60} derivatives in many fields of medicine, such as enzymatic inhibition, anti-HIV activity, neuroprotection, antibacterial applications, DNA cleavage, and photodynamic therapy. The functionalization of fullerenes is achieved through covalent bonding or physical interaction of foreign species. Although many prospective uses for fullerene-soluble complexes exist, their cytotoxicity effects on living cells should be brought into balance due to their sensitive function of surface derivatization.

In this aspect, ascorbic acid is timely: its combination with the fullerene C_{60} completely prevents the oxidative damage of the latter, leading to a protection of the cultured chromaffin cells against levodopa toxicity, thus suggesting the beneficial use of AsA-C_{60} (together with levodopa), as an efficient treatment of Parkinson's disease (see Chapter 14). It is also believed that AsA-C_{60} should be more effective in preventing oxidative damage than C_{60} alone, since both are potent antioxidants. Consequently, focusing on a detailed picture of the AsA in complex with C_{60} is of fundamental importance for biological applications.

In the first part of this chapter, we intend to give a complete description of ascorbic acid interaction with the fullerene C_{60}, including the effect of different spatial orientations of AsA relative to the C_{60} molecule [221]. With the help of classical molecular dynamics, the best molecular geometry corresponding to the strongest binding of AsA adsorbed on C_{60} is found. Molecular dynamics simulations are the most indicated computer software for this kind of biological systems because they involve a large number of atoms. The alternative ab initio methods in such cases require much more computational time, making the task impossible in several cases. Therefore, for reasons of computational economy, we ran our first-principles simulations using the classically optimized geometries. Quantum ab initio density functional theory (DFT) simulations, in both the local density and generalized gradient approximations, LDA and GGA, respectively, were carried out to estimate the AsA-C_{60} binding energy. Afterward, the electron transfer between the AsA and C_{60} molecules was calculated to assess the noncovalent nature of their interaction.

The second part of this chapter is devoted to the two-level adsorption of ibuprofen on C_{60} fullerene for transdermal delivery by using the same approach as presented for the ascorbic acid [222]. Ibuprofen (symbol: IBU; chemical formula: $C_{13}H_{18}O_2$) is a well-known pharmaceutical nonsteroidal anti-inflammatory drug (NSAID) with analgesic and antipyretic properties, which is frequently used and tolerated by patients for the treatment of painful and inflammatory conditions and available both by prescription and over the counter. Compared to other NSAIDs, it may have fewer side effects such as gastrointestinal bleeding, although it increases the risk of heart, kidney and liver failure. According to the World Health Organization (WHO), ibuprofen is an essential medicine for a basic healthcare system when used for acute pain, fever, and palliative care due to its low cost, safety, and efficacy. Ibuprofen appears to be slightly stronger than aspirin when treating soft tissue injuries, dental pain, and menstrual cramps, and it is equally effective for headaches, migraines, and fever reduction. Unlike to ibuprofen, however, aspirin is sometimes recommended to reduce the risk of heart attack or stroke.

Based on the motivation to investigate theoretically IBU-C_{60} systems to assess the feasibility of controlled delivery anti-inflammatory drugs through the skin, we present here the description of their noncovalent interaction after sampling different relative spatial orientations of IBU with respect to the C_{60} molecule. Again, with the help of classical molecular dynamics, the best molecular geometries corresponding to the strongest binding configurations of IBU adsorbed on C_{60} were found. After classically optimizing the system geometry, we have performed a new geometry optimization at the quantum DFT level of theory, selectively using the local density approximation (LDA) for the exchange-correlation energy and the B3LYP hybrid functional to estimate the properties of different IBU-C_{60} adsorbates.

Finally, in the third part of this chapter, we present the binding energy of the nonsteroidal anti-inflammatory ibuprofen to human serum albumin (HSA) [223]. The human serum albumin is a multifunctional protein abundant in the human blood plasma that is produced in the liver, with a molecular weight of approximately 66 kDa and the capacity to bind a large amount of chemical compounds. HSA functions primarily as a carrier protein for steroids, fatty acids, and thyroid hormones in the blood, and it plays a major role in stabilizing extracellular fluid volume by contributing to the oncotic pressure of plasma. Infection or inflammation in the human body, as well as disorders in the liver (cirrhosis or hepatitis) and kidneys, can result in a low albumin level. Due to its large binding capacity, the transport protein HSA helps in the solubilization and reduction of toxicity of drugs. It also protects drugs against oxidation, increases their half-lives, promotes their throughout the body, modulates their concentration at their targets, and improves the biological effects as well.

Once in the bloodstream, around 99% of IBU becomes strongly bound to HSA. As HSA is essential not only for the drug transport in the blood but also for the distribution throughout the human body, its high-level affinity for a drug molecule can hinder its access to the target site, decreasing its potency in vivo. Therefore, theoretical estimates of the IBUHSA binding affinity can be very useful to achieve an accurate understanding of this binding interaction, as albumin directly affects the pharmacokinetic and pharmacodynamic properties of ibuprofen, which is of prime importance for the design of new nonsteroidal, anti-inflammatory agents in future.

9.2 Ascorbic Acid

The purpose of this section is to give a complete description of ascorbic acid (AsA) interaction with the C_{60} buckminsterfullerene, including the effect of different spatial orientations of AsA relative to the C_{60} molecule. With the help of classical molecular dynamics, the best molecular geometry corresponding to the strongest binding of AsA adsorbed on C_{60} is found. Quantum ab initio DFT simulations, in both LDA and GGA, were carried out to estimate the AsA-C_{60} binding energy. Afterward, the electron transfer between the AsA and C_{60} molecules was calculated to assess the noncovalent nature of their interaction.

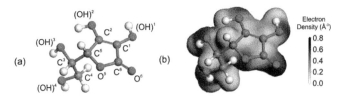

Figure 9.1 (a) Ascorbic acid ($C_6H_8O_6$) molecule with atom labels (b) electron density projected onto an electrostatic potential isosurface. After Santos et al. [221]

9.2.1 Computational Details

As depicted in Figure 9.1a, the ascorbic acid molecule is relatively small, with four –OH and two –CO groups that confer its antioxidant characteristics. Nine stable AsA structures were found experimentally in AsA crystals at low temperatures [224], while full optimizations for 36 AsA conformers have been reported through ab initio calculations using different theory levels and basis sets – RHF/6-31G, RHF/6-31G(d,p), RHF/6-311+G(d,p), and MP2/6-31(d,p) [225]. The conformations of protonated and deprotonated AsA, as well as that of oxidized AsA, were studied recently [226]. Figure 9.1b shows the calculated electron density projection onto an electrostatic potential isosurface of a possible conformation of isolated AsA. As expected, oxygen atoms present the highest electron density (darker regions) due to their stronger electronegativity in comparison to carbon and hydrogen.

In order to find the best geometry for the AsA-C_{60} adsorption, a classical annealing simulation was carried out using the Forcite Plus code. The universal force field was adopted to perform this simulation. The cutoff radius was chosen to be 18.5 Å. We carried out the annealing as follows: a total of 100 annealing cycles was simulated with an initial (midcycle) temperature of 200 K (300 K) and 50 heating ramps per cycle, with 100 dynamic steps per ramp. Using the NVT ensemble, we performed the molecular dynamics with a time step of 1 femtosecond (fs) and a Nose thermostat. After each cycle, the smallest energy configuration was optimized. Only the atoms of the AsA molecule were allowed to move during the calculations; the C_{60} atomic positions were kept fixed.

9.2.2 Optimized Geometries

The best geometry that we found after the annealing/geometry optimization procedure is shown in Figure 9.2. The non-hydrogen atoms belonging to the AsA molecule closest to the C_{60} center of mass are O_1 (6.37 Å), C_4 (6.74 Å), C_1 (6.80 Å), and O_6 (6.85 Å). The two hydrogen atoms bonded to the C_4 atom have the smallest

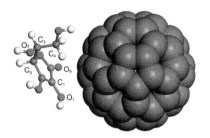

Figure 9.2 Atomic configuration with the lowest total energy for the AsA molecule adsorbed on C_{60} obtained after classical annealing and geometry optimization. O_1, O_6, C_1, and C_4 (O_3, C_3, and C_5) are the non-hydrogen atoms with the smallest (farthest) distances relative to the C_{60} center of mass. After Santos et al. [221]

distances to the C_{60} center of mass, 6.13 and 6.31 Å, respectively, followed by H_4 (6.41 Å) and H_1 (6.58 Å). On the other hand, the O_3, C_5, C_3, and O_5 are the farthest ones, with distances of 8.77, 8.17, 8.08, and 7.88 Å, respectively. To obtain the distance between a given atom and the approximately spherical surface of C_{60}, we should subtract 3.56 Å, the distance between any carbon atom belonging to the simulated C_{60} and the molecular center of mass.

Using the configuration of Figure 9.2, namely $\Gamma(0,0,0)$, we have built a set of inputs for a quantum first-principles geometry optimization. These inputs were generated by rotating the classically optimized $\Gamma(0,0,0)$ AsA molecule about three arbitrary orthogonal axes (labeled x, y, and z), passing through its center of mass, and keeping the C_{60} atoms fixed. Only rotations of $90°$ about each axis were allowed. For example, the configuration obtained by rotating the molecule by $90°$ around the x-axis was labeled $\Gamma(90,0,0)$. Only 10 inputs were chosen through this method namely $\Gamma(0,0,0)$, $\Gamma(90,0,0)$, $\Gamma(180,0,0)$, $\Gamma(270,0,0)$, $\Gamma(0,90,0)$, $\Gamma(0,180,0)$, $\Gamma(0,270,0)$, $\Gamma(0,0,90)$, $\Gamma(0,0,180)$, and $\Gamma(0,0,270)$. We have considered them as a representative sampling of the space of initial configurations to investigate the trends of AsA adsorption on C_{60} using the quantum ab initio approach.

All quantum calculations were carried out using the DMOL3 code in the DFT framework. Core and valence electrons were considered explicitly, and a double numerical basis set (DNP) was chosen to expand the electronic eigenstates with an orbital cutoff radius of 3.7 Å. The exchange-correlation potential was considered in both the LDA and GGA.

9.2.3 Adsorption Spectra

The results of the first-principles LDA-optimizations are shown in Figure 9.3, for the first five configurations: $\Gamma(0,0,0)$, $\Gamma(90,0,0)$, $\Gamma(180,0,0)$, $\Gamma(270,0,0)$, and

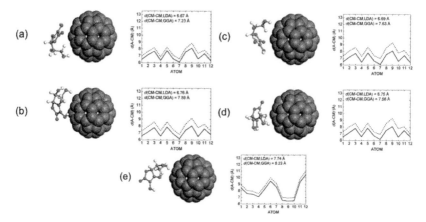

Figure 9.3 Left: LDA-optimized geometries obtained from the initial config-
urations: (a) $\Gamma(0,0,0)$; (b) $\Gamma(90,0,0)$; (c) $\Gamma(180,0,0)$; (d) $\Gamma(270,0,0)$; and (e)
$\Gamma(0,90,0)$. Right: Atomic distances d(A-CM) to the C_{60} center of mass (CM)
for each geometry obtained in the LDA (solid lines) and GGA (dashed lines)
approximations. Atoms 1–6 (7–12) correspond to C_1–C_6 (O_1–O_6). The d(CM-
CM) is the distance between the AsA and the C_{60} CMs. After Santos et al. [221]

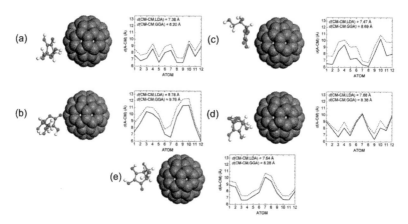

Figure 9.4 Same as in Figure 9.3 but for LDA-optimized geometries obtained
from the initial configurations: (a) $\Gamma(0,180,0)$; (b) $\Gamma(0,270,0)$; (c) $\Gamma(0,0,90)$;
(d) $\Gamma(0,0,180)$; and (e) $\Gamma(0,0,270)$. After Santos et al. [221]

$\Gamma(0,90,0)$. Figure 9.4 depicts the optimizations for the remaining configurations
$\Gamma(0,180,0)$, $\Gamma(0,270,0)$, $\Gamma(0,0,90)$, $\Gamma(0,0,180)$, and $\Gamma(0,0,270)$. On the left side of
each figure, it is possible to check the final molecular geometries obtained after the
LDA simulations. We do not show the GGA counterparts since they qualitatively
show the same general features of the LDA optimization, differing only by the interatomic and intermolecular distances. They tend to underestimate the strength of the
covalent and noncovalent bonds, leading to larger lengths in comparison to LDA.

Table 9.1 *Total energy of the AsA-C_{60} adsorbates configurations optimized using the LDA and GGA functionals. After Santos et al. [221].*

Starting geometry	Total energy (eV)	
	LDA	GGA
$\Gamma_{(0, 0, 0)}$	−80,145.7	−80,746.3
	Total energy relative to the $\Gamma_{(0,0,0)}$ (eV)	
	LDA	GGA
$\Gamma_{(90,0,0)}$	+0.52	+0.44
$\Gamma_{(180,0,0)}$	+0.45	+0.44
$\Gamma_{(270,0,0)}$	+0.52	+0.43
$\Gamma_{(0,90,0)}$	+0.70	+0.45
$\Gamma_{(0,180,0)}$	+0.88	+0.48
$\Gamma_{(0,270,0)}$	+0.74	+0.49
$\Gamma_{(0,0,90)}$	+0.67	+0.44
$\Gamma_{(0,0,180)}$	+0.42	+0.48
$\Gamma_{(0,0,270)}$	+0.72	+0.42

To give a more quantitative view of the optimized geometries, we have plotted a distance profile for each conformation of AsA adsorbed on C_{60}. The plots take into account only the non-hydrogen atoms (carbon and oxygen) of the AsA molecule, displaying their distances to the C_{60} center of mass (CM) and revealing the most relevant details of the atomic geometries of each AsA adsorbate. The distance between a given atom (A) and the C_{60} center of mass is denoted by d(A-CM), and the distance between the AsA and C_{60} CMs is d(CM-CM).

The total energy for all AsA-C_{60} configurations optimized using the LDA and GGA exchange-correlation functionals is presented in Table 9.1. The structure optimized from the geometry with minimal total energy after the classical annealing, $\Gamma(0,0,0)$, has the smallest total energy in both LDA and GGA first-principles calculations (for simplicity, we will use here the same Γ indexes to label structures from both classical and quantum results). The other geometries, however, have total energies very close to the Γ (0,0,0) case; they are larger by 0.74 eV (0.49 eV) at most for $\Gamma(0,270,0)$ if the LDA (GGA) exchange-correlation functional is used.

LDA (GGA) total energies for the other Γ adsorbates relative to $\Gamma(0,0,0)$ show a variation from 0.42 to 0.74 eV (0.42 to 0.49 eV). It is known that for various systems, like hydrogen-bonded complexes, the use of the LDA leads to better results in comparison to the GGA approximation when significant van der Waals forces and strong dipole-dipole forces are present. In these systems, introducing a dependency on the electron density derivatives leads to worse estimates for the interaction energies. In fact, it was shown, for example, that the LDA approach describes well the distance between the layers of graphite, rare-gas atoms adsorbed on graphite surfaces, the intermolecular distance in a face-centered cubic crystal of C_{60}, and the intermolecular interaction potentials of the methane dimer.

It is reasonable to assume that DFT-LDA calculations can describe nonbonding interactions in carbon systems such as AsA-C_{60} as well. Thus, the focus of our attention is mainly directed to the computationally less expensive LDA results, using the GGA data only to check for structural differences between the geometries predicted from both methods. In systems with strong charge transfer effects (not the one considered here), however, the introduction of contributions from electron density derivatives in the exchange-correlation functional improves the accuracy.

The lowest-energy geometry for both LDA and GGA calculations, $\Gamma(0,0,0)$, is qualitatively similar to the original classically optimized geometry, with the same pattern of atomic distances from the C_{60} center of mass, C_1, C_4, O_1, and O_6 (C_3, C_5, O_3, and O_5) being the nearest (farthest), as depicted in Figure 9.3a. As stated before, the GGA data follow very closely the LDA curves in general, with the most pronounced difference observed for the $\Gamma(0,0,90)$ (Figure 9.4c), followed by the $\Gamma(0,270,0)$ (Figure 9.4b) structure. The second-lowest LDA total energy conformation, $\Gamma(0,0,180)$ (Figure 9.4d), has C_3, C_5, O_3, and O_5 closest to the C_{60} center of mass and C_1, C_4, O_1, and O_6 as the outermost. The d(CM-CM) is maximum for the $\Gamma(0,270,0)$ configuration, with values equal to 8.78 Å (LDA) and 9.76 Å (GGA). It can be seen that $\Gamma(0,0,0)$, $\Gamma(90,0,0)$, $\Gamma(180,0,0)$, and $\Gamma(270,0,0)$ resemble one another very closely (Figure 9.3a–d). For the other Γ geometries, no similarities were found.

Looking now to the strength of the AsA-C_{60} interaction, we define the adsorption energy E_A as

$$E_A = E(\text{AsA} + C_{60}) - E(\text{AsA}) - E(C_{60}), \qquad (9.1)$$

where $E(\text{AsA} + C_{60})$ is the total energy for the AsA-C_{60} geometry, $E(\text{AsA})$ is the total energy of the AsA molecule, and $E(C_{60})$ is the total energy for the isolated C_{60}. The basis set superposition error was neglected due to the high quality of the basis set chosen. The LDA adsorption energy for each optimized structure was investigated by translating the AsA atoms (with no change of molecular geometry) along lines parallel to the axis formed by joining the AsA and C_{60} CMs, thus changing the distance d(CM-CM) between the AsA and C_{60} CMs.

The LDA adsorption energies E_A are presented in Figure 9.5. Some caution must be exerted in analyzing E_A for each case due to the trend of overbinding observed in the LDA of the exchange-correlation functional. One can see that the $\Gamma(0,0,0)$ geometry, again, has the smaller minimum at −0.54 eV in comparison with the other Γ structures, corresponding to the strongest binding.

The results for the $\Gamma(90,0,0)$, $\Gamma(180,0,0)$, and $\Gamma(270,0,0)$ are not shown in Figure 9.5 due to their structural similarity with $\Gamma(0,0,0)$. The second-lowest value of E_A is observed for the $\Gamma(0,0,90)$ structure (−0.31 eV), followed by $\Gamma(90,0,0)$ (−0.28 eV), $\Gamma(0,90,0)$ (−0.25 eV), $\Gamma(0,0,180)$ (−0.23 eV), $\Gamma(0,180,0)$ (−0.10 eV),

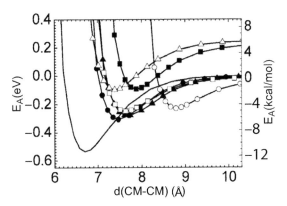

Figure 9.5 LDA adsorption energy of AsA on C_{60} in eV (left scale) and kcal/mol (right scale) as a function of the distance between the AsA and the C_{60} CMs, d(CM-CM) (in Å), for the geometries optimized from $\Gamma(0,0,0)$ (solid line only), $\Gamma(0,0,180)$ (solid squares), $\Gamma(0,0,90)$ (solid circles), $\Gamma(0,90,0)$ (solid triangles), $\Gamma(0,0,270)$ (open squares), $\Gamma(0,270,0)$ (open circles), $\Gamma(0,180,0)$ (open triangles). After Santos et al. [221]

and $\Gamma(0,270,0)$ (-0.10) eV. It can be noted that E_A does not have the same ordered minimal energy configuration as those observed for the total energy, $E(\Gamma(0,0,0)) <$ $E(\Gamma(0,270,0)) < E(\Gamma(0,0,270)) < E(\Gamma(0,0,90)) < E(\Gamma(0,90,0)) < E(\Gamma(0,0,180)) <$ $E(\Gamma(0,180,0))$.

9.3 Ibuprofen

In this section, we present results describing the noncovalent interaction of ibuprofen (IBU) with the C_{60} molecule after sampling their different relative spatial orientations. Our motivation is to investigate theoretically IBU@C_{60} systems to assess the feasibility of controlled delivery of the IBU anti-inflammatory drug through the skin by their association with C_{60}.

9.3.1 Classical Annealing Simulation

Pondus hydrogenii – quantity of hydrogen – is a scale used to specify how acidic (lower pH) or basic (higher pH) a water-based solution is. Since normal skin pH lies in the range 4.2–5.6 (somewhat acidic), the neutral IBU state proportion is higher than that of the charged carboxylic state, as shown in Figure 9.6a. Consequently, the first step of our calculations is the optimization of the geometry of an isolated uncharged ibuprofen molecule within the DFT approach. To perform this optimization, we have used the Gaussian 03 code, choosing the hybrid B3LYP exchange-correlation functional, which mixes Hartree-Fock and DFT exchange energy terms, and the 6-31G* basis set to expand the electronic orbitals.

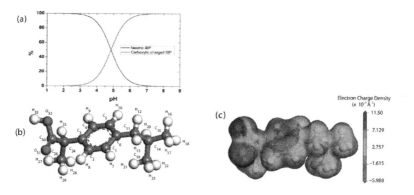

Figure 9.6 (a) The population of the neutral and charged carboxylic IBU states; (b) schematic representation of the most stable conformer of the ibuprofen molecule; (c) electron density projected onto an electrostatic potencial isosurface. After Hadad et al. [222]

Figure 9.6b shows the optimized geometry obtained for IBU with the atomic labels adopted here, while Figure 9.6c depicts a projection of the calculated electron density onto an electrostatic potential isosurface. As expected, oxygen atoms exhibit the highest amount of negative charge, with a secondary electron concentration at the benzenoid ring of IBU (see the darkest regions). Hydrogen atoms, in general, are positively charged.

This first geometry optimization also allowed us to calculate the vibrational frequencies and the infrared and Raman spectra for the IBU molecule discussed later on (Section 9.3.3). A single C_{60} molecule was also optimized, and its vibrational properties were obtained in the same way.

After optimizing both the isolated IBU molecule and the fullerene C_{60}, we used their geometries to build a set of 39 initial configurations for the IBU@C_{60} adsorbate, with different orientations of the IBU relative to the fullerene molecule. These initial configurations were then subjected to the same computational procedure as discussed for the ascorbic acid in the previous section.

Figure 9.7a shows the classical total energy for the optimized annealed structures of IBU@C_{60} versus the distance d (in Å) between the IBU and C_{60} centroids. From it, one can see a primary cluster of configurations for distances between 6.5 and 7.5 Å, with total energies in the 280–450 meV range. There is also a secondary cluster for distances between 8.5 and 10 Å, with total energy above 570 meV. We can associate these two clusters to two different sets of similar conformations of IBU adsorbed on C_{60}. They correspond to two possible adsorption "orbits" of this system, as the IBU molecules can take advantage of the spherical symmetry of C_{60} to rotate about it, keeping the distance between the centroids almost constant – i.e., two adsorption levels. A more detailed account of the adsorption energy as a

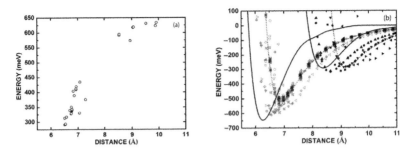

Figure 9.7 (a) Minimum energy (in meV) calculated via classical annealing simulation. (b) Adsorption energy (in meV) of ibuprofen on C_{60} as a function of the distance (in Å) between centroids, showing the LDA (represented by the solid lines) and the classical (represented by dotted lines) levels of calculation. After Hadad et al. [222]

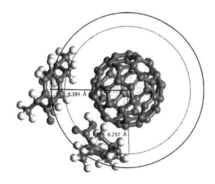

Figure 9.8 Two IBU@C_{60} adsorbate configurations A (B), whose centroid distances is 6.237 (8.384) Å, respectively. After Hadad et al. [222]

function of distance for each annealed IBU@C_{60} geometry is shown in Figure 9.7b, depicting the classically calculated $E_A(d)$, as given by Eq. (9.1), for some selected geometries of each configuration cluster. The solid and dotted lines show, respectively, the LDA and classical formation energies for the optimal configurations.

9.3.2 Adsorption Spectra

Aiming to investigate how a single C_{60} interacts and adsorbs several IBU molecules, we have performed a classical molecular dynamics (MD) simulation for a temperature $T = 300$ K. Here, we use the designation A (B) for the primary (secondary) lowest formation energy adsorbate and the LDA and CFF subscripts to indicate the LDA and classical force field calculated curves.

Figure 9.8 shows the A (centroid distance 6.237 Å) and B (centroid distance 8.384 Å) configurations of adsorbed IBU placed around the C_{60} molecule. The

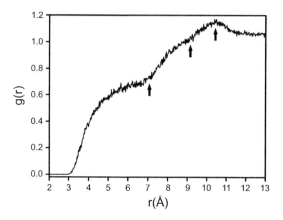

Figure 9.9 Pair correlation function g(r) for a C_{60} immersed in a box with a set of 60 IBU molecules, at room temperature. Observe the three distinct features (indicated by arrows) at about 7.3 Å, 9.5 Å, and 10.6 Å. After Hadad et al. [222]

A adsorbate has the isobutyl, propionic acid, and phenyl regions docked to the C_{60} surface. The B configuration has only its isobutyl region closer to the C_{60}, while the propionic acid and phenyl groups are moved away.

Figure 9.9 shows the pair correlation function, or radial distribution function, $g(r)$, obtained after the simulation of 60 IBU molecules interacting with C_{60} in a box with periodic boundary conditions. It is associated with the probability of finding an ibuprofen molecule at a certain distance r from the C_{60} centroid. Looking at the plotted curve, one can see three distinct features (indicated by arrows) at about 7.3 Å (looking like an inflection point), 9.5 Å (a very smooth secondary inflection point), and 10.6 Å (the maximum).

One can assign to each feature a distinct adsorption layer, with the first one (between 4 and 7 Å) corresponding to a set of IBU molecules in a configuration resembling the A adsorbate, depicted in Figure 9.8, with the propionic acid, phenyl, and isobutyl groups almost equally near to the fullerene molecule. The other features are related to the B adsorbate, as one can see from Figure 9.10 (a snapshot of the MD simulation).

Indeed, the IBU molecules in the first shell appear to be docked with the phenyl group parallel to the fullerene surface, while the IBU molecules in the second shell have (in general) their propionic acid regions moved away from the C_{60} and their isobutyl regions staying closer, as occurs for the B adsorbate shown in Figure 9.8.

Hence, it seems by our classical molecular dynamics simulations that the C_{60}, when solvated in an IBU solution, tends to organize the neighbor IBU molecules in shells or layers, which resemble the A and B adsorption geometries, even at room temperature. Three layers were obtained classically. The most internal

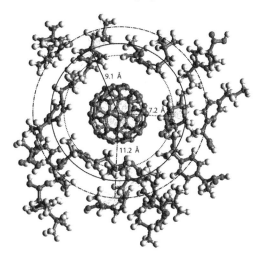

Figure 9.10 Snapshot of the molecular dynamics (MD) simulation showing a single C_{60} molecule surrounded by several IBU molecules. The distances correspond to the structural features showed in Figure 9.9. After Hadad et al. [222]

one is directly related to the A-IBU@C_{60} adsorbate configuration, obtained by using the DFT-LDA approach, and the two most external layers are related to the B-IBU@C_{60} adsorbate configuration, obtained by using the DFT-LDA approach as well. This is possible because, as shown in Figure 9.7, the first-level adsorption occurs in the 6.5–7.5 Å range, while the second level adsorption occurs in the 8.5–10.0 Å range. The existence of the two levels of IBU adsorption on C_{60} is very interesting since the delivery time of the drug should be more extended than in the case of a single-level adsorption.

9.3.3 Infrared and Raman Spectra

A single molecule of IBU has 93 normal modes, while C_{60} has 174 vibrational degrees of freedom with 46 distinct frequencies (due to symmetry). The number of infrared (Raman) first-order active frequencies for C_{60} is four (10), with 32 optically silent frequencies. The A and B adsorbates, on the other hand, have 273 distinct vibrational normal modes.

The left side of Figure 9.11a (9.11b) shows the infrared experimental (theoretical) spectrum for isolated ibuprofen from Reference [227], while the left side Figure 9.11c (9.11d) depicts the IR for B(A)-IBU@C_{60} adsorbate configuration. The left side of Figure 9.11e shows the infrared spectrum for the isolated fullerene C_{60}. As one can see, there are differences between the (a) and (b) curves, with the theoretical absorption bands being shifted toward higher (lower) frequencies in

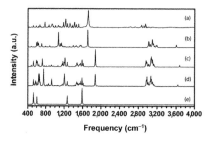
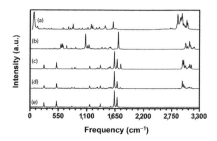

Figure 9.11 Infrared spectra (left side) and Raman spectra (right side) of ibuprofen: (a) experimental; (b) theoretical; (c) B-IBU@C$_{60}$ adsorbate configuration; (d) A-IBU@C$_{60}$ adsorbate configuration; (e) isolated fullerene C$_{60}$. After Hadad et al. [222]

the 2,400–4,000 (400–2,400) cm^{-1} range. This is expected, due to the limitations of the theoretical methodology.

The main absorption peaks of the theoretical spectrum for isolated IBU (in increasing order of frequency) occur at (from now on, all frequency units are in cm^{-1}) 622.13, assigned to a scissors movement of O$_{31}$=C$_{30}$–O$_{32}$ and C$_{30}$–C$_{24}$–C$_{26}$, together with a stretching of the bonds belonging to the phenyl group (in plane ring deformation); at 1,073.82, related to an asymmetric stretching of C$_{24}$–C$_{30}$–O$_{32}$ and a stretching of the C$_{24}$–C$_{26}$ bond, together with the in-plane CH phenyl bending and a twist of C$_{11}$H$_2$; at 1,709.52, associated with an asymmetric stretching of the C$_{24}$–C$_{30}$=O$_{31}$ bonds and the scissors movement of H$_{33}$–O$_{32}$–C$_{30}$ and C$_{30}$–C$_{24}$–H$_{25}$; at 3,099.06, assigned to C$_{19}$–H$_{22}$, C$_{15}$–H$_{16}$ stretching; and at 3,601.74, related to the stretching of the O–H bond. The last frequency is significantly higher than the value obtained by [227] from DFT computations, probably due to the differences in the basis set size and optimization strategy presented here.

The IR active modes of pure C$_{60}$, on the other hand, depicted in Figure 9.11e (left side), reveal four absorption bands: a peak at 529.78, which corresponds to a vibrational mode characterized by the radial breathing of carbon pentagons in an alternate fashion, experimentally assigned to 526; a second peak at 605.55, related to a radial breathing mode for which alternating hemispheres of the C$_{60}$ expand and contract (experimental value: 576); a third peak at 1,267.98, corresponding to a tangential mode with twisting carbon pentagons (experimental value: 1,182); and the fourth peak, at 1,586.20, corresponding to a tangential mode.

The interaction of IBU with C$_{60}$ leads to very similar spectra for both A and B adsorbate configurations, which is related to the weak character of their noncovalent bonding. Small differences in their vibrational spectra may arise due to the differences in relative orientation and shape of IBU in each system. One must also consider that small differences in the electron density distribution along the IBU molecule when interacting with C_{60} slightly modify the strength of the covalent bonds, shifting the normal-mode frequencies and their response to external electromagnetic fields.

In the A adsorbate configuration case, depicted in Figure 9.11d (left side), the main features of the IR spectrum occur for the following frequencies: 537.26 (mainly derived from the C_{60} mode at 529.78), 588.51 (related to the C_{60} mode at 605.55), 645.29 (CO–H out-of-plane bending, phenyl in-plane bending, C–C=O deformation), 719.84 (C_3–C_{24}–C_{26}, C_{11}–C_{14}–C_{19}, O=C-O-H deformations, phenyl CH out-of-plane bending), 745.95 (C_3–C_{24}–C_{30}=O deformation, phenyl CH out-of-plane bending, $C_{14}(C_{11}$–C_{15}–$C_{19})$ symmetric stretch), 870.98 (phenyl out-of-plane bending, CH_3 rocking), 1,166.18 (phenyl CH in-plane bending, C_{26}–H_3 rocking, C_3C_{24}–H bending, C–O stretch), 1,220.39 (phenyl CH in-plane bending, C_3–C_{24}, C–O stretch, C_6–H_3 rocking), 1,406.26 (C_{24}–C_{30}–O asymmetric stretching, C_3C_{24}–H bending, phenyl CH in-plane bending), 1,554.42 (phenyl C–C stretching, in-plane CH bending, C_6–C_{11}, C_3–C_{24} antisymmetric stretching), and 1,832.57 (C=O stretching, O–H bending, deriving from the isolated IBU mode at 1,709.52). Above 3,000, there are strong absorption bands for the A adsorbate configuration at 3,102.47, 3,108.73, and 3,123.61, all of them related to the stretching of CH_3 bonds, and also present in the isolated IBU spectrum. At 3,747.82, there is an absorption due to the stretching of the OH bond.

Looking now to the IR spectrum of the B adsorbate configuration, there is a peak at 538.42, which is equivalent (has a very similar vibrational eigenvector) to the 537.26 peak for the A adsorbate configuration case. One can find other matches for the absorption peaks at 588.47 (A: 588.51), 626.53 (CO–H out-of-plane bending, phenyl in-plane bending), 716.28 (phenyl out-of-plane bending, O=C–O–H deformation), 796.28 (phenyl out-of-plane CH bending, CH_3 rocking, O=C–O–H deformation), 1,086.35 (C_{24}–C_{26} stretching, $C_{26}H_3$ rocking, phenyl CH in-plane bending, $C_{11}H_2$ twisting), 1,173.86 (A: 1,166.18), 1,221.41 (A: 1,220.39), 1,409.41 (A: 1,406.26), 1,554.97 (A: 1,554.42), and 1,832.32 (A: 1,832.57).

Overall, the largest difference between frequencies of equivalent normal modes below 3,000 occurs for the normal mode at 442.96 for the A adsorbate (C_{24}–C_{30}–OH deformation, phenyl in-plane bending, C_{14} (C_{11}–C_{15}–C_{19}) symmetric deformation) and at 472.41 for the B system (C_{24}–C_{30}–OH deformation, phenyl in-plane bending), a difference of about 30. The second-largest difference occurs for the normal mode at 537.26 for the A system (C_{60} mode) and 510.51 for the B case (CO–H bending, phenyl out-of-plane bending), presenting a difference of about 27.

The right side of Figure 9.11 shows the Raman spectra of isolated IBU (experimental and calculated), A and B IBU@C_{60} adsorbate configurations, and molecular C_{60} using the same ordering of curves as in the IR case. A comparison between the experimental (Figure 9.11a, right) and the theoretical curves (Figure 9.11b, right) for isolated IBU shows that the bands below 1,800 span a wider frequency range, a feature that is not very clear when one looks only to the corresponding IR spectra. Computed modes with frequencies below 1,250 have Raman peaks shifted toward lower frequencies in contrast with the experimental data, while modes above this limit have Raman peaks shifted toward higher frequencies. For the single IBU molecule, a set of three peaks occur at 622.13 (also active in the IR spectrum), 640.82, and 672.69, which is related to the bending of bonds at the propionic acid group and, in the case of the last frequency, the bending of C–C bonds in the phenyl region. The peak at 709.53 is assigned to a normal mode, which corresponds to the bending of C–C bonds across the entire IBU molecule. The peak at 1,151.66 is assigned to the stretching of C–C bonds, while the peak at 1,709.52 (also IR active) originates from bond stretching in the propionic acid side. Scattering bands in the 3,000–3,200 range are mainly due to the stretching of C–H bonds of propionic acid and isobutyl.

The Raman active C_{60} normal modes (see Figure 9.11e, right side), on the other hand, occur theoretically (experimentally), at 269.81 (271), 516.15 (496), 1,155.28 (1,099), 1,363.68 (1,250), 1,635.45 (1,469), and 1,687.45 (1,575).

The Raman spectra of both A and B adsorbate configurations, shown on the right side of Figure 9.11c and d, respectively, share many features. Both have very intense peaks at frequencies corresponding to the isolated C_{60} normal modes: 266.47 (A), 266.62 (B), 269.81 (C_{60}); 496.88 (A), 496.81 (B), 516.15 (C_{60}); 1,503.37 (A), 1,503.44 (B), 1,634.45 (C_{60}); and 1,617.28 (A), 1,617.43 (B), 1,687.45 (C_{60}, very close to the strong Raman peak of isolated IBU at 1,709.52). The A adsorbate configuration has a Raman peak at 1,667.66 assigned to the asymmetric stretching of C–C bonds in the phenyl group, with the equivalent mode for the B configuration at 1,668.29. Both A and B adsorbates have Raman bands in the energy range between 3,000 and 3,200, with the same structure of modes, involving the stretching of CH bonds of IBU.

9.4 Human Serum Albumin

In this section, we deal with the binding of the nonsteroidal anti-inflammatory drug ibuprofen (IBU) to human serum albumin (HSA) by means of the DFT calculations within a fragmentation strategy, employing a dispersion corrected exchange-correlation functional. Our computer simulations presented here are a valuable approach for a better understanding of the binding mechanism of IBU@HSA, looking for a rational design and the development of novel IBU-derived drugs with improved potency.

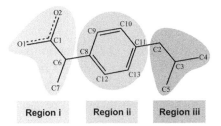

Figure 9.12 Chemical structure representation of the 2-(4-isobutylphenyl) propionic acid (IBU) molecule subdivided into three parts to help the analysis of its interactions with the HSA. After Dantas et al. [223]

9.4.1 Chemical Structure

As described in the previous section, ibuprofen has a flexible chemical structure. It tends to accumulate in appreciable quantities at inflamed compartments where anti-inflammatory/analgesic activity (synovial fluids, cerebrospinal fluid) is needed. To help the analysis of its interactions with the HSA, its chemical structure can be divided into three regions: the acidic side chain (region i), the central aryl moiety (region ii), and the hydrophobic terminal (region iii), as depicted in Figure 9.12.

Once in the bloodstream, IBU becomes strongly bound to HSA. Transcribing from the albumin gene (ALB), located at position q11-22 of chromosome 4, this single non-glycosylated polypeptide chain of 585 amino acid residues is formed by three homologous helical domains (I, II, and III), which are divided into six subdomains (IA, IB, IIA, IIB, IIIA, and IIIB), connected by random and extensive coils without beta-sheets elements [228].

A crystallographic study of HSA–hemin–myristate complex and five HSA–fatty acid complexes, formed using saturated medium-chain and long-chain fatty acids, revealed nine distinct binding locations in HSA. Site 1 (fatty acid FA1) is located in a D-shaped cavity at the center of the four-helix of subdomain IB. Site 2 (fatty acid FA2) is located between the subdomains IA and IIA. Sites 3 and 4 (fatty acids FA3 and FA4) lie in a large cavity in subdomain IIIA, forming the so-called Sudlow site II and recognize preferentially extended aromatic carboxylates. The site 5 (fatty acid FA5) is formed by a hydrophobic channel located in subdomain IIIB, while the binding site 6 (fatty acid FA6) is located at the interface between subdomains IIA and IIB and is occupied by both medium- and long-chain fatty acids, with no cluster of amino acids to stabilize carboxylates electrostatically. The binding site 7 (fatty acid FA7), or Sudlow site I, lies in a smaller hydrophobic cavity of subdomain IIA. Binding sites 8 and 9 (fatty acids FA8 and FA9) are located at the base and the top regions, respectively, of the gap between the subdomains IA–IB–IIA on one side, and IIB–IIIA–IIIB on the other.

Due to its large binding capacity, the transport protein HSA helps in the solubilization and reduction of toxicity of drugs. It also protects drugs against oxidation, increases their half-lives, promotes their distribution throughout the body, modulates their concentration at their targets, and improves the biological effects as well. Calculations of the intermolecular energies between the IBU ligand and the individual amino acid residues of HSA are a crucial step to achieve an accurate understanding of the binding interaction of the IBU@HSA system, as albumin affects directly the pharmacokinetic and pharmacodynamic properties of ibuprofen.

Details of the IBU@HSA binding interaction revealed by high-resolution crystal structure showed two major IBU binding regions, as depicted in Figure 9.13: Sudlow site 2, composed by the binding sites FA3 and FA4, located in a large cavity in the subdomain IIIA, and a secondary site located in a binding cleft that overlaps with FA6. These sites have different shapes, electrostatic potentials, and peculiarities with respect to the IBU binding. At the center of the FA3/FA4 cleft, the IBU molecule is oriented toward a polar patch formed by the Arg410, Tyr411, Gln414, and Ser489 amino acid residues at one end of the nonpolar pocket. For the binding site FA6, the drug enters through the 209–223 helix until reaching the interaction point with the side chains (amide groups) of Lys351 and Ser480 (Leu481 and Val482).

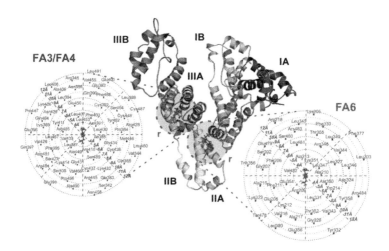

Figure 9.13 Chemical representation of the human serum albumin (HSA) showing the subdomains IA, IIA, IIIA, IB, IIB, and IIIB and the amino acid residues that form the two main binding sites of ibuprofen (FA3/FA4 and FA6). After Dantas et al. [223]

9.4.2 Interaction Energy Profile

The structural information of IBU complexed with HSA at the FA3/FA4 and FA6 binding pockets was obtained from the PDB file 2BXH containing X-ray diffraction data with a resolution of 2.7 Å [229]. The preparation of the molecular structure and the determination of the protonation state of IBU at physiological pH were accomplished using the MarvinSketch code and the protonation tool available in the Discovery Studio package. All heavy atoms are fixed at their X-ray crystal positions. Hydrogen atoms were added to the HBU–HSA complex, and only their respective atomic positions were optimized using the COMPASS (condensed-phase optimized molecular potentials for atomistic simulation studies) force field available in the forcite code. All calculations were converged to the total energy variation smaller than of 10^{-5} Ha, maximum force of 10^{-3} kcal $Å^{-1}mol^{-1}$, and maximum atomic displacement equal to 10^{-5} Å.

At physiological pH, IBU exists predominantly in the deprotonated (acidic) form due to its low (4.91) pKa value (pKa, the acid dissociation constant at logarithmic scale, is a quantitative measure of the strength of an acid in solution). Considering this state, its noncovalent interaction energies with amino acid residues of FA3/FA4 and FA6 within a pocket radius r up to 12 Å measured from the IBU centroid were calculated using the fragmentation molecular fractionation with conjugate caps (MFCC) methodology, discussed in Section 1.8 of this book. All DFT calculations were performed using the DMOL3 code adopting the GGA functional of Perdew, Burke, and Ernzerhof (PBE) [12], together with Grimme's long-range dispersion correction (GGA+D) [19]. A double numerical plus polarization (DNP) basis set was used to expand the Kohn-Sham orbitals for all electrons with unrestricted spin, which is comparable to Gaussian 6-31G**, a valence double-zeta polarized basis set.

An adequate modeling for the dielectric constant ϵ is important in the study of ligand-protein interactions (see Chapter 8), especially in the description of the electrostatic forces. Here we applied the COSMO (COnductor-like Screening MOdel) continuum solvation model, together with the linear-scaling quantum mechanical MFCC theory to calculate the electrostatic solvation energy of HSA, already proved to be an efficient method to study proteins in solution. Electrostatically embedded generalized molecular fractionation with conjugate caps (EE-GMFCC) is another useful approach to determine the electronic properties of the HSA, like the dipole moment, electron density of states, and the electrostatic potential.

To avoid the adoption of an arbitrary binding pocket size, which could risk missing important amino acid residues, we have performed a convergence study for the behavior of the total interaction energy of the IBU@HSA system as the radius of their interaction increases. When the variation of the total interaction energy (the sum of all individual residue–drug binding energies for a specific radius) is smaller than 10% between subsequent radius increase steps, we achieve the so-called Converged Binding Pocket Radius (CBPR). When this condition is reached, the total interaction energy of the IBU molecule complexed with each HSA site is obtained.

The assignment of hydrogen bonds (H-bond) is based on distance and angle criteria involving the H-bond donor (D) and the H-bond acceptor (A) atoms. A distance threshold of about 3.2 Å and a general hydrogen bond X–D–H–A–Y angle range between about 120° and 180° were chosen here. For the salt bridges, we have extremely well-defined geometric preferences. For the existence of hydrophobic contacts, three hydrophobic ligand atoms must lie in the range of the hydrophobic residue side chain. Finally, the distance between the centroid of the aromatic system and a charged atom should be less than 4.5 Å to the occurrence of a π-cation interaction [230].

Figure 9.14 depicts the behavior of the total interaction energy of IBU@ HSA (in kcal/mol) as a function of the binding pocket radius r (in Å), indicating a converged binding pocket radius (CBPR) of about 12 Å for a dielectric constant $\epsilon = 40$. Regions of steepest negative variation are caused by the incorporation of the residues Arg410, Ser489, and Lys414 (Lys351, Val482, Arg209, Leu481, and Ser480) to the binding pocket sphere as its radius crosses thresholds at 5, 6.5, and 8 Å (4, 4.5, 5, 7, and 8.5 Å) for the FA3/FA4 (FA6) pockets, respectively. Individual energy contributions of 65 (48) amino acid residues for $r = 8.5$ Å are enough to ensure total energy convergence with a value of -60.07 (-52.20) kcal/mol for the FA3/FA4 (FA6) binding site. Therefore, residues located at distances larger than 8.5 Å do not significantly contribute to the stability of the system. These results are in agreement with the hypothesis that the Sudlow site II is the main binding pocket for IBU in HSA [229].

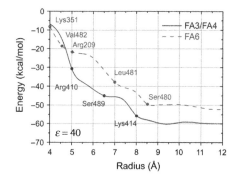

Figure 9.14 Variation of the interaction energy of IBU@HSA (in kcal/mol) as a function of the binding pocket radius r (in Å) calculated using GGA+D/PBE/DNP ($\epsilon = 40$) model. Amino acid residues responsible for the regions of steepest negative variation are highlighted. After Dantas et al. [223]

9.4.3 Binding Site and Residues Domain Graphic Panel

In the FA3/FA4 binding pockets, the carboxylate group, region of IBU, is oriented toward a basic polar patch formed by the residues (all interaction energy units are in kcal/mol) Arg410 (−16.50 interaction energy), Tyr411 (−1.38), Lys414 (−9.16) and Ser489 (−7.72), which is located at one end of a generally hydrophobic pocket at subdomain IIIA. In this site, seven nonpolar amino acid residues – namely Ile388 (−1.34), Phe395 (−1.41), Phe403 (−2.81), Leu407 (−2.72), Leu430 (−3.16), Val433 (−2.70), and Leu453 (−4.54) – surround the ii and iii hydrophobic regions of IBU, with smallest distances r = 2.67, 4.06, 2.61, 3.59, 2.83, 1.90, and 1.8 Å, respectively (see Figure 9.15a for details).

The key role played by the residue Arg410 in the IBU@HSA binding mechanism was also observed previously for ketoprofen, aromatic compounds with peripherally electronegative aspects, and curcuminoids [231]. Furthermore, it is believed that the carboxylic acid moiety of ligands has an important electrostatic interaction with the ionized guanidine group of Arg410. In fact, one can note that the Arg410 residue in our study is responsible for the largest attractive contribution (−16.50 kcal/mol) due to a salt bridge (2.0 Å) and π-cation (4.1 Å) interactions between the

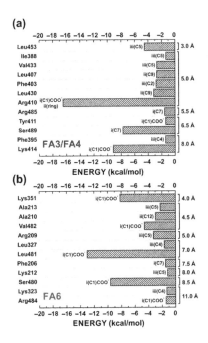

Figure 9.15 Graphic panel showing the most important FA3/FA4 (a) and FA6 (b) residues that contribute to the total binding energy of the IBU@HSA complex. The minimum distances between each residue and the ibuprofen molecule are also shown. After Dantas et al. [223]

guanidinium ion and the acid and aryl groups of IBU, respectively. According to [232], the π-cation of Arg10 has probably a greater stabilizing effect on the IBU than its salt bridge.

The Lys414 residue is also part of the polar patch of the FA2/FA6 binding site, but no single structural observations indicate the presence of dipole–dipole interactions. In our computer simulations, after the geometry optimization procedure, the anionic carboxylate region of ibuprofen interacts with the side chain Lys414 through a salt bridge (2.47 Å) with a pronounced attractive energy value of -9.16 kcal/mol.

Binding experiments using HSA mutants showed that Tyr411 is an important residue involved in the binding of curcuminoids, diazepam, and ketoprofen [231]. These ligands are bound at the center of the polar patch of FA3/FA4 site, forming H-bonds and hydrophobic interactions in respective order with the phenolic oxygen and aromatic ring of the residue Tyr411. In the case of IBU binding, region I of ligand forms an H-bond with the residue Tyr411 (1.88 Å H-bond length), with no relevant hydrophobic contacts. The polar patch of the FA2/FA6 site prevents the approximation of hydrophobic regions ii and iii of the IBU molecule.

Docking simulations suggest that there is an electrostatic attraction between oxygen atoms of the carboxylic acid of 11 profen drugs (including ibuprofen) and the side chains of residues Arg410 and Tyr411, with no role assigned to Ser489. In our optimized IBU@HSA complex, the deprotonated moiety of IBU stays close to the triad Arg410–Tyr411–Lys414 due to the highly attractive character of this region (-27.04 kcal/mol interaction energy for IBU), moving the drug away from the hydroxyl group of Ser489, which is located on the opposite side. The distance of about 4.6 Å between IBU (carboxylate) and Ser489 (hydroxyl) does not satisfy the limit for conventional H-bond. On the other hand, the Ser489 amino acid residue forms a typical H-bond with ligands such as diflunisal, indoxyl sulfate, and IBU itself, binding them near the polar patch in FA3/FA4. This uncertainty about the dominant intermolecular force between Ser489 and IBU, when one takes into account the spatial configuration of the optimized FA3/FA4 binding site complexed with IBU as well as the strong attraction of IBU towards Ser489 (-7.72 kcal/mol interaction energy), suggests the presence of an ion-dipole force.

Important residues for the binding of fatty acids to HSA – namely Ser342, Arg348, and Arg485 – interact weakly with IBU. However, for FA4, the carboxylate head groups of fatty acids are hydrogen-bonded to Arg410, Tyr411, and Ser489, which are the most important residues for IBU binding as well. Plasmatic fatty acids probably modulate competitively and allosterically IBU binding to HSA, which could explain the experimental dissociation equilibrium constant value of $10^{-6} M$ for IBU@HSA interaction system. This triad of residues is highly conserved in mammalian albumins, indicating the importance of intermolecular interactions at this location.

Docking simulations have shown that IBU has the most favorable binding profile among 11 profen drugs with different degrees of hydrophobicity. This occurs because only IBU has exclusively nonpolar moieties buried within the hydrophobic cleft of subdomain IIIA, giving a dominant role to London (dispersion) forces between the aryl and isobutyl group, regions ii and iii of IBU, and the amino acid residues Ile388, Phe395, Phe403, Leu407, Leu430, Val433, and Leu453, helping in the stabilization of the ibuprofen-albumin complex.

Crystallographic (structural) data suggest that IBU interacts mainly with electronegative portions of Lys351, Ser480, Leu481, and Val482 amino acids residues at the FA6 secondary site [229]. Indeed, as it can be seen from Figure 9.15b, these residues have strong attractive interactions (-8.17, -9.60, -13.02, and -4.62, respectively), although we cannot neglect the importance of the binding pocket residues Phe206 (-1.50), Arg209 (-3.29), Ala210 (-2.97), Lys212 (-1.18), Ala213 (-2.27), Lys323 (-1.30), Leu327 (-1.69), and Arg484 (-1.44), all units in kcal/mol.

The carboxylate group of fatty acid entering the FA6 site is recognized transiently by Arg209, Lys351, and Ser480 side chains, with no clear indication of anchor residues. The deprotonated carboxyl group, region i of IBU, by the way, is placed in the binding pocket formed by three nonpolar (Phe206, Leu481, Val482), two basic (Lys351 and Arg484), and one polar (Ser480) amino acid residues, with smallest distances of 2.58, 2.53, 1.54, 1.72, 5.68, and 3.38 Å, respectively. The anionic region i(C_1)COO$^-$ of IBU forms a salt bridge with Lys351 (cationic ammonium) and important hydrogen bonds with Ser480 (hydroxyl), Leu481 (amide), and Val482 (amide). The residues of this amphiphilic environment are responsible for most of the IBU–FA6 complex binding energy.

Region ii of the IBU molecule is surrounded by just one nonpolar (Ala210) amino acid, while five amino acid residues are close to region iii. Three of them are polar (Arg209, Lys212, and Lys323), and two are hydrophobic (Ala213 and Leu327). There is a cluster of nonpolar amino acid residues: Ala210 (-2.97 kcal/mol), Ala213 (-2.27 kcal/mol), and Leu327 (-1.69 kcal/mol), which stabilizes the IBU hydrophobic moieties through dispersion forces. Figure 9.16a (9.16b) depicts the most relevant amino acid residues involved in the IBU@HSA binding, considering the FA3/FA4 (FA6) binding pocket.

9.4.4 Electrostatic Potential Isosurfaces

Several studies suggest that the stability of IBU@HSA complexes is directly influenced not only by the amount and the degree of deprotonation of the hydroxyl groups [233] but also by the presence of the carboxylic acid moiety [234] as well as by the number of hydrogen bonds and the electrostatic character of the binding forces. With this in mind, our results unveil that the most important binding forces in the IBU@HSA system involve region i of IBU.

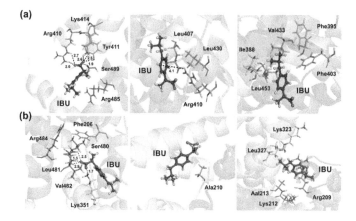

Figure 9.16 Arrangement of the most relevant amino acid residues involved in the IBU@HSA binding: (a) FA3/FA4 and (b) FA6 binding pockets. Drug-residue polar contacts are depicted by dashed lines. After Dantas et al. [223]

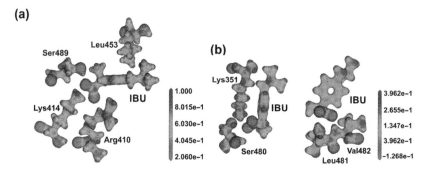

Figure 9.17 GGA+D/PBE/DNP/ϵ = 40 electrostatic potential isosurfaces of ibuprofen and the most attractive residues of sites (a) FA3/FA4 and (b) FA6. After Dantas et al. [223]

As it can be seen from the electrostatic potential isosurfaces plotted in Figure 9.17, positively charged regions of the residues Lys414, Arg410, and Ser489 (Lys351, Ser480, Leu481, and Val482) in the FA3/FA4 (FA6) binding site are near to the carboxyl group of the drug. The side chains (amide groups) of the residues Arg410, Lys414, and Lys351 (Leu481 and Val482), in contrast, present low charge densities centered at their basic nitrogen atoms, forming salt bridges (hydrogen bonds) with the IBU i(C_1)COO$^-$ group.

Notwithstanding the significant function of these polar contacts, recent investigations suggest that, within a family of compounds, the binding of IBU to HSA increases with the hydrophobicity of IBU [235]. Thus, we believe that controlling the number and the size of IBU nonpolar pharmacophores should be the main strategy for the development of new compounds with variable affinity for FA3/FA4

(FA6) pockets. As a matter of fact, the removal of the carboxylic acid moiety is not appropriate because it not only weakens the affinity of IBU for the FA3/FA4 (FA6) residues Arg410, Tyr411, Lys414, and Ser489 (Lys351, Ser480, Leu481, and Val482) of HSA but also disturbs the binding of the ligand to its functional molecular target, since this group is critical for therapeutic efficacy [236].

9.5 Conclusions

Many biological molecules noncovalently bound to the carbon fullerene C_{60} and nanotubes have simple electric charge distributions, with a negative and/or a positive center, as in the case of amino acids, for example. In this case, the adsorption energy as a function of intermolecular distance has a single minimum related to a unique spatial configuration of the adsorbed molecule on the carbon nanostructure, and it is straightforward to calculate the binding energy. This is not the case of both the ascorbic acid (AsA) and the nonsteroidal anti-inflammatory drug ibuprofen (IBU), which has four –OH groups plus two oxygen atoms in a ring, all of them contributing to the AsA–IBU@C_{60} noncovalent bonding in several ways.

In this system, for instance, the classical annealing simulation of a single AsA/IBU molecule interacting with C_{60}, as discussed in this chapter, is a very suitable method to obtain different initial geometries for studying AsA/IBU adsorption in the carbon fullerene C_{60}. This strategy could also predict the equilibrium structures for other complexes or molecules of interest with several negatively and/or positively charged centers.

The demonstration of noncovalent AsA/IBU functionalized C_{60} in biological molecules suggests also that the administration of this complex should be more effective in preventing the oxidative damage and toxicity of C_{60}. As a consequence, it is possible that the AsA/IBU-C_{60} associated, for example, with levodopa, reduces the neurotoxicity generated by isolated levodopa (see Chapter 13). In the latter case, new strategies for Parkinson's disease treatment by combining the clinical use of levodopa and potent antioxidants – that is, noncovalent AsA/IBU-functionalized C_{60} – should be envisaged.

In biological systems, the existence of water must definitively be taken into account. Complexes, as those presented in this chapter, interacting noncovalently with the fullerene C_{60} in vivo will have smaller adsorption energies in comparison to the vacuum case due to water effects, which could prevent drug binding. To avoid this outcome, one could attach organic functional groups covalently to the C_{60} surface to increase the strength of the AsA/IBU@C_{60} noncovalent interaction. As discussed in Chapter 7, molecular water stackings can contribute significantly to prevent (improve) directional electron (hole) transport in hydrated amino acid crystals and similar ordered biological molecules, due to a very large electron effective masses, and they can play a complex role on the carrier transport properties of biomolecular crystals.

Regarding the interaction of the IBU@HSA supramolecular, essential for drug transport in the blood and its distribution throughout the human body, theoretical estimation of its affinity can be very useful to screen drug candidates for acceptable levels of binding to HSA. Notwithstanding, a high-level affinity of HSA for a drug molecule can hinder its access to the target site, decreasing its potency in vivo. Therefore, the understanding of the interaction IBU@HSA, as discussed in this chapter, can be of prime importance for the design of new nonsteroidal, anti-inflammatory agents in the future.

10

Cholesterol-Lowering Drugs

10.1 Introduction

Hundreds of millions of adults have high cholesterol, which has generated a billion-dollar market of drugs devised to reduce and control the total serum cholesterol levels. Cholesterol is both good and bad. At normal levels, it is an essential substance for the body. However, if concentrations in the blood get too high, cholesterol becomes a silent danger that puts people at risk of heart attack. It has three primary functions, without which we could not survive, namely it contributes to the structure of cell walls, it makes up digestive bile acids in the intestine, and it allows the body to produce vitamin D as well as certain hormones.

Cholesterol is a basic structural component of all animal cell membranes, essential for the functioning of all human organs. As a lipid molecule, it is biosynthesized by all animal cells and transported around the body inside lipoprotein particles. As an oil-based substance, it does not mix with the blood, which is water based. Unfortunately, elevated cholesterol concentrations (hypercholesterolemia) can combine with other substances in the blood to form plaque, promoting the so-called atheroma development in arteries (atherosclerosis), which is potentially the main cause of coronary heart disease as well as other forms of cardiovascular disease [237].

Essentially, there are two different types of cholesterol: HDL (high-density lipoprotein) cholesterol and LDL (low-density lipoprotein) cholesterol. The former is called the "good" cholesterol because it carries cholesterol from other parts of the body back to the liver, which in turn then removes the cholesterol from the body. The latter is the "bad" cholesterol because a high LDL level leads to the buildup of plaque in the arteries [238]. A variety of things can raise the risk for high cholesterol – like age, heredity, and overweight, to cite just a few. There are usually no signs or symptoms that one has high cholesterol.

Regarded as gifts from nature, statin drugs, also known as HMG-CoA reductase inhibitors, are a class of lipid-lowering medications that reduce cardiovascular disease and mortality. Statins work in two ways: first, they block the action of the liver enzyme that is responsible for producing cholesterol. Second, they help the body to reabsorb the cholesterol that has built plaques in the artery walls. This reduces one's risk of blood vessel blockages and heart attacks. Be aware, however, that statins can help control one's cholesterol, but they do not cure it. One needs to keep taking medicines and get regular cholesterol checks to make sure that cholesterol levels are in a healthy range.

Statins are used for preventing and treating atherosclerosis, which causes chest pains, heart attacks, strokes, and even death in some individuals. So people with abnormally elevated cholesterol levels or a history of heart attacks should consider taking statins, despite their side effects – which include muscle pain, increased risk of diabetes mellitus, and abnormalities in the liver. Additionally, although this is a minor probability, they may lead to a rare severe adverse effects, particularly muscle damage.

Although the statins are arguably the most effective drug in lowering LDL cholesterol, the main carriers of bad cholesterol that plays a key role in the development of atherosclerosis and coronary heart disease, recently there also has been a renaissance of non-statin drugs. Among them, the most prescribed are ezetimibe and niacin, usually taken in conjunction with drugs with lower statin doses. Nevertheless, the discovery of statins has led to important progress in the primary and secondary prevention of coronary heart disease. Numerous clinical studies have correlated the reduction of blood cholesterol the reduction of the number of major coronary events, and the reduction general mortality in coronary patients with the use of statins [239].

Statins act by inhibiting the 3-hydroxy-3-methylglutaryl coenzyme A (HMG-CoA) reductase in the process of converting HMG-CoA to mevalonate, a committed step in the biosynthesis of cholesterol. It has been observed in clinical trials that this action decreases LDL cholesterol levels by 20%–60%, reducing coronary events by up to one-third over a five-year period. Statins also reduce HDL cholesterol levels on the order of 5%–8% [240]. All statins have the liver as the target organ. The percentage of the dose retained by the liver is greater than 70% (80%) for fluvastatin and lovastatin (simvastatin); there are no available data for atorvastatin and cerivastatin [241].

Patents covering the leading statins have expired recently (Lipitor/atorvastatin in 2010, Lescol/fluvastatin in 2011, Crestor/rosuvastatin in 2012), which increases the pressure for the development of new drugs for the hypolipidaemic market – in particular, new and more effective statins derivatives, which would open up new possibilities for the rational design and optimization of even better HGMR inhibitors.

In this chapter, we take full advantage of the published crystallographic data of the human HMG-CoA reductase (HMGR) complexed with statins (see next section for its crystallographic structure) to perform computer simulations to investigate the details of the binding interaction energies of the statins atorvastatin (A), rosu-vastatin (R), fluvastatin (F), cerivastatin (C), mevastatin (M), and simvastatin (S) to the HMGR enzyme [242]. Our computer simulation uses an ab initio quan-tum mechanical approach, based on the DFT and in the framework of the MFCC strategy [22].

Our purpose here is to elucidate why statins have differences in their efficiency to reduce cholesterol levels by obtaining and comparing the interaction energy between the HMGR residues and ligand atoms. The binding pocket size radius r, defined as the distance from the ligand (statin) to the centroid (HMGR), used to estimate the binding interaction energies, was varied from 2.5 to 12.0 Å, and a pro-file of the interaction energy with r was obtained for each HMGR-statin complex.

10.2 Crystallographic Data

Human HMG-CoA reductase (HMGR) consists of a single polypeptide chain of 888 amino acids. The amino-terminal 339 residues are membrane bound and reside in the endoplasmic reticulum membrane, while the catalytic activity of the protein resides in its cytoplasmic, soluble C-terminal portion (residues 460–888). A linker region (residues 340–459) connects the two portions of the protein [243].

X-ray diffraction data of 2.2 Å resolution shows that the catalytic portions of HMGR are composed of the residues 426–888, forming a tetramer whose quiet similar monomers are arranged in two dimers, each having two active sites, as shown in Figure 10.1. The HMGR monomers have am N-terminal (residues

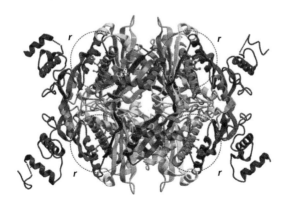

Figure 10.1 The HMGR tetramer bonded to four atorvastatins. The four similar binding pockets spheres (BPS), each one with radius r, are indicated by circles. After da Costa et al. [242]

460–527), a large L-domain (residues 528–590 and 694–872), and a small S-terminal (residues 592–682) domain.

The elucidated structures of HMGR co-crystallized with the six statins – compactin, Zocor (simvastatin), Lescol (fluvastatin), Baycol (cerivastatin), Lipitor (atorvastatin), and Crestor (rosuvastatin) – demonstrate that the HMGR-binding pocket is located between the L- and S-domains, suggesting that 21 residues from neighboring monomers (including those of the cys-loop) contribute to the binding – namely E559 (E = glutamic acid), C561 (C = cysteine), L562 (L = leucine), S565 (S = Serine), R568 (R = arginine), R590, V683 (V = valine), S684, N686 (N = asparagine), C688, D690 (D = aspartic acid), K691 (K = lysine), K692, K735, H752 (H = histidine), N755, D767, S852, L853, A856 (A = alanine), and L857 [243]. Besides, through an analysis of the distance and type of interaction between the residues and atoms of a statin, works based on HMGR–statin crystallographic data suggest and it is largely accepted that the most important residues in the binding pocket are S684, D690, K692 (polar interactions); E559, D767 (hydrogen bonds); and L562, V683, L853, A856, L857 (van der Waals contacts) [243].

Structural differences in statins may partially account for variations in potency of HMGR inhibition. The inhibitory concentration value, IC_{50}, measures how effective a drug is in inhibiting the biological function of its target or, more precisely, what is the concentration of drug required to block a biological process by half.

The range of inhibitory concentration values IC_{50} of some statins are as follows: atovartastin, 3.8–6.2 nM; rosuvastatin, 2.1–5.4 nM; cerivastatin, 2.8–10.0 nM; simvastatin, 4.3–18.0 nM; mevastatin, 23 nM; and fluvastatin, 3–28 nM. Investigations on the inhibition kinetics and the microcalorimetric analysis (including isothermal titration calorimetry) of pravastatin, fluvastatin, cerivastatin, atorvastatin, and rosuvastatin [244] showed binding enthalpies to the HMGR ranging between 0 (fluvastatin) and −9.3 kcal/mol (rosuvastatin) at 25°C. Rosuvastatin exhibited the strongest binding, showing Gibbs energy ΔG of about −12.3 kcal/mol, followed by cerivastatin, with $\Delta G \approx -11.3$ kcal/mol and atorvastatin ($\Delta G \approx -10.8$ kcal/mol).

10.2.1 Computational Details

The many roles of computation in drug discovery mainly focus on virtual screening, de novo design, evaluation of drug-likeness in the docking and molecular dynamics frameworks. For statins and other inhibitors of human HMGR, the computational modeling was performed mainly through docking.

In fact, a structure-based search for potential novel inhibitors of human HMGR was carried out by Zhang et al. [245] by combining CoMFA 3-D QSAR modeling and virtual screening, and finding a representative set of eight new promising compounds of non-statin-like structures. However, only atoms located within a radius of 6.5 Å from any atom of ligands of the protein ensemble structures were taken into account to build the active site.

Computer-aided molecular design tools – i.e., flexible docking, virtual screening in large databases, and molecular interaction fields – were employed by Da Silva et al. [246] to propose novel potential HMG-CoA reductase inhibitors that are promising for the treatment of hypercholesterolemia.

Molecular docking methods were used also to investigate 19 HGMR inhibitors to compare their hypolipidemic efficacies, and it was observed among that benzoic acid, catechin, hydroxycitric acid, and piperine have similar in silico properties to atorvastatin. However, a box of only $5 \text{Å} \times 5 \text{Å} \times 5 \text{Å}$ was created for the simulations, and the strength of local interactions was not estimated. On the other hand, quantum chemistry methods, often described as ab initio since they work without using empirical parameterizations, are being employed for the simulation of molecular systems with up to hundreds of atoms.

A detailed understanding of the ligand pathway actions leading to its bonding to HMGR residues in the binding pocket at the quantum biochemistry level of description is important and depends on the evaluation of the contributions of each amino acid residue to the total binding energy, allowing for the design of new ligand derivatives. Quantum mechanical (QM) methods are becoming popular in computational drug design and development mainly because high accuracy is required to estimate (relative) binding affinities, allowing one to compute the binding interaction at the atomic level.

However, as the macromolecules involved (in general, proteins) are very large, it is necessary to achieve a compromise between the computational cost of the calculations and the accuracy required to obtain trustworthy results. Recently, the aromatic interactions in the binding of ligands to a truncated HMGR was studied by Kee et al. [247] and showed that local density functional theory (DFT) methods match quantum chemistry second-order Moller-Plesset approach (MP2) energy values for aromatic binding better than hybrid or gradient-corrected DFT methods.

Taking into account all these facts and the need to achieve the best balance between computational cost per simulation run, we have opted for the LDA approach to estimate the strength of the drug–amino acid residue intermolecular interaction energies, which furnishes an improved picture of the general behavior tendencies instead of very precise values compared to experimental data.

A DNP basis set was chosen to expand the electronic Kohn-Sham orbitals, taking into account all electrons explicitly and with unrestricted spin. The DNP basis set is very accurate with neglectful basis set superposition error (BSSE) [134]. The orbital cutoff radius was 3.7 Å and the self-consistent field convergence threshold was, adjusted to 10^{-6} Ha. The hydrogen atomic positions of the system formed by the statin molecule and the HMG-CoA binding pocket optimized classically were optimized again using the DFT approach, with convergence tolerances set to 10^{-5} Ha for the maximum force per atom and 0.005 Å for the maximum atomic displacement.

10.3 Chemical Structure

The essential structural components of all statins are a dihydroxyheptanoic acid unit and a ring system with different substituents. The statin pharmacophore is a modified hydroxyglutaric acid component, which is structurally similar to the endogenous substrate HMG-CoA reductase, and the mevaldyl CoA transition state intermediate. The statin pharmacophore binds to the same active site as the substrate HMG-CoA and inhibits the HMGR enzyme.

The chemical structures of the statins differ with respect to their ring arrangements and substituents. These differences affect the pharmacological properties of the statins, such as their affinity for the active site of the HMGR, as well as the routes and modes of metabolic transformation and elimination.

The X-ray diffraction data of reference [243] provide the structures of HMG-CoA reductase complexed with statins with a resolution of 2.2 Å at high and neutral pH. These chemical structures were used as inputs for the calculations; the interaction energy of the active site was defined taking into account all amino acid residues inside a given radius.

Figure 10.2 depicts the structure and relevant functional groups of the statins atorvastatin (A), rosuvastatin (R), simvastatin (S), and fluvastatin (F), which are the main focus of the present study. The (**a**)-group (HMGR-like) is common to all statins and is directly related to their binding to the HGMR. Atorvastatin, rosuvastatin, and fluvastatin share the (**b**)-fluorophenyl group, while the remaining parts of their structures, including the (**c**)-phenyl and (**d**)-sulfonamide groups, are distinct. Simvastatin has butyril and decalin (**e**)-group making up its non-HMGR region, while fluvastatin has a characteristic (**f**)-benzene group.

The atomic positions of the non-hydrogen atoms at the binding pocket (including those belonging to the statin molecules) were kept fixed, while the hydrogen atomic positions were initially optimized by using the consistent valence force

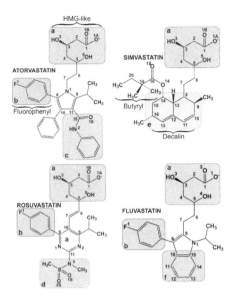

Figure 10.2 Atorvastatin (A), rosuvastatin (R), simvastatin (S), and fluvastatin (F) chemical structure, and their main functional groups, namely (**a**), (**b**), (**c**), (**d**), (**e**), and (**f**) groups. After da Costa et al. [242]

field (CVFF), which has parameters specific to amino acids. Afterward, simulations within the DFT formalism using the LDA for the exchange-correlation functional were carried out using the DMOL3 code [8].

It is well known that pure DFT methods are unable to accurately describe systems where noncovalent interactions, such as van der Waals forces, are relevant [248]. Besides, the LDA is not the best option to give a good description of hydrogen bonds. However, some DFT studies of layered crystals such as graphite as well as guanine hydrated crystals [249] have pointed out that the LDA gives reasonable values for atomic distances due to a cancellation of errors. Besides, Kee et al. [247] have investigated aromatic interactions in the binding of ligands in the active site of HGMR, showing that the LDA functional has a better agreement with the more sophisticated MP2 method in comparison to generalized gradient and hybrid functionals.

Figure 10.3 shows the binding sites of HMG-CoA reductase complexed with the chemical structures of the four representative statins: atorvastatin, rosuvastatin, simvastatin, and fluvastatin. The first (last) two are the most potent (least potent) statins according to our calculations; the corresponding most relevant amino acid residues of each binding sites are also shown.

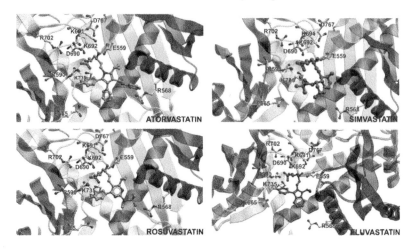

Figure 10.3 Binding sites taken from the chemical structures of the statins atorvastatin, rosuvastatin, simvastatin, and fluvastatin. The most important HMGR amino acid residues bonded to these drug molecules are shown: K735, R590, K692, K691, R702, and R568 attract the statins, and E665, D767, D690, and E559 repel them. After da Costa et al. [242]

10.4 Binding Interaction Energy Profiles

Binding interaction energies between each statin molecule and neighboring amino acid residues at the binding pocket of HMG-CoA reductase were calculated by using the MFCC strategy described in Section 1.8 [22]. The binding interaction energy between the statin molecule S and the amino acid residue R^i, $E(S - R^i)$, is given by

$$E(S - R^i) = E(S - C^{i-1}R^iC^{i+1}) - E(C^{i-1}R^iC^{i+1}) +$$
$$- E(S - C^{i-1}C^{i+1}) + E(C^{i-1}C^{i+1}), \tag{10.1}$$

where the C^i cap is obtained by attaching a carboxyl or amine group to the dangling bond of the residue R^i. On the right side of Eq. (10.1), $E(S - C^{i-1}R^iC^{i+1})$ is the total energy of the system formed by the statin and the capped residue; the $E(C^{i-1}R^iC^{i+1})$ term is the total energy of the capped residue alone; $E(S - C^{i-1}C^{i+1})$ is the total energy of the system formed by the statin molecule and the caps alone. Finally, $E(C^{i-1}C^{i+1})$ is the total energy of the system formed only by the molecular caps.

The total binding interaction energy $E(r)$ of each statin is obtained by adding up the binding energies with each one of the amino acid residues taken into account within the chosen binding pocket radius r. Each statin binds to HMGR in a specific way, which is determined by its own molecular structure and the interacting residues. It is not trivial to specify what the residues within the binding pocket

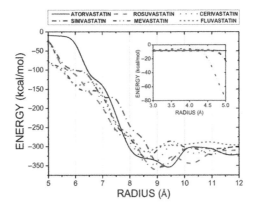

Figure 10.4 Variation of the binding interaction energy $E(r)$ (in kcal/mol) as a function of the binding pocket radius r (in Å) for atorvastatin (solid), fluvastatin (long-dashed), cerivastatin (dotted), mevastatin (dash-dotted), rosuvastatin (dashed-dotted-dotted), and simvastatin (small-dashed). The inset shows the interaction energy for the binding pocket radius smaller than 5 Å. After da Costa et al. [242]

sphere are using only crystallographic data or structural docking information. By increasing the binding pocket radius r, more and more residues are taken into account, changing the interaction energy. When r is small, one can expect a dominant contribution from near charged residues. At intermediate r sizes, the interaction energy probably will change more slowly and, finally, for large r values, the interaction energy must stabilize and converge. Indeed, such behavior is clearly depicted for different statins interacting with HMGR in Figure 10.4, where the interaction energy $E(r)$ (in kcal/mol) as a function of r (in Å) for atorvastatin (solid), fluvastatin (long-dashed), cerivastatin (dotted), mevastatin (dashed–dotted), rosuvastatin (dashed-dotted-dotted), and simvastatin (small-dashed) is displayed. Each energy curve is clearly distinct from the others, being exclusive to each statin.

Nevertheless, the MFCC strategy allows one to calculate the interaction energy between a given residue and its closer statin atom at the quantum level [22]. Here, we define the binding pocket sphere (BPS) as an imaginary sphere centered at the ligand centroid (see Figure 10.1), whose corresponding binding pocket radius r is defined as the distance between each statin centroid and the most distant HMGR residue binding to the ligand.

A quick look at Figure 10.4 shows that the interaction energy as a function of r stabilizes (varies by less than 10%) for all statins for $r \approx 11$ Å, which means that the HMGR-ligand-binding pocket size is underestimated in the classic work of Istvan et al. [243], which adopts $r \approx 9.5$ Å for atorvastatin. This reduced value of r excludes several charged residues (E850, D767, E655, R702, E528, and K662), which are relevant to evaluate the interaction energy at the binding site.

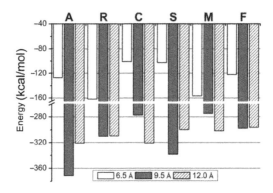

Figure 10.5 Total HMGR binding interaction energy of atorvastatin (A), rosuvastatin (R), cerivastatin (C), simvastatin (S), mevastatin (M), and fluvastatin (F) for the binding pocket radius r of 6.5 Å, 9.5 Å, and 13.5 Å, respectively. The 9.5 Å binding pocket for atorvastatin has nearly the same HMGR interacting residues proposed by Istvan et al. [243], except by the residues M657, G560, A751, M854, N658, and S661. After da Costa et al. [242]

The inset of Figure 10.4 shows that the interaction binding energy exhibits a sharp decrease starting at $r \approx 4.5$ Å for mevastatin and fluvastatin, and at $r \approx 4.7$ Å for simvastatin, rosuvastatin, and cerivastatin. For atorvastatin, the sharp decrease begins at $r \approx 5.7$ Å. There is a total minimum energy for $E(r)$ at $r \approx 8.6$ Å for mevastatin and fluvastatin, $r \approx 8.8$ Å for cerivastatin, $r \approx 9.2$ Å for simvastatin, and $r \approx 9.0$ Å for rosuvastatin. After this energy range, $E(r)$ tends to present small oscillations mainly due to contributions of charged amino acid residues that repel the ligands.

Figure 10.5 shows a comparison of the calculated interaction energies $E(r)$ for HMGR complexed with the statins of Figure 10.4 with $r = 6.5$, 9.5, and 12.0 Å, respectively. For the smallest binding pocket radius, the absolute value of the total HMGR interaction energy follows the order R > M > A > F > S > C, suggesting that rosuvastatin (cerivastatin) is the most (least) potent statin to inhibit HMGR if one assumes a direct correlation between statin potency and the strength of the total HMGR–statin interaction.

For the binding pocket with $r \approx 9.5$ Å, which is approximately the value used for atorvastatin in the work of Istvan and Deisenhofer [243], the absolute value of $E(r)$ obeys the sequence A > R > M > S > F > C, suggesting that atorvastatin (cerivastatin) is the most (least) effective statin to inhibit HMGR.

After the stabilization of the binding interaction energy $E(r)$ for $r > 11.0$ Å, it presents the following ordering: A > C > R > M > S > F, suggesting that atorvastatin and rosuvastatin are the most potent nontoxic statins, while simvastatin and fluvastatin are the least potent ones (cerivastatin, albeit having practically the same interaction energy of atorvastatin, is not considered here due to its toxicity).

The weakest bonding in the thermodynamic experiments was observed for fluvastatin, with the variation of the Gibbs energy $\Delta G \approx -9$ kcal/mol, while rosuvastatin has the highest affinity for HMGR [244]. The interaction energy varies from ≈ -320 kcal/mol for A and C, -310 kcal/mol for R, to -295 kcal/mol for fluvastatin.

The statins showing largest change when one switches from $r = 9.5$ Å to $r = 12.0$ Å are A ($E(r)$ increases by about 50 kcal/mol), C ($E(r)$ decreases by about 40 kcal/mol), S ($E(r)$ increases by about 35 kcal/mol), and M ($E(r)$ decreases by about 25 kcal/mol). Minimal changes of $E(r)$ are observed for R and F in the transition from $r = 9.5$ to 12.0 Å. These results are consistent with the results of clinical trials, which show that rosuvastatin and atorvastatin (simvastatin and fluvastatin) have the smallest (largest) maximum inhibitory concentration IC_{50} values among the nontoxic statins, 2.1 nM and 6.2 nM (18 nM and 28 nM), respectively.

10.4.1 The Graphic Panel BIRD

Figure 10.6 shows a graphic panel with the interaction energies between the four statins A, R, S, and F and the most important amino acid residues at the binding region of HMGR. This graphic panel is called BIRD – which is an acronym of the keywords binding site, interaction energy, and residues domain – and shows clearly

(a) The interaction energy (in kcal/mol) of the residue with the drug, depicted by the horizontal bars, from which one can assign quantitatively the role of each residue in the binding site – i.e., their effectiveness, whether attracting or repelling the drug.
(b) The most important residues contributing to the bonding on the left side.
(c) The region (boldface letters, identified in Figure 10.3) and the atoms of the drugs that are closer to each residue (placed on the left) at the binding site.

The binding energy of the amino acid residues interacting with the statin molecule inside the binding pocket will be defined here as the negative of the corresponding interaction energy calculated using the MFCC method. We can see from Figure 10.6 that six amino acid residues contribute strongly to the stabilization of these statins: R590, K691, K692, K735, R568, and R702, while four amino acid residues – D690, E559, E665, and D767 – display positive (repulsive) interaction energies.

In comparison, Istvan and Deisenhofer [243] point out that several polar interactions and hydrogen bonds are formed between the HMGR-like part, group (**a**) of statins and amino acid residues in the cis loop: S684 (serine), D690, K691, and K692. They also point out the formation of hydrogen bond networks involving

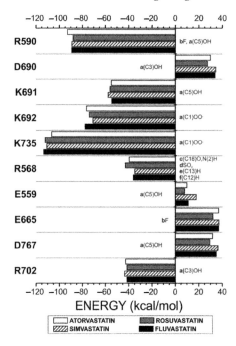

Figure 10.6 Binding site, interaction energy, and residues domain (BIRD) graphic
panel showing the most relevant residues of atorvastatin, rosuvastatin, simvastatin,
and fluvastatin, which contribute to the HMGR binding. After da Costa et al. [242]

E559, D767, and K735. Hydrophobic side chains of HMGR involving L562
(leucine), V683 (valine), L853, A856 (alanine), and L857 do participate in the van
der Waals interactions with the statin molecules but do not appear relevant in our
calculations, probably due to the fact that van der Waals forces cannot be described
in the pure DFT formalism. Simvastatin interacts through its decalin ring structure
with a helix of the enzyme, while rosuvastatin and atorvastatin have additional
binding interactions between their fluorophenyl groups and the R590 residue.

The K735 residue exhibits the strongest binding energy ($\approx$ 115 kcal/mol) to the
(**a**) group of F, R, S, and A, in decreasing order of intensity, through an ionic bond
involving the (C1)OO$^-$ terminal hydrophilic group – followed by the R590 residue,
with binding energy of about 90 kcal/mol for A, F, S, and R, also in decreasing order
of intensity.

If one looks at Figure 10.7 – which presents the main amino acid residues complexed with the statins A, R, S, and F and the main interatomic distances involved –
one can see that the residue K735 shows approximately the same interatomic
distances for the four statins investigated. The interaction with the residue R590,
on the other hand, involves the fluorophenyl (**b**) group of A, R, and F, which is
a bit closer in the case of A (2.6 Å), as shown in Figure 10.7. It seems that this

fluorine atom does not significantly change the binding energy in comparison to simvastatin, which interacts with the residue R590 through its butyril group and through its HMGR-like region – which is a bit closer to R590 when contrasted with R and F.

The third-most important residue, K692, has binding energy to statins between 75 and 77 kcal/mol, with binding intensity following the sequence F > A > R > S, and involves the (**a**) group. The smallest interatomic distance is observed, in this case, for fluvastatin, with 2.3 Å between an oxygen atom at F and a hydrogen atom belonging to the NH_3 group of the residue K692. The residue K691, by the way, interacts with the $(C_5)OH$ group via hydrogen bonding with binding energies between 54 kcal/mol and 58 kcal/mol, following the sequence of binding intensities S > R > A > F, with rosuvastatin and fluvastatin being the closer ones to the residue (1.6 Å and 1.7 Å, respectively, see Figure 10.7).

With a binding energy of about 41 kcal/mol, the residue R702 is bound more strongly to simvastatin, with interatomic distances of 6.4 and 6.8 Å between two

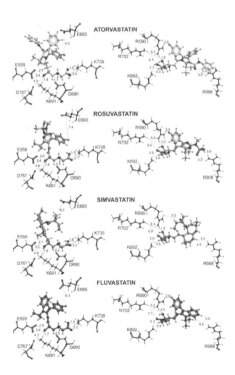

Figure 10.7 Distances between HMGR residues and the statins atorvastatin, rosuvastatin, simvastatin, and fluvastatin, respectively. Their interaction energies are the most important contributions to the total binding energy of the HMGR–statin complex. After da Costa et al. [242]

oxygen atoms at the (**a**) group and a hydrogen atom belonging to one of the NH_2 groups of the amino acid residue.

Finally, the residue R568 binds with different functional groups of A, group (**c**); R, group (**d**), which contains SO_2; S, group (**e**), decalin with two rings; and F, group (**f**), benzene ring, which leads to a noticeable variation of the interaction energy. In this case, R exhibits the strongest binding (≈ 43), followed by A (≈ 39), F (≈ 37), and S (≈ 35), all units in kcal/mol. Considering the interatomic distances, R is closer to the residue R568, with distances of about 4 Å, followed by S, with distances between 4.6 and 5.5 Å. The most distant ligand with respect to the residue R568 is F, the second-weakest interaction with R568 of the four statins investigated, for which the interatomic spacing stays between 5.9 and 6.6 Å.

All in all, in region (**a**), the four statins show a variation of the interaction energy among themselves always smaller than 10 kcal/mol for the attracting residues R590, K691, K692, K735, and R702.

The most important repulsive residues in the binding pocket are, in decreasing order of significance, E665, D767, D690, and E559. The residue E665 shows interaction energies between 32 and 37 kcal/mol, with A, S, and F having practically the same interaction energy. On the other hand, R has the smallest energy because the residue E665 is far away, inside the spherical shell between 11.0 and 11.5 Å centered at its centroid. Besides, the interaction with the residue E665 involves the fluorophenyl group of A, R, and F and the butyryl group of S.

Regarding the residue D767, we have interaction energies following the sequence $S > F > A > R$, varying from 36 (S) to 29 kcal/mol (R), which is interesting because S has the largest interatomic distance to E665 (4.9 Å). Note that A has practically the same interatomic distance value measured for S.

It is also worth noting that by extending the BPS radius to include both residues D767 and R702, the increase of binding energy due to the interaction with the residue R702 is not matched by the decrease promoted by the positive interaction energy of the residue D767 (the net gain in the binding energy due to the contributions of the residues D767 and R502 is 5 kcal/mol on average).

Next in the sequence, we see that the residue D690 repels the four statin molecules by interacting with the HMGR-like (a)-group, with the largest calculated repulsion being assigned to S, followed by F, A, and R, in decreasing order. Interaction energies for this residue vary from 35 (S) down to 27 kcal/mol (R). Atorvastatin has the smallest interatomic spacing to the residue D690, 1.7 Å, while R and F has the largest interatomic distances, of about 2.5 Å.

Finally, the residue E559 exhibits the largest variation of the interaction energy, starting with S, with 18 kcal/mol, decreasing to 12 kcal/mol (F), 10 kcal/mol (A), and 8 kcal/mol (R).

The preceding results point to severe limitations on previous modeling of statins. In the case of the CoMFA 3D QSAR modeling and virtual screening of Zhang et al. [245] previously mentioned, only ligand atoms located within a 6.5 Å radius from any HMGR residue in the binding pocket were taken into account. Consequently, the significant contribution of the residues E665, R702, and R568 to the statin binding was disregarded.

On the other hand, in their work on the use of virtual screening, flexible docking, and molecular interaction fields to design novel HMG-CoA reductase inhibitors for the treatment of hypercholesterolemia, Da Silva et al. [246] have considered the residues K691, D690, N755, R590, E559, N684, K735, L853, V683, and L857 as the main interaction residues with cerivastatin. However, they seem to have overestimated the role played by the residues S684, V683, L857, L853, E559, and N755, and they completely missed the residues K692, R568, E655, D767, and R702, which do contribute substantially to the binding energy.

According to Kee et al. [247], K691, K692, D690, and S684 are the most important residues binding statins to the HMGR active site, and that the unexploited protein–ligand interaction between the residue Y479 (Y = tyrosine) and HMGR can be important in the design of future statin drugs. This means that the strongest binding residues K735 and R590 were neglected, as well as R568, E559, E665, D767, and R702. Moreover, the residue Y479 does not belong to the 12.0 Å binding pocket radius, which indicates that it will not be important in the design of future statin drugs, in striking contrast with the conclusion of Kee et al. [247].

10.5 Conclusions

Recently, a set of blockbuster drugs, including Pfizer's Lipitor (atorvastatin), expired their US patents, leading to plummeting sales due to the competition with generic versions and the loss of billions of dollars of revenues from the largest pharmaceutical companies [250].

In order to conciliate the production of more efficient medications and the necessity of decreasing their cost of development, we suggested in this chapter that the use of relatively cheap computational simulations at the quantum level can be very promising as a valuable tool to understand and develop new drugs.

We also reached the conclusion that it is necessary to perform a careful convergence study to check which amino acid residues must be taken into account to provide a good description of the binding interaction. In the case of statins, amino acid residues within a 12 Å distance from each statin molecule centroid are required to achieve total interaction energy convergence, which means that previous works have underestimated the size of the binding site and the role of specific residues (like the residues E665, D767, and R702).

According to the calculations we carried out, atorvastatin has the largest total binding energy, 320 kcal/mol, while rosuvastatin has a binding energy of 310 kcal/mol, and simvastatin (Zocor) and fluvastatin (Lescol) have total binding energies of about 290 kcal/mol. Such a difference is mainly caused by the decrease of the repulsive interactions between the HGMR pocket site and the molecules of A and R. This suggests that the inhibitory potency of statins is mainly dictated by the strength of their attraction to the binding pocket, notwithstanding the presence of other factors that could affect the efficiency of these molecules beyond binding energy.

The main advantage of the methodology proposed here is the possibility of evaluating which amino acid residues contribute more intensely to the stabilization of the statin–HMGR complex, which can be very helpful for purposes of drug design. For example, the MFCC-based approach allows us to infer that modifications in region (**a**) (the HMG-like region), which define a molecule as a statin, are not required to improve the binding interaction, as regions (**c**), (**d**), and (**e**) are also important in this aspect.

The binding energy analysis of statins complexed with HGMR presented here has a good correlation with thermodynamic studies [244] and clinical trial data [251], as they point to rosuvastatin (Crestor) and atorvastatin (Lipitor) as the best available statins on the market.

11

Collagen-Based Biomaterials

11.1 Introduction

Biomaterial science is an expanding area, which encompasses a wide range of medical knowledge. Collagen, a well-known protein, is regarded as one of the most useful biomaterials in the medical, dental, and pharmacological fields. It is an essential component of most tissues in the human body, carrying out important functions such as the preservation of their structure, shape, and physical characteristics. Its excellent biocompatibility and safety due to its biological characteristics made it the primary resource in medical applications. Besides, it is the most abundant family of extracellular matrix proteins expressed in connective tissues of animals [252].

A single collagen molecule is a structural insoluble fibrillar protein used to set up larger collagen aggregates that perform unique physiological functions in bones, skin, cartilage, ligaments, and tendons. Also, there has been increasing appreciation of the biological importance of collagen in many cellular processes, such as adhesion, proliferation and migration, matrix degradation/remodeling, tissue regeneration, and homeostasis. Furthermore, due to a number of biological properties such as high biocompatibility, rare adverse reactions, and weak antigenicity, collagens have been widely used as a natural material for diverse biomedical applications, including their clinical use as biomimetic scaffolds and drug-delivery systems. Therefore, the understanding of how these properties are derived from collagen's fundamental structural units requires a comprehensive knowledge of the mechanisms underlying its structure and stability [253].

As a matter of fact, about 25% of the protein mass of the human body is made from collagen, having at least 45 distinct collagen genes accounting for 28 protein isoforms being identified in vertebrates [254]. Their main differences are related to the number and ordering of amino acid residues, as well as the way by which they are associated with each other, forming a molecular cable network that strengthens the tendons and sheets that support the skin and the internal organs. Thermal stability, mechanical strength, and the ability to be engaged in specific interactions with other biomolecules make collagen fibers important for so many biological organisms. Biomedical applications of collagen require improved methods to synthesize its triple-helix feature, essential for the development of artificial structures displaying its natural properties.

Collagen, as said before, has a triple-helix structure consisting of three α chains, each chain showing one or more collagenous domains characterized by a repetition of a X–Y–Gly motif (or triplet), where the X and Y amino acid residues are often proline (Pro) and hydroxyproline (Hyp), frequently classified as imino acids [255]. Seven types of collagen (I, II, III, V, XI, XXIV, and XXVII) are assembled in stable fibrils, forming a complex 3-D fibrous superstructure. They are composed of individual helices of three polypeptide chains/strands twisted in a right-handed manner and held together by a ladder of intermolecular backbone hydrogen bonds between adjacent strands. Variations among the collagen family members include differences in assembly of basic polypeptide chain, different lengths of the triple helix, and differences in termination of the helical domain [256].

Type I collagen is the most common in mammals, forming the structure of skin, teeth, bones, corneas, and tendons, with approximately 340 amino acid triplets giving rise to a twisted, rodlike conformation with length of about 300 nm. It is an extracellular matrix protein, with an essential role in maintaining the architecture of multicellular organisms, as well as important industrial uses such as implants, sutures, leather, and prostheses. Structurally, type III collagen is usually assembled as a homotrimeric member composed of three α_1 (III) chains resembling other fibril collagens and found in association with type I collagen in areas where extensibility is necessary, such as the skin, the placenta, and blood vessel walls. Fibril-forming collagens have a unique cleavage site for matrix metalloproteinases, which is located at about three-quarters of the molecular length from the short amino acid terminal (the so-called N-terminal). In the aorta, it is essential for the maintenance of the mechanical strength of the extracellular matrix and to withstand the high blood pressure produced by the heartbeat.

It was suggested that an amino acid deficient region (the so-called carboxyl-terminal or C-terminal for short) near the collagenase cleavage site in type III collagen is important for binding and cleavage by this enzyme. The T3-785 peptide was designed to model this region, a collagen-like triple-helical peptide that provided the first visualization of how the sequence of collagen defines distinctive local conformational variations in triple-helical structure, having the unusual feature of exhibiting several consecutive amino acid triplets with no proline or hydroxyproline residues [257].

Circular dichroism studies showed the T3-785 peptide as a trimer, with an equilibrium melting temperature of $\approx$300 K, below the melting temperature of

333 K for (Pro–Hyp–Gly)$_{10}$. Besides, the T3-785 molecule is rod-shaped, forming a collagen triple helix with three staggered chains. Its residue sequence is formed by three Pro–Hyp–Gly amino acids repetitions (the N-terminal zone) followed by the amino acids sequence Ile–Thr–Gly–Ala–Arg–Gly–Leu–Ala–Gly (Central zone), ending with four more repetitions of the amino acids Pro–Hyp–Gly (C-terminal zone). The Central region corresponds to the collagenase cleavage site.

Using classical modeling techniques, the folding mechanism of collagen-like peptides was explored, yielding insights into the folding pathway of native collagen and the formation of misfolded structures. It was also shown that atomistic simulations can lead to reliable standard free energy values in aqueous solutions for the transition from dissociated monomers to triple-helical collagen model peptides. The role of hydration in collagen triple-helix stabilization was investigated by De Simone et al. [258], with their results showing that the mechanism of triple helix stabilization is sequence dependent. On the other hand, the relationship between interchain salt bridge formation and triple-helical stability was studied using detailed molecular simulations, revealing that not all salt bridges have the same importance for the stabilization process. Finally, a new modeling approach was recently proposed to replicate the supermolecular arrangement of collagen proteins using periodic boundary conditions, with a good agreement with experimental observations, corroborating theories about the fibril's structure [259], and giving rise to a novel computational model approach to enhance self-assembly and biofunctionalization of collagen peptides [260].

At the quantum level, the role of various collagen triplets influencing the stability of collagen was investigated by using density functional theory (DFT) computational model, employing the B3LYP exchange-correlation functional and 6-31G* basis set calculations of relative and solvation free energies of collagen triplets, revealing that the collagen-like conformation is energetically more stable than its extended one [261]. Suitable ab initio models to explore the stability of the implicit and explicit hydration network of collagen at the B3LYP/6-31G(d) level of the theory were developed, whose collagen and β-sheet forming sequences GGG and AAG (PPG and POG) triplets destabilize (stabilize) the collagen triple helix. The semiempirical Parameterization Method 6 (PM6) model and the CAM-B3LYP functional were carried out to optimize the geometry of close-packed (CP) motif for collagen and the more established 7/2 structure, suggesting a possible biological function for molecular hydrogen localized in the cavity of the CP structure. Employing ONIOM (Our own N-layered Integrated molecular Orbital and molecular Mechanics) and AM1 (Austin Model 1) calculations, Tsai et al. evaluated the effect of mutations in collagen-like triple helices, demonstrating the importance of the glycine residue in the repeating triad X–Y–Gly [262].

However, the number of atoms that can be treated by quantum mechanics is still small. Thus, the idea of representing the total energy of macromolecular structures as a combination of fragmented energies has received increasing interest, and indeed, the ability to perform fragment-based quantum mechanical calculations on large biomolecules could be a very useful approach to predict their energetic properties. Recently, a computer simulation study to describe, at the quantum level, the noncovalent interaction energies among the amino acid residues of T3-785 tropocollagen triple-helical structure found that proline can stabilize the collagen triplet only when other residues are in the polyproline II conformation [263]. DFT and ONIOM calculations at the B3LYP/D95(d,p) level and the semiempirical AM1 Hamiltonian were carried out to optimize the geometry of triple-helical collagen-like structures [262], demonstrating that the glycine residue can act as an D amino acid in the repeating triad X–Y–Gly.

The purpose of this chapter is to analyze the conformational stability of the collagen-like peptide T3-785 (PDB: 1BKV) [264] by using molecular quantum chemistry calculations within the MFCC scheme [22]. The MFCC approach is a route to accurately investigate large biological systems with low computational cost, which already has employed in Chapter 10 to describe ligand–protein interactions at the quantum level. It was also used to investigate dengue viral infection [265]. Taking into account the entire system, the system ABC composed by the residues of the chains A, B, and C, residue–residue interaction energy calculations have been undertaken in a peer-to-peer way between the system ABC with those of B (subsystem AB), C (subsystem BC), and A (subsystem CA), respectively. We identified the interchain binding interactions involved in the stability of the structural zones (N-terminal, Central and C-terminal) and triplets (Pro–Hyp–Gly, Ile–Thr–Gly, Ala–Arg–Gly, and Leu–Ala–Gly) of T3-785. Finally, the importance of the amino acid residues Hyp, Gly, Ile, Thr, Ala, Arg, and Leu to collagen's triple-helix stability were evaluated. All of these quantum binding energy calculations contribute to the development and synthesis of artificial collagenous materials with high stability for biomedicine and nanobiotechnology applications.

We also present an adequate description of the integrin $\alpha_2\beta_1$-collagen triple-helix complex interactions. An improvement in the understanding of its physical, chemical, and biological properties is necessary to address some of the drawbacks in collagen-based applications [266]. Despite its significant therapeutic potential, only a few peptide-based agents have been used to regulate the $\alpha_2\beta_1$ integrin functions by targeting the α_2I domain. The greatest difficulty for the development of these drugs lies in the complexity of the integrins, since they act in several cellular types. Also, despite the simplicity of the collagen molecule, the outline of its structural stability is still one of the most difficult challenges today, as corroborated by a number of investigations (see [253] and the references therein). Furthermore, collagen interaction comprises one of the most complex cell adhesion systems already described, the challenge being the development of drugs with high specificity that avoid the redundancy of these molecules. Thus, the understanding of the molecular basis related to the integrin–collagen complex is quite important for the improvement of new compounds targeting this system.

11.2 Chemical Structure of the Collagen-Like Peptide T3-785

The crystallographic structure of the collagen-like peptide T3-T85 was obtained by X-ray diffraction technique with a resolution of 2.0 Å and is stored in the Protein Data Bank (PDB) files under the code 1BKV [257]. It was used to obtain the atomic positions for the calculations describing the interaction energies between the chains of tropocollagen. It also allows one to visualize how the primary sequence of amino acid residues making up collagen fibers produces distinctive local conformational variations in the triple-helical structure.The protonation state of the system at physiological pH was obtained using the PDB2PQR server, an automated pipeline for the setup, execution, and analysis of the Poisson-Boltzmann electrostatics calculations [267]. Biomolecular structures determined by crystallography rarely include hydrogens explicitly because they have a very low electronic density of state. Thus, while the atomic positions of the non-hydrogen atoms were kept fixed, the hydrogen atomic positions were added to the structure and optimized by using the parameterization tool for organic molecules CHARMM22 (Chemistry at Harvard Macromolecular Mechanics 22 forcefield) [268]. The calculations were carried out with convergence tolerances set to 10^{-5} kcal/mol (total energy variation), 10^{-3} kcal/(mol Å) (RMS gradient), and 10^{-5}Å (maximum atomic displacement).

In order to achieve the structural stability of the triple-helix collagen promoted by interactions with extended hydration network, all water molecules forming hydrogen bonds with a particular residue were included for completeness in the fragments where it appears. The analysis of the binding scenario was based on the following criteria:

(a) Conventional and water-mediated hydrogen (H)-bonds between the inter-chain residues within the collagen-like peptide T3-783 were defined using geometric and energetic criteria derived after a search of 77,378 binding sites from 25,096 PDB entries.

(b) H-bond donor (D) and H-bond acceptor (A) atoms, distance, d_{D-A}, range from 2.6 to 3.1 Å.

(c) The acceptor–hydrogen–donor angle, θ_{A-H-D}, does not less than 120°.

Besides classical hydrogen bonds, there is unequivocal evidence for the existence of non-conventional H-bonds between CH donor groups and oxygen acceptors in

13 well-defined protein structures [269]. Here, non-conventional H-bonds occurred when $d_{D-A} \leq 3.5$ Å and $\theta_{A-H-D} \geq 120°$.

Finally, ion-induced dipole interaction, a long-range interaction that may exist between ions and molecules with no permanent dipoles, occurring when an external electric field temporarily distorts the electron cloud of a neutral/nonpolar molecule until it forms an induced dipole moments, was implemented mainly following the measures found in Reference [270]. For the existence of hydrophobic (induced dipole–induced dipole) contacts, three hydrophobic R^i (R^j) atoms, namely carbon atoms with accessible surface and halogens, must lie in the range of the hydrophobic R^j (R^i) side chain. The maximum distance is set to the sum of the van der Waals radii of atoms in question with a tolerance of 0.8 Å.

Figure 11.1 depicts the T3-785 peptide sequence from three different outlooks. In Figure 11.1a, we show the amino acid sequence that composes each single chain of this collagen-like structure, each one being comprised by three Pro–Hyp–Gly amino acid repetitions (N-terminal zone), followed by the residues Ile–Thr–Gly–Ala–Arg–Gly–Leu–Ala–Gly (Central zone), and ending with four repetitions of Pro–Hyp–Gly (C-terminal zone). These three repetitions are also shown in

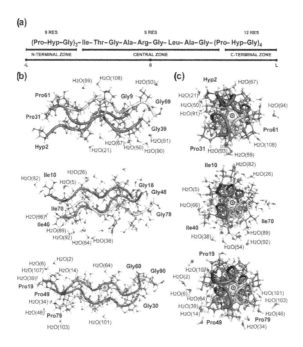

Figure 11.1 (a) Schematic, (b) transversal, and (c) longitudinal representation of the three zones that structurally constitute the collagen-like triple-helical peptide T3-785. The three monomers were designed by the letters A (dark gray), B (gray), and C (light gray). After Oliveira et al. [264]

Figure 11.1b and 11.1c, displaying a transversal and longitudinal molecular overview of the crystallographic structure, respectively. As it can be seen, the homotrimer is formed by A- (dark gray), B- (gray), and C- (light gray) chains, fragmented in the N-terminal zone (top), Central zone (middle), and C-terminal zone (bottom). Variations in the helical symmetry among its different zones indicate that the triple-helical twist is sequence dependent. The 10/3 (7/2) helical pitch of Pro/Hyp-poor regions (Pro–Hyp–Gly triplet) could play a role in the interaction of collagenous domains with other biomolecules (nucleation of collagen fibrils).

Despite its simplified sequence and regular structure, the definition of the molecular basis of the collagen triple-helix stability has been hitherto proved to be a difficult task. Several computational studies have been performed to describe the stability of the collagen molecule [271]. Molecular dynamics simulations were also employed to study the hydration structure effect on the preservation of collagen structure and the change in the stability of a collagen-like fibril segment, as a result of interaction with water and gallic acid [272]. The interaction of two pentameric collagen-like bundles was investigated using molecular dynamics simulations [273], and the stability of the triple-helical collagen models was also assessed.

11.3 Energetic Description

After the determination of the suitable chemical structure of the collagen-like peptide T3-785, quantum simulation within the DFT formalism together with the Local Density Approximation (LDA) approach based on the exchange-correlation functional of Perdew and Wang (DFT/LDA-PWC) [10] were carried out by using the DMOL^3 code. As in Chapter 10, a double numerical plus polarization (DNP) basis set was chosen to expand the electronic Kohn-Sham orbitals taking into account all electrons with unrestricted spin explicitly, being 10–100 times faster and in good agreement with those obtained using the large-scale Gaussian basis sets, such as 6-311+G(3df,2pd) [134]. The same orbital cutoff radius (self-consistent field convergence threshold) 3.7 Å (10^{-6} Ha) was considered in order to reduce the computation time with little impact on the accuracy of the results. The hydrogen atomic positions of the peptide T3-785 optimized classically were optimized again using the DFT approach. Besides, we adopted a reliable semiempirical correction on the description of dispersive, covalent, and ionic bonds within the improved density functional theory (DFT+D) proposed by Ortmann et al. [274].

Based on these quantum chemistry computations and using the formalism of the MFCC method described in Section 1.8, we calculated the interaction energies among the amino acid residues belonging to different chains of the T3-785 peptide, forming three subsystems representing interactions between two chains (AB, BC,

Figure 11.2 Structural representation of the collagen-like peptide T3-785 according to the temperature factors (B-factor) signed for all atoms. Dark gray/light gray color change represents the range of the B-factor from 10 to 100 Å^2. After Oliveira et al. [264]

and AC; see Figure 11.1). For each amino acid residue of interest R^i, we draw an imaginary sphere with radius equal to 8.0 Å and evaluate the interaction energy $EI(R^i\text{-}R^j)$ with each residue R^j, considering at least one atom inside the sphere. The residues are identified by their names and corresponding chain.

Due to the intrinsic complexity of the collagen molecule, reductionist approaches through the use of peptide models have been widely applied to obtain information on the structure of this complex system. Following this line, the T3-785 molecule, a collagen-like triple-helical peptide, was chosen as the collagen-model peptide for studying the conformational stability of human collagen.

It is important to notice that the Pro–Hyp–Gly triplets of the T3-785 collagen-like peptide are highly inaccurate, as evidenced by low crystallographic quality, particularly poor electron density map, and lower average temperature factor of its atoms (see Figure 11.2). Therefore, N-terminal (C-terminal) residues Pro31, Pro61, and Hyp2 (Gly30, Gly60, and Gly90) in chains A, B, and C, respectively, and, hence, the sequences Hyp2–Gly3, Pro31–Hyp32–Gly33, and Pro61–Hyp62–Gly63 (Pro28–Hyp29–Gly30, Pro58–Hyp59–Gly60 and Pro88–Hyp89–Gly90) were discarded from our analysis in the same way as it was done by other studies [257].

11.4 Interaction Binding Energies

The absolute interaction binding energies of the collagen-like peptide T3-785 zones are summarized in Table 11.1, which shows the binding energies of the AB (−625.67), BC (−566.12), and CA (−566.31) subsystems, as well as of the entire ABC system. Energies are given in kcal/mol and percentage (from now on in this chapter, all energy units are given in kcal/mol). At the AB (BC; CA) subsystem column, each zone of A (B; C) chain interact with B (C; A) chain. Accordingly, the three chains preserve collagen triple-helical conformation with a pronounced attractive energy value of −1758.10, in which the Central zone (N-terminal; C-terminal) is accountable for 42.09% (32.61%; 25.29%) of this binding.

The regions N-terminal and C-terminal are composed by three and four Pro–Hyp–Gly triplets, respectively, and this difference accounts for about 8% of their

Table 11.1 *Absolute interaction binding energies of the collagen-like peptide T3-785. After Oliveira et al. [264].*

	Subsystem AB		Subsystem BC		Subsystem CA		System ABC	
N-Terminal	–129.8	20.7%	–171.9	30.4%	–143.1	25.3%	–444.7	25.3%
Central	–300.1	48.0%	–187.4	33.1%	–252.5	44.6%	–740.0	42.1%
C-Terminal	–195.8	31.3%	–206.9	36.5%	–170.7	30.2%	–573.4	32.6%
TOTAL	–625.7	100.0%	–566.1	100.0%	–566.3	100.0%	–1758.1	100.0%

Table 11.2 *Average interaction binding energies of the triplets in the collagen-like peptideT3-785. After Oliveira et al. [264].*

	Subsystem AB		Subsystem BC		Subsystem CA		System ABC	
Pro–Hyp–Gly	–65.1	17.3%	–75.7	28.8%	–62.8	19.9%	–67.9	21.3%
Ile–Thr–Gly	–74.6	19.8%	–61.9	23.5%	–59.2	18.8%	–65.2	20.5%
Ala–Arg–Gly	–167.6	44.5%	–68.1	25.9%	–154.8	49.1%	–130.2	40.9%
Leu–Ala–Gly	–69.6	18.5%	–57.4	21.8%	–38.5	12.2%	–55.2	17.3%

interaction binding energies. For the AB (CA) subsystem, the order of energetic importance in these three zones follows the sequence: Central > C-terminal > N-terminal, with the negative energy value of -300.13, -195.77, and -129.77 (-252.51, -170.75, and -143.05), respectively. On the other hand, the BC subsystem shows an inversion in the order of the binding energy between the Central and C-terminal zones, with the second becoming more attractive (-206.90) than the first one (-187.40), as a result of the small intensity interactions between the amino acid Arg44 and the residues from the C-chain.

11.4.1 Binding Energies of the Triplets

The average binding energies of the triplets Pro–Hyp–Gly, Ile–Thr–Gly, Ala–Arg–Gly, and Leu–Ala–Gly of the A (B; C) chain interacting with B (C; A) within subsystem AB (BC; CA), as well as of the entire ABC system, a part of the collagen-like peptide T3-785, are depicted in Table 11.2. Energies are given in kcal/mol and percentage. For instance, the energy value of -65.11 calculated for the Pro–Hyp–Gly triplet in AB subsystem actually represent the average value of the binding energies of the 15 amino acid residues that compose the seven Pro–Hyp–Gly sequences of the A-chain – namely the residues Pro4, Hyp5, Gly6, Pro7, Hyp8, Gly9, Pro19, Hyp20, Gly21, Pro22, Hyp23, Gly24, Pro25, Hyp26, and Gly27, due to their interactions with the B-chain. Likewise, the (average) energetic values assigned in the last column of Table 11.2 represent the attractive contribution of the four triplet types for the T3-785 peptide (ABC system) stability. Therefore, one can see that Ala13–Arg14–Gly15 (A-chain), Ala43–Arg44–Gly45 (B-chain), and Ala73–Arg74–Gly75 (C-chain) triplets stand out from the others with a strong average attractive energy of -130.15, almost twice as large as any other one. Nevertheless, Pro–Hyp–Gly, Ile–Thr–Gly, and Leu–Ala–Gly triplets present a very close attractive contribution (-67.87, -65.23, and -55.12, respectively).

11.4.2 Binding Energies of the Residues

The energetic affinities between the specific amino acids of chains A–B, B–C, and A–C, as well as the entire ABC system, from a perspective of the relative (average) binding energies of the residues Pro, Hyp, Gly, Ile, Thr, Ala, Arg, and Leu of A (B; C), a part of the collagen-like peptide T3-785, are shown in Table 11.3. Energies are again given in kcal/mol and percentage. Analyzing the average energy of residue–residue interactions of each amino acid that composes the T3-785 collagen-like peptide, the energetic importance follows the order: Arg (39.5%) > Hyp (17.60%) > Thr (16.43%) > Ala (11.13) > Gly (7.80%) > Pro (3.50%) > Leu (2.17%) > Ile (1.87%). These results are consistent with those obtained by Persikov et al. [275] through circular dichroism (CD) spectroscopy of the collagen-model Ac–(Gly–Pro–Hyp)$_3$–Gly–X–Y–(GlyPro–Hyp)$_4$–Gly–Gly–CONH$_2$, asserting that the same residue could display different propensities related to the nonequivalence of these positions in terms of the interchain interactions and the solvent exposure.

As one can see from Table 11.3, the amino acid arginine is the most relevant to conformational stability of the T3-785 peptide, accountable for 39.50% of the sum of average total interaction energy per amino acid residue. The residues Arg14,

Table 11.3 *Average interaction binding energies of the residues in the collagen-like peptide T3-785. After Oliveira et al. [264].*

	Subsystem AB		Subsystem BC		Subsystem CA		System ABC	
Pro	−7.7	2.9%	−10.9	5.8%	−5.9	2.4%	−8.2	3.5%
Hyp	−38.7	14.4%	−48.3	25.5%	−36.8	15.0%	−41.2	17.6%
Gly	−19.0	7.1%	−17.0	9.0%	−18.8	7.7%	−18.3	7.8%
Ile	−6.1	2.3%	0.4	−0.2%	−7.4	3.0%	−4.4	1.9%
Thr	−40.1	14.9%	−48.1	25.3%	−27.4	11.2%	−38.5	16.4%
Ala	−39.1	14.6%	−20.3	10.7%	−18.8	7.7%	−26.1	11.1%
Arg	−110.2	41.1%	−42.5	22.4%	−124.9	51.0%	−92.5	39.5%
Leu	−7.3	2.7%	−3.0	1.6%	−4.9	2.0%	−5.1	2.2%

Arg44, and Arg74 are strongly attracted to the B, C, and A strands, respectively, with the highest attractive binding energies (-110.23, -42.46, and -124.94, respectively). Although Arg44 is still one of the main residues in the BC subsystem, it has a much lower interaction energy compared to its chemical partners because it has just a single strong interaction with the residue Ala73 (-33.32), while the residue Arg14 (Arg74) shows a high attraction to the Ala43 and Leu46 ones (Leu16, Ala17, and Pro19 ones) with energy values of -42.57 and -54.76 (-50.47, -38.27 and -41.63), respectively. Analyzing their conformational and solvation aspects, one can determine what Arg amino acid is able to interact not only with their side-chain functional groups but also with other nearby amino acid residues. Interestingly, the most significant energetically Arg-residue systems tend to pick up more surrounding water.

11.4.3 Hydrogen Bond Pattern

The T3-785 peptide shows a typical Rick and Crick hydrogen bonding pattern [276], where N–H groups of the Gly act as donor and the C=O (carbonyl) groups of the Xaa residues act as acceptors. These interchain H-bond, Gly:NH–OC:Xaa, present average distance (angle) approximately equal to 2.88 Å (166°). Specifically, 25 out of the 156 Gly–Xaa interactions analyzed (excluding terminal glycines) showed this pattern, mainly those with attractive residues binding energies: Gly6–Pro34 (-4.08), Gly9–Pro37 (-6.54), Gly12–Ile40 (-5.63), Ala43–Gly15 (-1.26), Gly18, Leu46 (-7.63), Gly21–Pro49 (-7.79), Gly24–Pro52 (-11.65), and Gly27–Pro55 (-8.94) in the subsystem AB; Gly36–Pro64 (-5.80), Gly39–Pro67 (-8.88), Gly42–Ile70 (-10.70), Gly45–Ala73 (-4.73), Gly48–Leu76 (-9.58), Gly51–Pro79 (-4.16), Gly54–Pro82 (-6.23), and Gly57–Pro85 (-5.10) in the subsystem BC; finally, Gly66–Pro7 (-5.20), Gly69–Ile10 (-6.80), Gly72–Ale13 (-10.80), Gly75–Leu16 (-6.57), Gly78–Pro19 (-3.96), Gly81–Pro22 (-5.73), Gly84–Pro25 (-4.87), and Gly87–Pro28 (-4.16) in the subsystem CA. The average value of these interaction binding energies is three times higher than that without the H-bond.

As shown in Figures 11.3 and 11.4a, a second network of repetitive hydrogen bonds is observed in the imino acid–free zone. Here, water molecules connecting glycine carbonyl and Xaa amides are located on adjacent chains, Xaa:NH–HOH–OC:Gly, creating a set of water-mediated interchain H-bonds, as it was proposed in Reference [277]. These intermolecular contacts are observed in Gly9–Ile40 (-10.92), Gly12–Ala43 (-16.01), and Gly15–Leu46 (-9.66) in subsystem AB; Gly39–Ile70 (-8.24) and Gly45–Leu76 (-8.18) in subsystem BC; and Gly66–Ile10 (-12.17), Gly69–Ala13 (-7.56), and Gly72–Leu16 (-7.95) in subsystem CA. Observe that the Ala73:NH–HOH:OC–Gly42 H-bond does not exist since

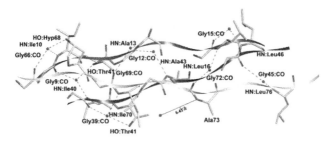

Figure 11.3 Three-dimensional spatial visualization of the interchain water-mediated hydrogen bonds among the amino acid residues of the collagen-like peptide T3-785. After Oliveira et al. [264]

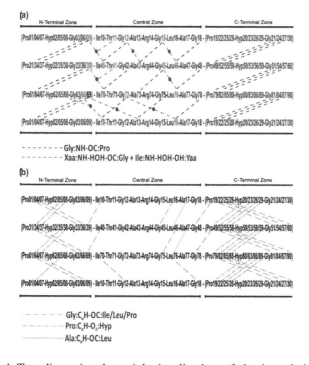

Figure 11.4 Two-dimensional spatial visualization of the interchain hydrogen bond patterns, namely (a) conventional (b) and non-conventional H-bonds of the collagen-like peptide T3-785. After Oliveira et al. [264]

the nearest water molecule in Ala73:NH is 6.43 Å away, explaining the weak interaction (−1.01) calculated for the Gly42–Ala73 residue pair. Those strong attractive interactions confirm the hypothesis that this second network of hydrogen bonds strengthens the triple-helical conformation in imino acid–poor regions of collagen [255] and refutes the idea that regions of triple-helical conformation without imino acids will be more flexible and dynamic than those where all X/Y positions are occupied by Pro/Hyp residues.

In three situations, water molecules involved in this second set of bonds make an additional contact with the side chain (hydroxyl group) of Yaa residues, namely Ile:NH–HOH–OH:Yaa. In our collagen model, one water molecule connects Ile40 (Ile70) with Gly9 and Thr11 (Gly39 and Thr41), the latter within the same chain. The importance of Ile40:NH–HOH–OH:Thr11 (Ile70:NH–HOH–OH–Thr41) contact is expressed by the energetic value equal to −31.29 (−43.86), third- (first-) most important of subsystem AB (BC). At least, the water molecule between the residues Ile10 and Gly66 is also in contact with Hyp68, which directly affects the Hyp68–Ile10 interaction, the second-most attractive in subsystem CA (−47.50).

In addition, we evaluated the non-conventional hydrogen bonds, Gly:C_αH–OC:Xaa (Ile/Leu/Pro) (Figure 11.4b). We found six contacts involving Ile/Leu and Gly residues, namely Gly39:C_αH–OC:Ile10 (−5.96), Gly45:C_αH–OC:Leu16 (−3.88), Gly69:C_αH–OC:Ile40 (−3.83), Gly75:C_αH–OC:Leu46 (−4.42), Gly12: C_α H–OC:Ile70 (−3.29), and Gly18:C_αH–OC:Leu76 (−7.98). In addition, 15 Pro-Gly pairs also present these non-covalent forces, namely Pro4–Gly33 (−5.09), Pro7–Gly36 (−6.12), Pro19–Gly48 (−8.03), Pro22–Gly51 (−4.79), Pro25–Gly54 (−6.58), Pro34–Gly63 (−4.81), Pro37–Gly66 (−4.48), Pro49–Gly78 (−6.13), Pro52–Gly81 (−5.41), Pro55–Gly84 (−7.00), Pro64–Gly6 (−6.61), Pro67–Gly9 (−3.00), Pro79–Gly21 (−4.72), and Pro82–Gly24 (−7.39) as well as Pro85–Gly27 (−3.91).

We also highlight the existence of 14 important non-conventional H-bonds, namely Pro:C_γH–O$_\delta$:Hyp, involving residues of glycine or proline with other proline or hidroxyproline of adjacent chains, respectively: Hyp5–Pro34 (−20.66), Hyp8–Pro37 (−27.54), Hyp20–Pro49 (−41.01), Hyp23–Pro52 (−11.77), Hyp26–Pro55 (−14.34), Hyp35–Pro64 (−16.54), Hyp38–Pro67 (−22.42), Hyp50–Pro79 (−20.62), Hyp53–Pro82 (−19.19), Hyp56–Pro85 (−25.15), Hyp65–Pro7 (−17.52), Hyp80–Pro22 (−26.48), Hyp83–Pro25 (−13.86), and Hyp86–Pro28 (−10.81). Finally, we distinguished two Ala:C_αH–OC:Leu contacts in Ala17–Leu46 (−23.05) and Ala47–Leu76 (−13.76) interactions.

The existence and importance of this non-conventional H-bond for collagen triple-helix stability was initially discussed by Bella and Berman [278]. It is believed that C_αH–OC network interaction acts to align the three chains and thereby cooperatively decreases the total energy of the biological system. In our study, the hydrogen bonds – whether conventional, water-mediated, or non-conventional types – are present in 62 residue–residue interactions, which together account approximately for 44% of the total interaction energy between the collagen strands. It is therefore likely that these H-bond patterns contribute to the local stability of the triple helix and help to maintain the triple-helical conformation, confirming the water-bonded model for collagen.

11.5 Graphical Panel of the Most Relevant Interactions

The thermal stability of collagen-like models with many Gly–Pro–Yaa triplets in the Central zone decreases gradually with a number of arginine residues in the Yaa position. However, when Glu and Arg amino acids occupy Xaa and Yaa, respectively, interchain electrostatic interactions increase the stability of the triple-helix heterotrimer. In fact, conformational studies in collagen-like peptides indicate that Arg residues are entropically favorable in this situation. In our model, the residues Arg14, Arg44, and Arg74 are in the Yaa position of their respective triplets.

Figure 11.5 depicts the interaction energy (in kcal/mol) of the residues Arg14, Arg44, and Arg74 at the AB, BC, and CA subsystems, respectively. The residue Arg14 (Arg44) interacts attractively with Ala43, Gly45, Leu46, and Ala47 (Ala73, Gly75, Leu76, and Ala77) residues, with energy values equal to −42.57, −8.50, −54.76, and −12.28 (−33.32, −4.65, −11.35, and −2.54), respectively. Similarly, the residue Arg74 binds with the A-chain residues Gly15 (−1.45), Leu16 (−50.47), Ala17 (−38.27), Gly18 (−7.85), Pro19 (−41.63), and Hyp20 (−12.98).

Additionally, the trans-side-chain conformation adopted for Arg14 (Arg74) residue and characterized by the dihedral angle 162.82° (−160.75°) is responsible for a close approach of its guanidine group with Leu46 (Leu16), strengthening

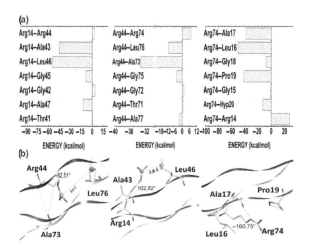

Figure 11.5 (a) Graphical panel and (b) spatial arrangement of the most relevant interactions involving the Arg residues 14, 44, and 74 in subsystem AB, BC, and CA, respectively. Interchain intermolecular contacts are depicted by black dashed line (H-bond), gray solid line (non-conventional H-bond), and dotted line (ion-induced dipole). After Oliveira et al. [264]

an ion-induced dipole interaction type between them. Furthermore, the bifurcated non-conventional H-bond between Arg14 and Ala43 (Arg14:C_αH–OC:Ala43) as well as the conventional H-bond formed between Arg74 and Ala17 (Arg74:N$_\varepsilon$H–OC:Ala17) justify the energy values of -42.57 and -38.17, respectively, with a single oxygen acceptor simultaneously participating in the two hydrogen bonds. Likewise, the high ion-induced dipole interaction between the residues Arg74 and Pro19 is also favored by the trans-conformation ensuring parallel orientation of their side chains.

On the other hand, Arg44 acquires gauche (+) conformation determined by the dihedral angle equal to $-82.51°$. In this state, there is a small displacement of its protonated guanidine moiety in the opposite direction of their C-chain interacting residues, with the exception of the Ala73 residue. In fact, the distance between guanidine (Arg44) and isobutyl centroids (Leu76) is approximately 5.6 Å, while the Arg14–Leu46 system does not reach 3.8 Å (see Figure 11.5). Concerning the Arg44–Ala73 interaction, there is one H-bond (Arg44:C_αH–OC:Ala73) very similar to those observed in the Arg14–Ala43 pair, yielding a very close interaction energy. Finally, the repulsion involving the binding of residues Arg74–Arg14, Arg44–Arg74, and Arg14–Arg44 is related to strong charge–charge interactions among the guanidine functional groups of the same charge.

Threonine residues account for an average interaction energy of -38.50, mainly due to the binding energy contribution of the residue Thr11 (Thr41; Thr71) interacting with the B- (C-; A-) chain with an energy contribution of -40.05 (-48.07; -27.39). Thus, the threonine residues are the third-most relevant amino acid class to maintain the conformational integrity of the T3-785 peptide. Among all threonine–residue interactions, the two most effective are those in which water molecules mediate the hydrogen bond between the polar group of Thr11 and Thr41 with the α-amino group of Ile40 and Ile70, respectively, particularly the bindings between the residues Thr11–Ile40 (-31.39) and Thr41–Ile70 (-43.86) (see Figure 11.6). A third binding interaction, Thr71–Ala13 (-28.77), is characterized by a non-conventional H-bond between Thr71:C_αH and Ala13:CO.

In collagenous domain characterized by a repetition of Xaa–Yaa–Gly triplets, the Xaa and Yaa positions are frequently occupied by proline (Pro, 28.1%) and hydroxyproline (Hyp, 38.1%) amino acid residues respectively, which together represent approximately 22% of all amino acids of collagen strands. The abundance of these residues decreases the entropic cost for collagen folding, although increasing dramatically the thermal stability of the triple helices. It is known that appropriate ring pucker, enforced by a stereoelectronic or steric effect, preorganizes the torsion angles to those required for the triple-helix formation [253].

In our calculations, within a radius of 8 Å, Pro4, Pro7, Pro19, Pro22, and Pro25 (A-chain); Pro34, Pro37, Pro49, Pro52, and Pro55 (B-chain); and Pro64, Pro67,

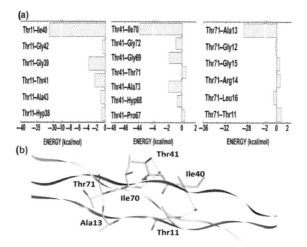

Figure 11.6 Graphical panel (a) and spatial arrangement (b) of the most relevant interactions involving the Thr residues 11, 41 and 71 in subsystem AB, BC and CA respectively. Interchain intermolecular contacts are depicted by black dashed line (hydrogen bond) and gray solid line (non-conventional hydrogen bond). After Oliveira et al. [264]

Pro79, Pro82, and Pro85 (C-chain) interact with many residues of the B-, C-, and A-chains, with a total attractive energy value of -6.58, -1.50, -18.39, -6.26, -5.82; -18.66, -4.78, -6.89, -16.67, -7.75; and -3.33, -5.32, -7.64, -5.67, -7.73, respectively. Consequently, the average interaction energy of prolines from the A-chain with residues from the B-chain (Pro-AB) is equal to -7.71 (2.88%). Similarly, Pro–BC and Pro–CA have binding interaction energy equal to -10.95 (5.77%) and -5.96 (2.42%), respectively (see Table 11.3).

Although it was proposed that the van de Waals interactions between the proline residues are important for stabilization of the triple-helix structures [279], nevertheless, among the 35 Pro–Pro interactions of the collagen-like model T3-785 assessed here, just a few have some degree of attractiveness. As it can be seen from Figure 11.7, the pyrrolidine rings of the interchain parallel prolines are relatively distant from each other (distance C_γ–$C_\gamma > 7.9$ Å), preventing strong Pro–Pro interactions. This is in good agreement with the parameters of the crystal structure of (Pro–Pro–Gly)$_9$ at 1.0 Å resolution [280].

Regarding the Hyp5, Hyp8, Hyp20, Hyp23, and Hyp26 (Hyp35, Hyp38, Hyp50, Hyp53, and Hyp56; Hyp65, Hyp68, Hyp80, Hyp83, and Hyp86) residues within the A- (B-; C-) chain, they are attracted to the B- (C-; A-) chain in a stronger way, specifically with binding energies equal to -31.13, -45.25, -60.68, -29.22, -27.05 (-27.17, -85.63, -44.69, -43.32, -40.69; -32.57, -52.65, -55.36, -21.11, -22.06). As indicated by the average interaction binding energies

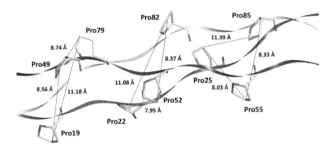

Figure 11.7 Distances (in Å) of the proline amino acid pairs present in the collagen-like peptide T3-785. After Oliveira et al. [264]

of −38.66 (Hyp–AB), −48.30 (Hyp–BC), and −36.76 (Hyp–CA), hydroxyprolines are energetically important in interstrand interaction, mainly as a result of the association of non-conventional H-bonds (see more details in the Section 11.4.3). In fact, our calculations show that the largest residue–residue attractions involving proline and hydroxyproline residues occur in the presence of non-conventional hydrogen bonding. The carbonyl groups of prolines in A (B; C) make a contact of this nature with the $C_\alpha H$ protons of glycines in B (C; A), specifically by the residues pairs: Pro4–Gly33, Pro7–GlyY36, Pro19–Gly48, Pro22–Gly51, Pro25–Gly54 (Pro34–Gly63, Pro37–Gly65, Pro49–Gly78, Pro52–Gly81, Pro55–Gly84; Pro64–Gly6, Pro67–Gly9, Pro79–Gly21, Pro82–Gly24, Pro85–Gly27).

Unlike the other amino acid residues, the alanines one are in both positions of Gly–Xaa–Yaa triplet. In the Xaa (Yaa) location, Ala13, Ala43, and Ala73 (Ala17, Ala47, and Ala77) interact with average energies of −14.56 (−37.58). Although not conformationally distinct from each other, the distances of Ala(Xaa)–residue interchain interactions are, on average, 28% greater than those observed in Ala (Yaa)–residue. Due to the nearness of interacting amino acid residues of different chains, Ala in Yaa position presents more favorable binding energies, namely Ala17–Leu46 (−23.05), Ala17–Pro49 (−11.68), Ala47–Leu76 (−13.76), Ala47–Pro79 (−16.39), and Ala77–Pro19 (−11.74) (see Figure 11.8). Considering the chemical nature of the amino acids involved, these contacts have probably a strong hydrophobic character (induced dipole–induced dipole forces). Additionally, Ala17–Leu46 and Ala47–Leu76 also make non-conventional hydrogen bonds.

Regardless of their structural and functional peculiarities, many types of collagen are formed by Xaa–Yaa–Gly triplet repetitions. The packing of the triple-helical coiled-coil structure requires Gly in every third position. Because of its compact structure, assembly of the triple-helix puts this residue at the interior of the helix, and the side chain of the Gly is small enough to fit into the center of the helix. Although the Gly residues do not present a charged (polar) side chain yielding an

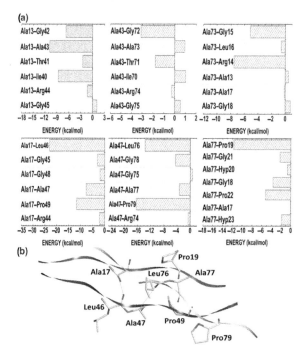

Figure 11.8 (a) Graphical panel and (b) spatial arrangement of the most relevant interactions involving the Ala residues in Xaa and Yaa positions of the Xaa–Yaa–Gly motif. Interchain intermolecular contacts are depicted by the gray solid line (non-conventional H-bond). After Oliveira et al. [264]

ion–dipole (dipole–dipole) interaction, like the Arg (Hyp and Thr) ones do, they can perform strong intermolecular hydrogen bonds. In fact, in the T3-785 peptide each Gly triplets present an average binding energy value of -18.27 (-17.02; -18.81) for the AB (BC; CA) subsystem. Therefore, the CO carbonyl (NH and $C_\alpha H$) moiety act as an acceptor (donor) element in a well-defined H-bond patterns, which are essential for the consistency of the triple-helix collagen.

According to their distribution in the host–guest triple-helical peptides, Ile and Leu amino acids residues occupy the Xaa position in 11.2% of cases, with similar average binding energies equal to -4.38 and -5.08, respectively. The weak attraction between the residues Ile10–Ile40 (-1.15), Leu16–Leu46 (-0.34), Leu16–Ala47 (-0.49), Leu16–Gly48 (-0.33), Leu46–Leu76 (-2.40), Leu46–Ala77 (-0.15), Leu46–Gly78 (-0.26), Ile70–Ile10 (-0.62), Ile70–Ala13 (-1.58), Leu76–Ala17 (-1.76), and Leu76–Pro22 (-0.82), suggest the presence of an induced dipole–induced dipole interaction among their nonpolar side-chain atoms (see Figure 11.9).

More significantly, attractions occur when these hydrophobic contacts are replaced by ion–induced dipole forces in Arg–residue contacts, water-mediated

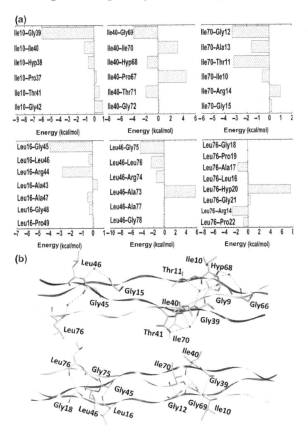

Figure 11.9 (a) Graphical panel and (b) spatial arrangement of the most relevant interactions involving Leu and Ile residues of the collagen-like peptide T3-785. Interchain intermolecular contacts are depicted by black dashed line (hydrogen bond) and gray solid line (non-conventional H-bond). After Oliveira et al. [264]

H-bonds (Lys9–Ile40: -10.92, Thr11–Ile40: -31.39, Gly15–Leu46: -9.66, Gly39–Ile70: -8.24, Thr41–Ile70: -43.86, Gly45–Leu76: -8.18, Gly66–Ile10: -12.17, and Hyp68–Ile10: -47.50), or even non-conventional H-bonds Gly39–Ile10 (-5.96), Gly45–Leu16: -3.88, Gly69–Ile40: -3.83, Gly75–Leu46: -4.20, Gly12–Ile70: -3.29, and Gly18–Leu76: -7.98), consistent with the hydration network description of the collagen-like triple-helical peptides with (Leu–Hyp–Gly)$_n$ and (Gly–Leu–Leu)$_n$ repeat triplets [281].

11.6 Integrin–Collagen Triple-Helix Complex Interaction

The control of cellular interactions in multicellular organisms is fundamental for many biological processes – such as embryogenesis, immune responses, tissue integrity and homeostasis, to cite just a few. Several interactions occurring among cell–cell and cell–matrix components are performed by integrating specific ligands to cell adhesion receptors. Among them, the best known are the members of the family of integrins, a type I heterodimeric transmembrane glycoprotein, which function as the major metazoan receptor for cell adhesion and transmission of bidirectional signals. They play a central role in physical support functions for signal transduction, actin cytoskeleton assembly, gene expression, and cell functions – including cell adhesion, migration proliferation, differentiation, and apoptosis [282].

Integrins are composed of two subunits, namely α (containing ≈ 750 residues) and β (with $\approx 1,000$ residues), not covalently associated, jointly forming 24 distinct heterodimeric structures with different binding and distribution properties in various tissues [283]. These subunits show a large extracellular domain folded into N-terminal globular portions, which are combined to create the ligand-binding surface. This globular portion is followed by a type I transmembrane domain, a single-stranded structure of about 25–29 amino acids residues forming a coiled α-helix, and a short C-terminal cytoplasmic tail region (10–70 amino acids), which binds to cytoskeletal elements through cytoplasmic adapter proteins. They can also bind to a number of proteins of the extracellular matrix (ECM), which are found on the surface of other cells during different physiological contexts, making them a target to therapies for a large amount of diseases, such as thrombosis, infection, immune system disorders, osteoporosis, and cancer.

Among their heterodimeric structures, the $\alpha_2\beta_1$ integrin, a collagen receptor, has been involved in platelet adhesion, epithelial differentiation, morphogenesis, cancer, wound healing, angiogenesis, inflammation, and immunity. In vitro studies have shown that the deletion of the α_2 structure causes a reduction in platelet response, blocking the binding to type I collagen, as well as the destabilization of thrombus formation. Patients either with reduced levels of $\alpha_2\beta_1$ integrin expression on platelets or with the presence of auto-antibodies had platelet activation impaired by collagen but not by other molecules. Studies related to metastasis in tumor cells have revealed that the overexpression of the $\alpha_2\beta_1$ integrin may be related to the acceleration of metastasis in experimental models of melanoma and cancer of the stomach. Three decades after the identification of the VLA-2 cell surface protein in T cells, a recent study demonstrated that the $\alpha_2\beta_1$ integrin helps the interleukin 7 (IL-7) receptor on Th17 lymphocytes to mediate bone loss [284]; therefore, it is a pharmacological target against osteoporosis. Another analysis from cells derived of hepatic lesion inferred that the stimulation of $\alpha_2\beta_1$ integrin resulted in the formation of multicellular networks, increasing the fibrogenesis process in liver cells [285].

Taking into account the previous comments, there recently has been a great deal of interest in the investigation of the pharmacological targeting of $\alpha_2\beta_1$ integrin, mainly for the treatment of thrombosis and angiogenesis. Structural analysis of the α_2 subunit extracellular portion reveals the existence of seven homologous repeats in its N-terminal portion that folds into seven blades on a β-helix, where an I-domain comprising about 200 residues is inserted. Within this I-domain, one can find a metal-ion dependent adhesion site (MIDAS), playing a crucial role in the integrin–ligand interaction process. Extensive investigations have elucidated the structural motifs of this domain at the α_2 subunit, required for the ligands binding, inferring that the GFOGER (glycine–phenylalanine–hydroxyproline–glycine–glutamate–arginine) sequence is indeed an adhesive motif found in types I, II, and XI of the collagen structure, being also capable of binding to $\alpha_2\beta_1$ integrin without its prior activation.

11.7 Structural Representation

Crystallographic data of the integrin α_2I-domain co-crystallized with the synthetic triple-helix collagen structure containing GFOGER motif was obtained from the PDB database (www.rcsb.org) under the code (resolution) 1DZI (2.1 Å) [286]. For the purposes of the analysis performed here, it was necessary to modify the coordination metal of the 1DZI crystallographic structure from Co^{2+} to Mg^{2+} because, although the presence of the Co^{2+} ion has been previously shown to intensify the integrin–collagen binding mechanism [287], its toxicity is harmful for human beings. Besides, the magnesium ion is reported as the most effective bivalent cation present in the MIDAS region of integrin's I-domain.

Afterward, a structure optimization was performed using the hybrid QM/MM method in two layers through the ONIOM approach, available in the G09 code of the Gaussian software, to better deal with the size of the structure and to guarantee the accuracy of the results by the use of quantum methods. Through this method, the system can be subdivided into an infinite number of layers, and different levels of theory can be applied individually to each layer. For the layers described by a QM theory, we used the Minnesota hybrid exchange-correlation functional M06-2X [288] and the 6-311+G(d,p) basis set to expand the electronic orbitals. The Assisted Model Building with Energy Refinement (AMBER) force field was used to perform the MM calculations within an electronic incorporation scheme [289]. The M06-2X functional has shown good performance in several studies to describe noncovalent interactions, isomerization energies, and thermochemical properties in a number of systems, as well as accuracy in calculating hybrid QM/MM geometry optimization – which has been presented by metallopeptidase, nucleotide-nanotube, and thymidylate synthase catalyse systems, to cite just a few taking into account the 6-311 and 6-31 basis set.

Thus, within this hybrid layout, it was defined that the magnesium ion Mg^{2+} and the amino acids localized within a radius of 8.0 Å around it were free to

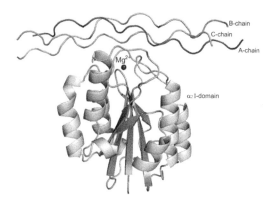

Figure 11.10 Structural representation of the integrin α_2I-domain co-crystallized with the synthetic triple-helix collagen structure containing GFOGER motif (PDB ID: 1DZI) in each of its monomers: A-chain (dark gray), B-chain (gray) and C-chain (light gray). Also shown is the magnesium ion Mg^{2+} in the MIDAS motif of the integrin–collagen interaction. After Bezerra et al. [266]

move, while the ion and the amino acids from the integrin and the collagen were screened within a radius of 5.0 Å around the coordinate metal. The hydrogen atoms that have not been optimized during the ONIOM process had their atomic positions optimized using the COMPASS force field available in the Forcite code, with the same convergence tolerances used in Section 11.2. The binding energy of the minimized collagen-α_2I-domain integrin complex (see Figure 11.10) was calculated by again using an MFCC-based scheme, as in the previous section, to study the stability of the collagen triple-helix structure.

It is worth mentioning that the whole integrin α_2I-domain is composed of almost 200 amino acids, making the calculation of all pair interactions in the collagen–integrin complex, at a quantum level, very tricky and almost impossible by computer simulation. One way to overcome this situation is to consider a significant number of amino acid residues of the collagen molecule inside an imaginary sphere with a pocket radius (r) equal to 8.0 Å, and evaluate their binding energy with each residue of the integrin (R_j) showing at least one atom inside the sphere. We consider, as a threshold radius, the pocket radius $r = 8.0$ Å based on the study of protein–protein interactions [290].

All water molecules forming hydrogen bonds with a particular residue were included for completeness in their respective fragments. Conventional and water-mediated H-bonds were defined using geometric and energetic criteria derived after a search of 77,378 binding sites from 25,096 PDB entries, considering the distance between the H-bond donor (D) and the H-bond acceptor (A) atoms, d_{D-A}, ranging from 2.6 to 3.1 Å, and the acceptor–hydrogen–donor angle, θ_{A-H-D}, not falling below $120°$.

The magnesium ion Mg^{2+} in the MIDAS motif was taken into account in the calculation procedures attached to the nearest residue. As it was previously shown, the glutamic acid from the collagen middle chain (Glu33) is an important residue to its coordination in the MIDAS. Thus, two models have been used in this work to represent the magnesium ion position during the calculations, namely (i) attached to the glutamic acid residue Glu33 of the collagen (MG1 model) and (ii) associated to the closest residue of integrin, Thr221 (MG2 model). DFT calculations for the MFCC procedure were performed using the $DMOL^3$ code, adopting the GGA-PBE functional [12] together with Grimme's semiempirical correction of the dispersive, covalent, and ionic bonds [19].

11.8 Interaction Energy Profiles

The mechanisms by which the $\alpha_2\beta_1$ integrin plays its key role in the human physiology as well as in disease development is now performed here through a computer simulation based on the MFCC approach, aiming to quantify the residue–residue interaction energies of the amino acids composing a synthetic collagen structure, with the single strains formed by the sequence Ac–[Gly–Pro–Hyp]$_2$–Gly–Phe–Hyp–Gly–Glu–Arg–[Gly–Pro–Hyp]$_3$–NH$_2$, and the integrin α_2 (I-domain).

The calculation of the interaction energies was performed in two moments: first with the metal Mg^{2+} associated with the glutamic acid Glu33 (MG1) of the collagen and then with the amino acid threonine Thr221 (MG2) of the integrin. Considering a pocket radius $r = 8.0$ Å, 238 pair interactions were identified; among them, 89 (104;45) were found between the integrin and the collagen A-chain (B-chain; C-chain).

Table 11.4 depicts the results obtained through the sum of individual energies of each amino acid from the triple-helical collagen model. From there, one can see that the total binding energy of the integrin–collagen complex is -728.28 (-335.76) kcal/mol for MG1 (MG2). The A-chain (C-chain) comprising the amino acids Gly1–Hyp21 (Gly45–Hyp65) contributed with -266.08 and -191.09 (-7.65 and -6.44) kcal/mol to the total binding energy when the calculation was performed to the MG1 and MG2 models, respectively. Meanwhile, for the interactions that occur between the B-chain (comprising the amino acids Gly23–Hyp43) and the integrin residues, when the magnesium ion Mg^{2+} was associated with the amino acid Glu33 (Thr221), the total energy value was -454.59 (-138.23) kcal/mol, the difference being due to a "positive zone" created when the B-chain is coupled to its glutamic acid (Glu33), attracting negatively charged amino acids from the integrin. Observe also that the last GPO repetition does not present any contact with the integrin in all collagen strains.

Table 11.4 *Total binding energy of each amino acid from the triple-helical collagen structures. After Bezerra et al. [266].*

	Total binding energy (kcal mol⁻¹)					
	A-chain		B-chain		C-chain	
Amino-acid sequence	MG1	MG2	MG1	MG2	MG1	MG2
Gly	—	—	—	—	—	—
Pro	—	—	—	—	—	—
Hyp	—	—	—	—	0.93	0.93
Gly	−0.01	−0.01	0.20	0.20	0.34	0.34
Pro	−2.63	−2.63	−0.23	−0.23	−0.34	−0.34
Hyp	−11.32	−11.32	−0.54	−0.54	0.37	0.37
Gly	−1.95	−1.95	−1.27	−1.27	0.29	0.29
Phe	−5.10	−5.10	−11.03	−11.03	−0.45	−0.45
Hyp	−29.33	−29.33	−2.93	−2.93	−1.0	−1.00
Gly	−2.93	−2.93	−3.75	−3.36	−1.99	−1.99
Glu	3.64	3.64	−289.53	−16.04	30.47	32.27
Arg	−217.08	−142.00	−133.78	−85.41	−31.65	−32.91
Gly	0.94	1.04	2.43	3.44	4.03	3.97
Pro	−0.69	−0.69	−3.70	−3.86	−4.98	−4.36
Hyp	0.04	0.04	−9.20	−15.93	−0.05	−0.04
Gly	0.14	0.14	0.31	0.31	−0.87	−0.87
Pro	—	—	—	—	−2.84	2.84
Hyp	—	—	−1.58	−1.58	0.18	0.18
Gly	—	—	—	—	—	—
Pro	—	—	—	—	—	—
Hyp	—	—	—	—	—	—
Total binding energy	−266.08	−191.09	−454.60	−138.23	−7.56	−6.44

11.8.1 Gly–Phe–Hyp–Gly–Glu–Arg (GFOGER) Motif

Let us first consider the α_2I-domain pair interactions with the collagen A-chain. We have analyzed 89 pair interactions, and the most relevant ones are depicted in Figure 11.11. Although the number (positions) of water molecules (amino acids) may change in the actual dynamics process, it should be mentioned that a number of MD computer simulation works indicates their lower mobilities at the binding interface when compared to other amino acids and water molecules on the protein surface, mainly those showing high interaction energy. Thus, the use of the ONIOM geometry minimization reinforces the position of their amino acids close to the strongest, stable, and, consequently, less mobile residues. As one can see from Figure 11.11a, the energetic behavior almost does not change when the magnesium ion Mg^{2+} is associated with either the glutamic acid Glu33 (MG1 model) or the threonine Thr221 (MG2 model), excluding the pair interaction Arg12–Thr221. From these data, it can be seen that the GFOGER sequence of the collagen A-chain, represented by the Gly7–Phe8–Hyp9–Gly10–Glu11–Arg12 sequence, has great relevance for binding to the α_2I-domain of integrin, since the two collagen amino acids that have the pair interaction showing the highest energetic values, Glu11 and Arg12, belong to this sequence.

The energetic result of the pair Glu11–Arg288 depicts one of the highest values found for the interaction of the A-chain amino acids with integrin, accounting for −35.19 kcal/mol. It is clearly a result of an electrostatic interaction between the negatively (positively) charged side chain of glutamic acid (arginine), 8.3 Å apart from each other. On the other side, the interaction Glu11–Glu256 shows a repulsive energy of 41.69 kcal/mol. Although their negatively charged carboxyl groups are

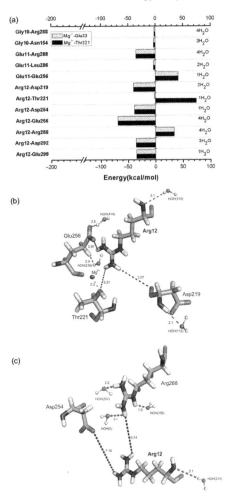

Figure 11.11 The most relevant interactions involving the α_2I-domain with the collagen A-chain: (a) graphical panel (model M1–shadow area; model M2–black area); (b) and (c) main residue interactions. Potential direct hydrogen bonds, water hydrogen bonds, salt bridge, metal–residue, and charge–charge interactions are indicated by the dashed lines. After Bezerra et al. [266]

quite distant (approximately 9.0 Å), this high energy is seen as an effect of these charged side chains that are facing one another in the crystallographic structure. Taken together, these two interactions (the attractive and the repulsive) are mainly responsible for the final slightly repulsive interaction between Glu11 and the α_2I-domain ($\approx$3.00 kcal/mol). Besides, the energetic value of this interaction was the same in both the MG1 and MG2 models, regardless the residue that interacts with the ion, which means that this connection is little influenced by the MIDAS region.

The residue Arg12 presents seven significant interactions with the α_2I-domain residues – five attractive and two repulsive. The attractive interactions (all energy interactions, from now on, are given in kcal/mol unit) occur between the residues Arg12–Glu256 (−68.37) > Arg12–Asp219 (−40.16) > Arg12–Asp254 (−38.27) > Arg12–Asp292 (−35.48) > Arg12–Glu299 (−34.71). Three interactions between the arginine and the aspartic acid/glutamic acid are highlighted by the attractive behavior presented by arginine positively charged side chain (guanidinium group) and these acidic residue's negatively charged side chain (COO⁻), which interact forming a salt bridge, allowing the occurrence of this high electrostatic attractiveness among the residues involved. In addition, the presence of water molecules among them allows the occurrence of hydrogen bonds and the increase of their affinity.

Likewise, the repulsive interactions are Arg12–Thr221 (74.34) > Arg12–Arg288 (34.40). The repulsion between the arginines is explained by the fact that the interaction has a strong charge–charge interaction between the functional guanidinium group of the same charge. The high repulsion shown in the interaction Arg12–Thr221 in MG2 is related to its direct interaction with the Mg^{2+} coordination meta, which is also the sole residue–residue interaction in the A-chain depicting different values, as far as the MG1 and MG2 models are concerned. We observed that when this same interaction is calculated with the ion Mg^{2+} attached to Glu33, the energy found is attractive (−1.81 kcal/mol), but of low intensity, stressing the major role played by the Mg^{2+} ion for the repulsion. Figures 11.1b and 11.1c represent the major interactions carried out by the amino acid Arg12 and residues of the integrin. From there, it is also possible to see that the residue Arg12 is mainly surrounded by integrin's negatively charged residues, though positively charged (Arg288) and polar uncharged (Thr221) ones are within the binding pocket radius (8.0 Å).

Although the interactions shown in the preceding paragraphs are seen as the most intense ones, it is important to take into account other amino acids that also contribute to the stability of the complex. Among the residues found in the collagen A-chain, Hyp9 and Phe8 represent the second- and the third-highest total binding energies −29.32 and −5.10, respectively. The residue Phe8 is involved in an interaction with Tyr157 (−3.50), forming a π–π stacking interaction, where their rings are positioned in a face-by-face configuration at a distance of 5.0 Å, and with Tyr285 (−1.42) in a non-conventional hydrogen bond, in which the carbonyl oxygen of the tyrosine main chain is at a distance of 2.40 Å to the phenylalanine C–H (ring).

Meanwhile, the residue Hyp9 makes direct H-bonds with the residues Asn154 (−2.63; 2.20 Å) and Ile156 (−5.02; 2.57 Å), beyond a non-conventional H-bond with Tyr157 (−5.11; 2.49 Å) and some hydrophobic contacts with Ser155 (−3.80). It is important to notice that all direct hydrogen bonds are made between the hydroxyl side chain of hydroxyproline and the carbonyl oxygen of the Ans154 and Ile156 main chain. Besides, the non-conventional H-bond seen between the residues Hyp9 and Tyr157 is also formed with the hydroxyl of Hyp.

We now turn to the α_2I-domain interactions with the collagen B-chain. We observed 104 pairs interaction between the amino acids, Gly29–Phe30–Hyp31–Gly32–Glu33–Arg34, and the integrin α_2I-domain, whose most relevant ones are presented in Figure 11.12. It is possible to see that the interactions among these amino acids and the residues of the integrin, different from the previous case, show considerable energetic variations as far as the association of the magnesium ion Mg^{2+} is concerned (MG1 and MG2 models). Thus, as shown in Figure 11.12a,

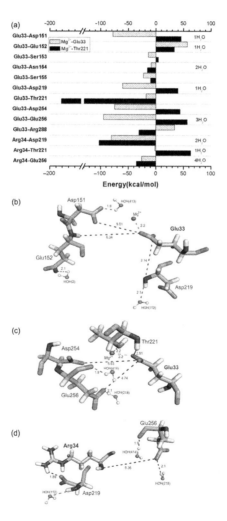

Figure 11.12 Same as in Figure 11.11 but for the collagen B-chain. After Bezerra et al. [266]

when we consider Mg^{2+} interacting directly with the residue Thr221 (MG2 model), we obtain seven repulsive interactions and six attractive ones. The positive energy values (repulsion) are presented in the following interactions: Arg34–Thr221 (63.29) > Glu33–Glu256 (57.17) > Glu33–Asp151 (46.05) > Glu33–Asp254 (44.16) > Glu33–Asp219 (40.24) > Glu33–Glu152 (33.44) > Glu33–Ser153 (4.65), while the attractive ones are Glu33–Thr221 (−181.277) > Arg34–Asp219 (−102.36) > Arg34–Glu256 (−35.20) > Glu33–Arg288 (−30.62) > Glu33–Asn154 (−15.41) > Glu33–Ser155 (−9.74).

On the other hand, when the coordination metal Mg^{2+} is interacting directly with the amino acid Glu33 (MG1 model), we obtain two repulsive interactions and 11 attractive interactions. The repulsive energy values are presented in the following interactions: Glu33–Glu152 (57.24) > Glu33–Arg288 (34.22). The attractive interactions are Glu33–Glu256 (−94.64) > Arg34–Asp219 (−80.63) > Glu33–Asp151 (−77.75) > Glu33–Asp254 (−74.49) > Glu33–Asp219 (−60.38) > Arg34–Glu256 (−25.68) > Glu33–Ser154 (−22.55) > Glu33–Thr221 (−17.64) > Glu33–Ser153 (−13.76) > Glu33–Asn154 (−8.56) > Arg34–Thr211 (−1.46).

The amino acid residue Glu33 presented 10 interactions with significant energy values. Three of them occurred between Glu33 and the aspartic acids (Asp219, Asp151, and Asp254). When the magnesium ion Mg^{2+} is directly coordinating Thr221, all interactions of the residue Glu33 with the aspartic acid are repulsive since both amino acids are negatively charged, promoting an electrostatic repulsion between them. However, when the coordination metal is associated with the residue Glu33, the same interactions become attractive. Bella and Berman [278] found that only a negatively charged residue makes direct contact with the metal. Thus, when a carboxyl group of the ligand is getting closer, there is a rearrangement around the metal site, which can only be achieved by additional conformational changes extending along the I-domain. Furthermore, the integrin I-domain residues that define the sequence DXSXS (Asp151–Glu152–Ser153–Asn154–Ser155) keeps the same coordination as the metal ion, regardless of whether or not the integrin is bound to a ligand [286].

The most important interactions promoted between the residue Glu33 and the glutamic acid of the integrin occur between the pair interactions Glu33–Glu152 and Glu33–Glu256. In the MG2 model the values for both interactions are repulsive (33.44 and 57.17, respectively), as shown previously, due to the electrostatic repulsive forces occurring between the charged side chains of the glutamic acid residues. When compared to the energetic values considering the MG1 model, there is a small increase in the repulsion between Glu33–Glu152 (57.24) and the appearance of a high attractiveness between the Glu33–Glu256 residues (−94.64), this attractiveness being due to the system Glu33–Mg^{2+} presenting a positively charged surface. In addition, the amino acid Glu256 presents mediated connections between the water molecules with the metal [286].

Other relevant interactions occur between the residue Glu33 and the serines Ser153 and Ser155, also with a significant difference when one compares the MG1 and MG2 models. After the binding of the collagen with the integrin, the ion promotes a rearrangement in order to maintain its direct interaction with the integrin residue; i.e., it interacts indirectly through hydrogen bonds with the serine, which explains the increase of the attraction interaction between Glu33 and these serines in the MG1 model [286]. The residue Glu33 also has important interactions with the residues Asn154, Arg288, and Thr221. The interaction with Asn154 is attractive and has hydrogen bonds, regardless of the interaction of the metal, being reduced when the metal interacts with the residue Glu33. The interaction energy is attractive (−30.62) when the metal interacts with the integrin residue and repulsive (34.22) when the metal interacts with the residue Glu33. The attraction is a consequence of the charges of these amino acids, while the repulsion is due to the fact that the glutamic acid has its conformational state altered by a direct bond to the coordinating metal.

The highest energy value found among all interactions occurred between the residues Glu33 and Thr221 (−181.27) when the calculation is performed with the ion metal interacting with Thr221 (MG2 model). The threonine Thr221 residue belongs to the MIDAS region of the integrin, whose binding site is essential for the binding between the collagen molecule and the $\alpha_2\beta_1$ integrin. Thus, the metal must be associated with the residue Thr221 so that the integrin becomes active and can be attached to the collagen. Threonine does not have a negative formal charge, which should leave the metal ion strongly electrophilic and capable of forming a strong bond with Glu33. After binding to the ligand, the ion will coordinate the glutamic acid and will interact with the threonine through hydrogen bonds.

Another residue of the collagen that stands out because of its importance is Arg34 (crucial for the collagen binding with high affinity to the integrins) interacting with the residues Asp219, Thr221, and Glu256. The highly attractive bond between Arg34 and Asp219 is because its side chains are, respectively, positive and negative, thereupon forming a salt bridge of great importance for the complex. The same occurs for the interaction with the residue Glu256 allied to hydrogen bonds mediated by the presence of water. Figure 11.13b,c,d depict the major interactions performed between the collagen B-chain and the integrin.

Following the amino acid sequence in the collagen B-chain, Phe30 contributes with −11.03 to its total binding energy. Individually, this phenylalanine shows two strong interactions, namely Phe30–Asn154 (−2.58) and Phe30–Gln215 (−5.00). These energies are related to the non-conventional hydrogen bonds, with the former

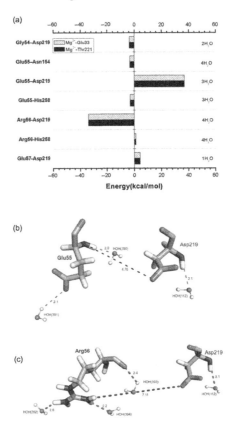

Figure 11.13 Same as in Figure 11.11 but for the collagen C-chain. After Bezerra et al. [266]

being created between two C–H groups of Phe side chains and the oxygen atoms of Asn154 at the distance of 2.48 Å and 2.87 Å, respectively.

Meanwhile, the residue Phe30 made non-conventional H-bonds with two water molecules that are hydrogen-bonded to the residue Gln215, yielding a water-mediated hydrogen bond. It is important to notice that both pair residues also present a number of hydrophobic contacts, which increase the interaction energy. Different from the residues Hyp9 and Hyp31, they do not show the same energetic pattern and number of interactions, as they are only involved in seven pairs, presenting a total binding energy of −2.93 related to a direct hydrogen bond with Asn154 (−4.65) through its carbonyl oxygen and the amine side chain of the asparagine from a distance of 2.75 Å.

Finally, we consider the α_2I-domain interactions with the collagen C-chain Gly51–Phe52–Hyp53–Gly54–Glu55–Arg56, which has 45 interactions with residues of the integrin. The interactions between the residues Glu54–Asp219 (36.98) and Arg56–Asp219 (−33.88) were the only ones that showed considerable importance for the C-chain, as depicted in Figure 11.13. One can note that, as in the A-chain case, there is no difference in values of energy taking into account the coordination of the metal ion models (MG1 and MG2). The negative charges of the side chains of the glutamic and aspartic acids explain the repulsion found. The interaction between the arginine and the aspartic acid is due to the formation of a salt bridge between the polar chains of the residues. In addition, the presence of water molecules in the interactions allows the emergence of hydrogen bonds creating a microsolvation environment (see Figure 11.13b and 11.13c).

11.8.2 Gly–Pro–Hyp (GPO) Motif

In the collagen domain characterized by a repetition of Gly–X–Y triplets, the X and Y positions are frequently occupied by proline (Pro, 28.1%) and hydroxyproline (Hyp, 38.1%) amino acid residues, respectively, which together represent approximately 22% of all amino acids of the collagen strands. The abundance of these residues decreases the entropic cost for collagen folding, although it dramatically increases the thermal stability of the triple helix. It is known that appropriate ring pucker, enforced by a stereo electronic or steric effect, pre-organizes the torsion angles required for the triple-helix formation [253].

As depicted before, Gly–Pro–Hyp repetitions do not contribute strongly to the binding mechanism of integrin α_2 and collagen when compared to the Gly–Phe–Hyp–Gly–Glu–Arg motif. As one can see from Figure 11.14, the GPO (Gly–Pro–Hyp) repetitions are commonly positioned far from integrin, excluding some of the [Gly–Pro–Hyp]$_3$ in the end of the strands, mainly in the collagen B-chain.

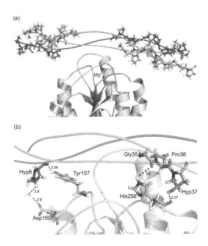

Figure 11.14 (a) The GPO (Gly–Pro–Hyp) motif in the collagen chains: A-chain (dark gray), B-chain, and C-chain (gray): (a) structural representation; (b) main intermolecular interactions. After Bezerra et al. [266]

The second-strongest binding energy found to this motif was -11.318 kcal/mol related to the amino acid Hyp6 (collagen A-chain), which is mainly related to a direct hydrogen bond with the residue Tyr157 (-4.726 kcal/mol; 2.04 Å), as well as the solvent that creates a network of water-mediated hydrogen bonds with the residue Asp160 (-5.574 kcal/mol). The amino acid His258 (integrin) is seen making direct H-bonds with the residues Hyp37 (-5.078 kcal/mol) and Gly35 (-4.12 kcal/mol) at the distances 2.27 Å and 3.20 Å, respectively, while it is involved in hydrophobic interactions with Pro36 (-3.285 kcal/mol). It is important to notice that these two residues are members of the first repetition after the Gly–Phe–Hyp–Gly–Glu–Arg motif in the collagen B-chain, which is the collagen closest region to the integrin receptor in the crystallographic structure analyzed here. This proximity gives to this GPO motif the possibility of making direct polar interactions with the α_2I-domain; none of the other [Gly–Pro–Hyp]$_3$ (end of the strain) repetitions have made this kind of contacts. Therefore, as it is shown in Table 11.4, there is no other individual strong interaction between this motif in all collagens' strains and the integrin.

It has been shown that hydroxyproline residues in collagen work in the maintenance of its structure by the formation of water bridges to other collagen amino acids, mainly to the glycine Gly. Also, the integrin-type collagen receptors cannot bind to the basic triple-helical Gly–Pro–Hyp sequence, but they require specific motifs formed by residues in two or all three collagens' chains, in agreement with the energetic description presented here and in the previous section (type III collagen-like peptide T3-785), in which this motif is mainly related to the collagen stabilization.

11.9 Conclusions

In this chapter, we presented quantum chemistry calculations considering initially the electronic structure of type III collagen-like peptide T3-785, in order to identify the nature of the binding interactions that stabilize its 3-D structure. The interchain interaction energies of amino acids that compose this collagen reveal the most energetically important regions (triplets) in this system, obeying the sequence: N-terminal zone < C-terminal zone < Central zone (Leu–Ala–Gly < Ile–Thr–Gly < Pro–Hyp–Gly < Ala–Arg–Gly). In order to evaluate the individual importance of each amino acid type in this peptide, we compared their average interaction binding energies, giving the following order: Arg > Hyp > Thr > Ala > Gly > Pro > Leu > Ile.

We also highlighted the importance of intermolecular interactions to keep the conformational stability of collagen. Figure 11.15 depicts a condensed graphical view of the most (dark gray, on the left) and fewest (gray, on the right) energetically

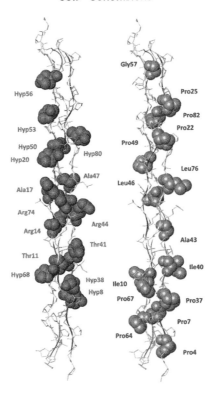

Figure 11.15 Pictorial view of the 15 most (dark gray, on the left) and fewest (gray, on the right) energetically significant amino acids within the peptide T3-785. After Oliveira et al. [264]

significant amino acids within the peptide T3-785. Only two (Ala1 and Ala47) of the top 15 amino acid residues do not present a polar side chain, although they perform a dipole–dipole interaction (Ala47: C_αH–OC: Leu76). On the other hand, the residues with lowest affinity are essentially nonpolar, as is the case of C-/N-terminal prolines. The aliphatic amino acid residues Ala, Leu, Ile, and Gly jointly account for 22.97% of the sum of average interaction energy per amino acid residue, corroborating the hypothesis that the process of self-organization in triple helix is also directed by hydrophobic interactions [281].

The strong attractive character observed in the amino acid residues Arg14, Arg44, and Arg74 indicates the importance of ion-induced dipole force for triple-helix stability. Besides them, the conventional H-bonds (Gly:NH–OC: Xaa) are present in stronger interactions involving Gly residues. Non-conventional (Gly:C_αH–OC:Xaa) and water-mediated H-bonds (Xaa:NH–HOH–OC:Gly) involving the carbonyl, amine, and hydroxyl of residues Gly, Ile, and Thr are important for their interchain interactions. The binding energy values calculated for residue–residue pairs with mediator water molecules suggest that the solvent network plays an important role in the stabilization of the collagen model T3-785. As a matter of fact, water could determine the subtle balance among driving forces of different physical natures, thus broadening the array of available binding mechanisms in biomolecular associations.

Regarding the integrin $\alpha_2\beta_1$-collagen triple-helix complex interaction also discussed in this chapter, we presented their electronic structure in order to identify the nature of the most important binding interactions in the complex. The results obtained demonstrated not only the importance of the MIDAS motif, located in the α_2I-domain, but also the binding interactions of the coordination metal Mg^{2+} with the glutamic acid Glu33 (MG1 model) of the collagen and with the amino acid threonine Thr221 (model MG2) of the integrin.

We found that the collagen A-chain is influenced by the coordination ion Mg^{2+}, mainly in its amino acid residues Glu11 and Arg12, while the collagen C-chain has no metal ion contribution in the collagen–integrin binding. In the interactions between the A-chain and the integrin, the Arg12 arginine residue and the aspartic acid residues (Asp219, Asp254, and Asp292) exhibit high attractive affinity due to the presence of salt bridges. With respect to the C-chain, significant attractive interactions occur between the residues Arg56–Asp219.

Several of these amino acids are of fundamental importance for the formation of the integrin–collagen complex, and mutations in any of them may lead to serious diseases for humans. The residues composing the collagen A-chain are also important to the maintenance of collagen–integrin interaction, while the collagen C-chain amino acids show a small participation in this complex. Besides, our results show that the metal ion Mg^{2+} and the residue Glu33 (collagen) play a central role in the binding with the receptor; the affinity of this interaction probably reflects the exceptionally strong bonds formed by the metal bridge.

The collagen B-chain, on the other hand, is the most important chain interaction between the triple-helix collagen and the α_2 I-domain of the integrin. The attractive interactions occur between the residues Glu33–Asp219, Glu33–Asp254, Glu33–Glu256, Arg34–Asp219 (Glu33–Thr221, Arg34–Asp219, Arg34–Glu256), when the metal is bound to the glutamic acid Glu33 (threonine Thr221). The arginines present in this chain have a significant importance because they form salt bridges with the integrin negatively charged amino acid residues, mainly those forming the MIDAS motif. Mutations in this arginine can negatively influence the interaction between the macromolecules, as well as the absence of the residue Thr221 for integrin structure. In addition, we must also consider the fact that the Central region of the collagen B-chain is closer to the integrin site analyzed here.

The results generated in this chapter provide detailed knowledge about the stability and interaction of collagens, and this will certainly help researchers working in the area of de novo protein design, mainly those related to the synthesis of model collagen-like peptides.

12

Antimigraine Drugs

12.1 Introduction

Migraine is an episodic syndrome consisting of a variety of clinical features that result from dysfunction of the sympathetic nervous system. It is considered one of the most common neurological diseases. The causes that lead to it are still not definitely clarified. It is a complex condition with a wide variety of symptoms, the main feature usually being a painful headache, that affects many people worldwide. There is currently no cure for migraine, although it may be related to a mix of environmental and genetic factors, and the treatments available are varied and differ from person to person.

A migraine attack can be very frightening with severe throbbing pain or a pulsing sensation, usually on just one side of the head – often accompanied by nausea; vomiting; and extreme sensitivity to light, sound, and smell. It usually comes on gradually and is generally made worse by physical activity. It is a severe and disabling neurological disorder that typically begins in adolescence or early adulthood, showing higher incidence among women in reproductive years, peaking in their fourth decade of life. It usually lasts from four hours to three days and has an enormous impact on the work, family, and social lives of the sufferers, significantly diminishing their quality of life. Up to one-third of sufferers experience an aura – a spectacular and sometimes frightening focal neurological disturbance manifesting as visual, sensory, or motor symptoms that signals that the headache will soon occur.

Migraine ranks as the third most common (and costly) disabling neurological disease (disorder). Despite that, unfortunately, migraine remains a poorly understood disease that is often undiagnosed and undertreated.

The pathophysiology of migraine is still unclear. In recent decades, important strides have been made to understanding it, giving rise to the vascular hypothesis, which dominated the later part of the twentieth century. According to this theory, migraine is a consequence of a vascular dysregulation, with the aura preceding headache thought to result from hypoxemia related to transient vasoconstriction and migraine pain from rebound vasodilation.

Current thinking has moved away from vascular dysregulation as a primary cause of migraine, asserting that neurovascular dysfunction is the possible primary driver of the not-yet-clear pathophysiology of the disorder, by means of the activation of the trigeminovascular system, cortical spreading depression, and neuronal sensitization [291]. As a consequence of these findings, a number of drug classes and molecular targets clinically used in therapeutic settings to control acute attacks and as preventive measures were released, such as nonsteroidal anti-inflammatory drugs (ibuprofen; see Chapter 9), triptans, opioid analgesics, ergot alkaloids, antiemetics, antiepileptic drugs, and antidepressants. Currently, among these classes of drugs, ergot alkaloids and triptans are two of the most widely used for the treatment of acute migraine, with its action being related to vasoconstriction process or/and decreasing the release of serotonin neurotransmitter through the binding and activation of serotonin receptors [292].

Serotonin is a monoamine neurotransmitter primarily found in the gastrointestinal tract, blood platelets, and central nervous system of animals, including humans. Drugs altering the serotonergic system are clinically used for treating depression, anxiety, psychosis, nausea, and vomiting, to cite just a few. Since the early days of headache research, serotonin is seen as an important target to migraine pathogenesis, mainly those related to their receptors 5-HT_{1B} and 5-HT_{1D} [293].

In this context, computer simulation approaches may be a useful tool to provide a guide to search new specific antimigraine drugs, to identify efficient ligands not only to the serotonin receptors but also to other cell surface receptors like the G-protein-coupled receptors (GPCRs), the largest and most diverse group of membrane receptors in eukaryotes [294]. The most common and computationally cheaper method is based on the molecular mechanics simulation, which unfortunately releases a limited amount of information about the interaction between specific residues of the receptor and the different ligands, although it is particularly useful for the design of some drugs.

However, the use of quantum biochemistry for in silico drug design recently has been gaining more attention due to its high accuracy, despite a higher computational cost. In this sense, fragmentation methods have been developed to make the macromolecules computationally less expensive [24]. Among these methods, the molecular fractionation with conjugated caps (MFCC) approach, largely employed in previous chapters, has been widely used, particularly to calculate the interaction energy between the amino acid fragments of the receptor and the ligands with great success.

Taking into account the preceding remarks, it is the aim of this chapter to investigate the binding energy profile of the drug dihydroergotamine (DHE), one of the most widely prescribed antimigraine drugs, considered the classical drug for the treatment of migraine and cluster headache, and the individual residues of the serotonin $5\text{-HT}_{1B}\text{R}$ through quantum biochemistry techniques based on the MFCC scheme within the density function theory (DFT) framework [295]. The individual contribution of each amino acid residues involved in the DHE-serotonin receptor binding was calculated using the X-ray structure of the 5-HT_{1B} receptor co-crystallized with dihydroergotamine (PDB ID: 4IAQ) [296]. To improve our calculations, we implement an electrostatic embedding scheme, here called electrostatic embedding MFCC (EE-MFCC) scheme.

Our main goal is to depict the DHE-$5\text{-HT}_{1B}\text{R}$ binding through a detailed energetic profile of the interactions between the residues in the binding site and the antimigraine drug, essential not only to provide information about adjustments to improve the effectiveness of new drugs specific to $5\text{-HT}_{1B}\text{R}$ therapy but also to allow a better understanding of the GPCR's binding features. A comparison of our results obtaining by using the EE-MFCC approach with those considering the ordinary MFCC scheme is also done for completeness.

12.2 Serotonin Receptors and the Antimigraine Drugs

Serotonin (5-hydroxytryptamine) is an ubiquitous monoamine acting as one of the neurotransmitters at the synapses of nerve cells. It acts through several receptor types and subtypes in different and complex processes. Among them, the 5-HT_{1B} receptor, located on the human chromosome 6q13 (HTR1B gene), comprises approximately 390 amino acids assembled as seven highly conserved transmembrane helices (TMH1–THM7) connected by extracellular and intracellular loops [297], as shown in Figure 12.1. It is a G-protein-coupled receptor (GPCR), coupled to the G-protein alpha subunits $G\alpha_i$ or $G\alpha_0$, leading to the inhibition of adenylyl cyclase and the decrease of the intracellular cyclic adenosine monophosphate (cAMP) levels.

The $5\text{-HT}_{1B}\text{R}$ is widely distributed throughout the brain – mainly on the frontal cortex, striatum, and hippocampus – where it has been related to the modulation of the synaptic release of serotonin and other neurotransmitters, indicative of its role as terminal auto- and heteroreceptor, respectively. Moreover, there is evidence that $5\text{-HT}_{1B}\text{R}$ may be involved in the mechanisms underlying schizophrenia, aggression, depression, attention-deficit/hyperactivity disorder, and drug abuse. Nevertheless, expression of $5\text{-HT}_{1B}\text{R}$ is prevalent in human cerebral artery smooth muscle cells, where vasospasm has been implicated in the pathogenesis of migraine [298].

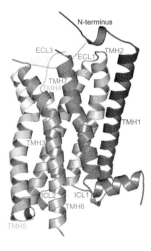

Figure 12.1 Serotonin receptor 5-HT$_{1B}$ structure composed of seven transmembrane helices (TMH), two intracellular, and three extracellular loops (ICL and ECL, respectively). After Lima Neto et al. [295]

Therefore, 5-HT$_{1B}$R is an attractive target for the development of rational therapies not only to migraine but also to some other neurological dysfunctions.

Activation of 5-HT$_{1B}$ receptors in cerebral arteries by agonists causes vasoconstriction and has been posited as a therapeutic explanation for some antimigraine drugs [292]. Among them, dihydroergotamine (DHE; (5'α,10α)-5'-Benzyl-12'-hydroxy-2'-methyl-3',6',18-trioxo-9,10-dihydroergotaman) is one of the most widely prescribed drug (see Figure 12.2). Marketed under the trade names D.H.E. 45 and Migranal, DHE is a synthetic ergot alkaloid, or ergoline, derived from hydrogenation of the double bond between atoms C10=C11 of ergotamine (ERG), which acts as a selective 5-HT$_{1B}$R agonist, where it exerts its antimigraine effects [299]. However, after discovering triptans in the 1990s, the clinical employment of dihydroergotamine underwent a reduction due to the acute side effects related to its activity at other monoaminergic receptors [300]. Nonetheless, dihydroergotamine continues to be one of the most widely used antimigraine drug.

The recent determination of the first high-resolution crystal structure of the human 5-HT$_{1B}$ receptor co-crystallized with agonists ergotamine and dihydroergotamine [296] was a first step toward a deeper understanding of the ligand-binding mechanism in a 5-HTR receptor. In spite of that, the structural and sequence similarities among the members of these GPCRs prevent obtaining selective agonists and/or antagonists to a single receptor. Furthermore, evidence suggests that conformations induced by noncovalent interactions between ligand and amino acids from the GPCRs can give rise to different biological responses in the same

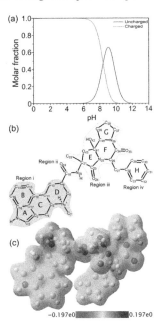

Figure 12.2 Protonation state of dihydroergotamine (DHE). (a) The molar fraction curves as a function of the pH. (b) Atom labeling and different regions of DHE in protonated form. (c) DFT electron density projected onto an electrostatic potential isosurface of charged ligand. After Lima Neto et al. [295]

receptor, which cannot be seen by a single snapshot from the crystallographic structure. Thus, techniques that can help the characterization of the ligands-5-HTR interactions are quite welcome in drug design aiming new selective ligands to serotonin receptors [301].

12.3 Drug–Receptor Complex Data

To perform the computer simulation we use the first X-ray crystallographic structure of human serotonin 5-HT$_{1B}$ receptor in complex with dihydroergotamine (PDB ID: 4IAQ) at 2.80 Å of resolution [296]. Also, we accomplish the study of protonation state set up at physiological pH of DHE and 5-HT$_{1B}$R amino acids using the MarvinSketch code version 5.5.0.1 (Marvin Beans Suite, ChemAxon) and PROPKA 3.1 package, respectively.

12.3.1 Classical Calculation

We add the atoms not resolved by X-ray diffraction and therefore absent in the crystallographic files, such as the hydrogens and some amino acid side chains, to the structure and submit them to a classical geometry optimization using the classical force field CHARMm, fixing the other atoms. This force field is especially parameterized for organic molecules, which brings to the calculations a higher accuracy. We set the same convergence tolerances as in Chapter 11. Afterward, the atomic positions and charges were stored to carrying out the subsequent steps.

Interaction energy between DHE and 5-HT$_{1B}$R is calculated either through the MFCC-based formalism or better, by including the surrounding electrostatic interactions in all steps through the use of point charges, the so-called electrostatic embedding MFCC (EE-MFCC) scheme. Here, we label the ligand dihydroergotamine as L, and the residue interacting with it as R_i, with i denoting the index of the ith amino acid residue. The C_i (C_i^*) cap is formed from the neighboring residue covalently bonded to the amine (carboxyl) group of the residue R_i along the protein chain, providing a better description of its electronic environment. Also, C_{PC} represents the set of all crystallographic amino acids, without the reference residue (R_i) and caps (C_i, C_i^*), where each atom is replaced by the corresponding atomic charge (obtained classically after optimization process) in the corresponding atomic site (point charges). For these fragmented and electrostatically embedded structures, the interaction energy between the ligand and the individual fragments, $EI_{EE-MFCC}(L - R_i)$, is calculated according to

$$EI_{EE-MFCC}(L - R_i) = E(L - C_i R_i C_i^* - C_{PC}) - E(C_i R_i C_i^* - C_{PC})$$
$$- E(L - C_i C_i^* - C_{PC}) + E(C_i C_i^* - C_{PC}), \quad (12.1)$$

where the first term, $E(L - C_i R_i C_i^* - C_{PC})$, is the total energy of the system formed by the ligand and the capped residue; the second term, $E(C_i R_i C_i^* - C_{PC})$, is the total energy of the residue with the caps; the third term, $E(L - C_i C_i^* - C_{PC})$, is the total energy of the system formed by the caps and the ligand; while the fourth and final term, $E(C_i C_i^* - C_{PC})$, is the energy of the caps with dangling bonds hydrogenated. For each term, the presence of the background charges distribution of the remaining residues (excluding the ligand, the reference residue, and the caps at each one) are included in the calculation through the influence of the long-range interactions of the environment in the ligand–residue binding energy as point charges. This makes the electrostatic effect of the environment more realistic than the usual continuum models due to its capacity to reproduce more accurately the heterogeneity of the protein environment. Here, the influence of the electrostatic interactions of the surrounding environment is obtained by comparing EE-MFCC results with those found using the MFCC scheme without point charges.

12.3.2 Quantum Calculation

After fragmentation considering the MFCC and EE-MFCC schemes, energetic calculations for each ligand–residue interaction at the binding site were performed using the Gaussian G09 code, within the DFT formalism, with the GGA functional B97D, including Grimme's dispersion terms [19] and COSMO continuum solvation model.

To expand the Khon-Sham orbitals for all electrons, we selected the 6-311+G(d,p) basis set for the calculations. It is important to notice that DFT calculations were carried out for each term of Eq. (12.1). Electrostatic potential isosurface of the charged ligand has been calculated using the electrostatic embedding scheme, where explicit molecules are surrounded by the full set of point charges.

To avoid missing important interactions, a convergence study of the total binding energy as a function of the ligand-binding pocket radius was performed, limiting the number of amino acid residues to be analyzed. To this end, we added the individual interaction energy of those amino acid residues within imaginary spheres with a pocket radius r centered at the ligand, considering $r = R/2$, $R = 1, 2, 3, 4, \ldots,$ N_n, N_n being the next natural numbers in the sequence. The binding pocket radius r achieved convergence when the energy variation in subsequent radius is smaller than 10%.

12.4 Interaction Energy of the Amino Acid Fragments

One of the main targets in the pharmaceutical industry is the GPCR family, amounting to almost 30% of every drug sold on the market. Among them, serotonergic systems represent a large share of the top-selling drugs of the past decade [302]. Since off-target interaction is the major cause of side effects and failures in clinical trials of serotonergic ligands, the identification of selective ligands is a critical aspect of the drug discovery process.

In this context, the recent publication of 5-HT$_{1B}$R co-crystallized with dihydroergotamine and ergotamine has motivated structure-based drug research through in silico simulations, which have been quite helpful to guide the development of most selective compounds [296]. Nevertheless, there are a few quantum mechanics–based studies about this family to analyze and compare many GPCRs. Among them, the use of fragment molecular orbital (FMO) simulations within a second-order Møller-Plesset perturbation theory (MP2) formalism at GPCR crystallographic structures depicted the relevance of quantum calculations coupled to fragmentation methods to unveil their binding features, including a map of relevant interactions within a pocket radio $r = 4.5$ Å from each ligand [303]. As far as we know, there is no other quantum mechanical study relating dihydroergotamine binding at 5-HT$_{1B}$R, despite its high pharmaceutical relevance. As an improvement of the previously described quantum mechanical techniques, we present here an MFCC-based approach to depict the binding features of the DHE-5-HT$_{1B}$R

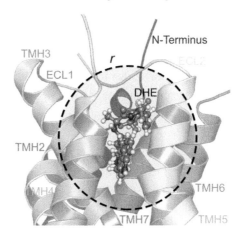

Figure 12.3 Arrangement of the dihydroergotamine (DHE) surrounded by 5-HT$_{1B}$R helices and loops. The binding pocket sphere with radius (r) is also shown in the picture as a circle around it. After Lima Neto et al. [295]

complex including the surrounding electrostatic environment through the use of amino acid point charges, the so-called EE-MFCC model.

12.4.1 Protonation State

To start our in silico calculations, we obtain the protonation state of dihydroergotamine (Figure 12.3). As showed in Figure 12.2a, there is a prevalence of DHE protonated state (96.12%–90.77%) at 7.0–7.4 physiological pH range. We adjust the molecular structure of ligand to a charged form ($+1$) by adding a single hydrogen atom to the amine group at N6 of DHE. It is in agreement with the analysis of the DHE's crystallographic structure and is a common behavior showed by serotonin and other ergolines, such as the ergotamine and the lysergic acid diethylamide (LSD) [304], all of them presenting serotonergic activity.

We subdivide the ligand into four regions to facilitate further investigation, namely (i) the ergoline moiety (A–D rings); (ii) the peptide link; (iii) rings E–G; and (iv) the phenyl H ring (see Figure 12.2b).

The electron density distribution of dihydroergotamine in the charged state is depicted in Figure 12.2c, where it can be seen that the entire molecule has a most positive feature (dark gray shade), being accentuated on rings A–D, an effect of positively charge N6 atom, which is close to N1 from indole moiety. Despite that, the rings E–H influence the charge distribution by giving a characteristic closer to neutral (light gray shade) to regions (iii) and (iv), mainly due to the oxygens in rings E and F.

12.4.2 Interaction Energy

Similar to serotonin and other ergot alkaloids, DHE occupies a large and deep ligand-binding cavity of 5-HT$_{1B}$R, defined by residues from helices III, V, VI, VII, and ECL2, comprising the orthosteric binding pocket, besides an extended binding pocket close to the extracellular entrance. DHE adopts a binding mode with the ergoline ring system (rings A–D) occupying the orthosteric binding pocket and the cyclic tripeptide moiety inside regions (iii) and (iv), bound to the upper extended binding pocket, as shown in Figure 12.3 [296]. Also, the positively charged amine (N6) has been proved to be fundamental for the formation of a salt bridge between a number of ligands and carboxylate moiety of an aspartic acid of the orthosteric site, which works as a docking point to charged amine of many ligands in 5-HT$_{1B}$R and other GPCRs [305].

Through the use of EE-MFCC scheme, here justified because it can give a deeper understanding of the binding features related to the DHE-5-HT$_{1B}$R complex by inspecting the binding energy of each amino acid individually, one can depict the interaction energy of 108 residue–DHE complexes, where the majority of them are nonpolar (aliphatic and aromatic) amino acids within a pocket radius 6.5 Å from the ligands. This characteristic is reflected on the individual interaction energies that are found between −5.00 and 5.00 kcal/mol.

Therefore, we select the strongest interactions with ranging values less than −6.00 kcal/mol (strongly attractive) and greater than 6.00 kcal/mol (strongly repulsive), plotting them in Figure 12.4. It shows 13 significant residues as the most important from the EE-MFCC calculations, namely (all units in kcal/mol) D129 (−93.92), V201 (−7.38), F330 (−8.44), F351 (−7.86), L126 (−7.96), I130 (−7.43), V200 (−6.69), D352 (−38.59), T355 (−6.42), E198 (−20.69), D204 (−17.32), R114 (17.48), and D123 (−21.44). Among them, only the residue R114 is repulsive, which reflects the amino acid distribution within the protein structure since it is the only one to have a positive charge from its guanidine group.

It is well known that hydrogen bonds (HB) and salt bridges (SB) are very important to the biological recognition mechanism of ligands. For aminergic GPCR (class A) members, it is shown that the most important feature in the ligand-binding modes is an SB between the protonated amine group of a ligand and an aspartic acid from TMH3 (orthosteric binding pocket), which for 5-HT$_{1B}$R is the residue D129. From Figure 12.5a, we can observe that it is the only residue making an electrostatic interaction with (i)(N6)H. This salt bridge is formed between its carboxylic acid group and the charged amine from DHE, almost face-to-face in a distance of 1.95 Å. In addition, one can see that there are a small number of HBs between DHE and 5-HT$_{1B}$, accounting only for the residues V201 and T134 at a distance of 2.02 Å and 2.20 Å ((iii)(C23)O24 and (i)(N1)H), respectively.

Recently, Wang and co-workers conducted a radioligand competition assays with [^{3}H]-LSD on mutated ligand-binding pocket 5-HT$_{1B}$R, which indicated the complete abolition of LSD and ERG bindings after substitution of the residue D129 with alanine [296]. Besides, it was also shown that GPCRs, with the amino acid Asp at the same position, interacting with ligands displaying common charged amino group, present the highest interaction energy after FMO calculations [306]. Similarly, Zanatta and co-workers used MFCC scheme to calculate the binding energy between dopamine D3 receptor (D$_3$R) and eticlopride [303], and between haloperidol [305] and risperidone [307], finding that the aspartic acid at the same position (showed as the residue D110) is the most energetic residue for all ligands. Thus, it was already expected that D129 is one of the most energetically relevant amino acids.

On the other hand, V201 is a residue that composes the extracellular loop 2 (ECL2), a region of the extended binding pocket. Throughout the evaluation of a number of GPCRs crystallographic structures, it is believed that ECL2 functions as a "lid" by closing the ligand in the binding site. Accordingly, it is expected that residues from ECL2 could help in modulating the ligand behavior.

In addition to hydrogen bond, the residue V201 is also involved in hydrophobic contacts with (iv)(C30)H (1.94 Å), which increase its binding energy. Since the residue T134 has a slightly lower energetic value than our reference (-5.84 kcal/mo), we decided not to show it in Figure 12.4. However, its importance should be mentioned here, once it plays a relevant position at the receptor and is conserved among the GPCR members, being also part of TMH3 that occupies a region of orthosteric binding pocket close to the nitrogen atom at indole ring, (i)(N1)H, acting as a HB acceptor.

Among the GPCRs with crystallographic structure, only human histamine H1 receptor (H$_1$R), human 5-HT$_{1B}$R, and human 5-HT$_{2B}$R present a threonine in the same position, although its mutation to alanine reduce the binding affinity of ERG, LSD and 5-HT to 5-HT$_{1B}$R, which also occurs to the residue V201 [296]. Our energetic results are in agreement with major electrostatic interaction contribution found in ERG-5-HT$_{1B}$R complex [306]. Hence, one should pay attention to these residues during the development of new drugs.

Additionally, H-bonds and salt bridges hydrophobic interactions (HIs) show also a high relevance for protein stability and maintenance of protein–ligand complexes. In the case of transmembrane proteins, such as GPCRs, HIs appear to be an important factor to preserve its structure within the membrane. Also, hydrophobic interactions help to stabilize some ligands inside the binding pocket r. Here, most of the studied residues form HIs with dihydroergotamine, as mentioned before. The residues F330 (TMH6) and I130 (TMH3) are involved in hydrophobic contacts, non-conventional H-bonds, and π interaction with region (i) of DHE

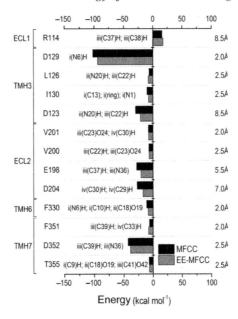

Figure 12.4 Graphic panel showing the most relevant residues that contribute to the DHE–5-HT$_{1B}$ complex. The minimal distances between each residue and the ligand, as well as the region and the atoms of the ligand interacting to each residue at the binding site, are also shown. We depict the results by using both the EE-MFCC and MFCC models, in gray and black, respectively. After Lima Neto et al. [295]

(Figure 12.5a). A cation–π interaction is made between the aromatic ring of F330 and DHE [i(N6)] in a distance of 4.50 Å, while a non-conventional H-bond is created with (ii)(C18)O19. Meanwhile, I130 shows a σ–π interaction with A ring of DHE (2.42 Å) and another non-conventional H-bond with (i)(N1) in a distance of 2.91 Å.

Mutation studies have shown that these amino acids are very important to ligands, recognition, since the residues F330A completely and I130A abolished the binding of radioligands [296]. Furthermore, the results from the FMO/MP2 calculation in ERG-5-HT$_{1B}$R show the same relevance to these two residues and confirm the presence of dispersion energy as the main binding energetic contributor. The residue F330, together with its neighbor F331 (TMH6; Figure 12.5b), with interaction energy equal to -4.15 kcal/mol, is likely involved in the activation process on 5-HT$_{2A}$R, especially as it pertains to the movement of helix 6, which has been related to the activation of β_2 receptor, although the mutation F331A does not cause a relevant impairment on ergolines binding, in agreement with the lower interaction energy found in this work.

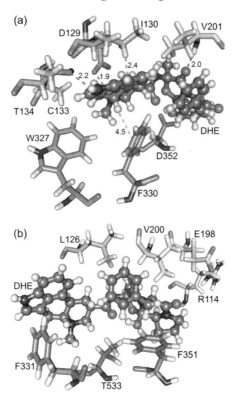

Figure 12.5 Detailed view of the most relevant amino acids involved in the binding of DHE with 5-HT$_{1B}$. Potential hydrogen bonds, cation–π and σ–π interactions are indicated by the dark gray, gray, and light gray dashed lines, respectively. After Lima Neto et al. [295]

Together with the residues F330, I130, and F331, the residues C133 (TMH3) and W327 (TMH6) form a narrow hydrophobic cleft, which occupies the orthosteric binding pocket and packs tightly against the nearly planar ergoline ring system (see Figure 12.5a and b). These two residues were shown to be highly important because mutation to alanine created a complete impairment that blocks the binding of ergolines and 5-HT [296].

The residue C133 is an amino acid that is highly conserved between serotonin receptors, being present in all of them except 5-HT$_2$R and 5-HT$_4$R [302]. It interacts with (i)(C2)H and (i)(C4) from a distance of 2.53 Å and 2.33 Å, respectively. Besides, mutation on serine occupying this position on 5-HT$_{2A}$R to alanine affected the affinity of ligands such as 5-HT and tryptamine. Also, the residue W327 occupies a region that could be a sodium binding site. It interacts with (i)(C4) and (i)(7)H from a distance of 2.77 Å and 2.00 Å, respectively. Despite that, we found a small binding energy for C133 (0.41 kcal/mol) and W327 (−3.20 kcal/mol), which could indicate, mainly for the former, a major structural role for direct protein–ligand complex formation in DHE-5-HT$_{1B}$R.

Most of the main energetic results found by EE-MFCC are related to charged amino acids, such as D129, D352 (Figure 12.5a), E198, R114 (Figure 12.5b), D204, and D123. Among them, only the residues D352 and D129 had their importance shown by experimental or theoretical studies [296]. Furthermore, all of them have their binding energies related to an effect of electrostatic attraction or repulsion induced by positively charged distribution of the ligands, as it was shown to D352 [306].

The residues R114, D123, E198, D204, and D352 occupy a position close to the extracellular portion of the protein, interacting primarily with regions (ii)(N20), (iii)(N36), and (iv)(C29)H at varying distances (see Figure 12.4). On the other side, the residues L126 (Figure 12.5a), V200, T355, and F351 (Figure 12.5b) are involved in hydrophobic contacts and weak H-bond with regions (ii) and (iii), showing their relevance depicted by crystallographic, mutation, and computational studies with 5-HT, LSD, and other antimigraine drugs, such as ERG and sumatriptan [294, 296, 302, 306]. The residue L126 is interacting with (ii)(N20)H and (iii)(C22)H at distances 2.88 Å and 2.21 Å, respectively.

Meanwhile, V200 (T351) makes contact with region (iii)(C22)H and (iii)(C23) O24 ((iii)(C39)H and (iv)(C33)H)) at distances 2.32 Å and 2.67 Å (1.85 Å and 2.20 Å), respectively. Furthermore, the residue T355 is in close contact with regions (i)(C9)H, (ii)(C18)O19, and (iii)(C41)O42, at distances 2.02 Å, 2.72 Å and 2.87 Å, respectively.

Overall, among the amino acids analyzed here, we found that the binding energies follow the decreasing residue sequence: D129 > D352 > D123 > E198 > D204 > F330 > L126 > F351 > I130 > V201 > V200 > T355 > R114. Except for R114, D123, E198, and D204, all of these amino acids are consistent with published analyses of crystallographic structures, computational, and mutation studies. Adding to this, some residues that are not among the most energetic ones, have their binding energies and interactions depicted due to the relevance showed in other studies. For completeness, we display in Figure 12.6 the electrostatic potential isosurface with projected electron densities for DHE bound to some of the most important residues at the binding pocket site.

12.5 Total Binding Energy

To evaluate the binding interactions through the fragment-based quantum mechanics method, it is important to take into account every significant attractive and repulsive amino acid residue that can influence this mechanism. Therefore, instead of taking an arbitrary region of the binding site, we performed a search for an optimal

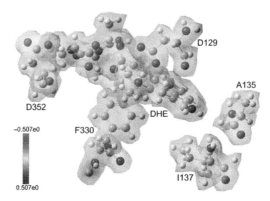

Figure 12.6 Electrostatic potential surface of DHE interacting with some 5-HT$_{1B}$R amino acids. After Lima Neto et al. [295]

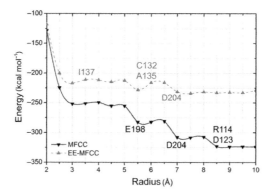

Figure 12.7 The total interaction energy as a function of the binding pocket radius r calculated using the GGA functional B97D in the MFCC and EE-MFCC schemes. Amino acid residues responsible for the regions of steepest negative and positive variation are highlighted. We represent the results from the EE-MFCC and MFCC models in gray and black, respectively. After Lima Neto et al. [295]

binding pocket radius (r) in which a variation less than 10% of the sequential pocket radius could be observed after a radius increase. For this task, the binding pocket radius r is varied from 2.0 Å (-117.91 kcal/mol) to 10 Å (-228.46 kcal/mol) in order to determine the best value of r, found to be 7.0 Å corresponding to an energy of -230.93 kcal/mol, from which the convergence was achieved.

According to Figure 12.7, there is a sharp increase in the total binding energy (all energy units in kcal/mol) between $r = 2.0$ Å (-200.30 for 2.5 Å) and 3.0 Å (-216.88), due to the attractive binding energies of the residues closer to the DHE, making H-bonds and hydrophobic interactions. Afterward, a slight decrease of the total binding energy is seen at the pocket radius $r = 3.5$ Å (-210.87), due to the repulsive energy of the residue I137. A small stability is seen among the radius

$r = 4.0$ Å (-211.36), 4.5 Å (-214.39), and 5.0 Å (-212.26), but it is broken by the attractive energy of residue E198, which increases the total energy at $r = 5.5$ Å (-227.90). Residues C132 and A135 are responsible for reducing the total binding energy at pocket radius $r = 6.0$ Å (-210.87). At $r = 7.0$ Å (-230.93), the total binding energy is decreased by residue D204, and small shifts are seen until the pocket radius r reaches 10.0 Å.

12.5.1 Energetic Description of Ligand Regions

The structure of some serotonergic ligands share similarities in composition and conformation, which becomes easy to understand when off-target effects are frequent [300, 302]. In this context, we depicted the energetic aspect and the number of contacts made by individual DHE regions in Figure 12.8.

We selected each region based on the structural relevance shown by the crystallographic structure of DHE [304]. Energetically, regions (iii) and (i) are the first- and second-most relevant – accounting for -116.12 and -103.75, respectively – followed by regions (ii) (-44.26) and (iv) (-36.56). This sequence changes when we compare the number of contacts, namely region (i) (59) > region (iii) (36) > region (iv) (16) > region (ii) (4). This kind of information could be useful for drug development since it can shed some light into the main molecular groups and regions of the ligands.

Ligand Region	Number of Contacts	Total Energy (kcal mol⁻¹)	
		MFCC	EE-MFCC
Region i 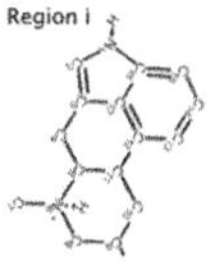	59	-160.23	-103.76
Region ii 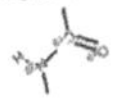	4	-54.01	-44.26
Region iii 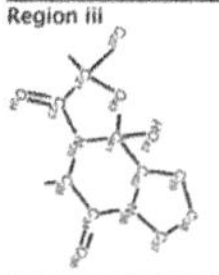	36	-143.61	-116.12
Region iv 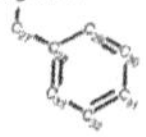	16	-47.88	-36.56

Figure 12.8 Binding energies and number of contacts of the individual regions of DHE. Energetic results are shown as related to the EE-MFCC and MFCC calculations. After Lima Neto et al. [295]

Table 12.1 *Interaction energies (MFCC and EE-MFCC) per serotonin receptor secondary structure. After Lima Neto et al. [295].*

Protein Segment	Total Energy (kcal mol⁻¹)	
	MFCC	EE-MFCC
N-terminus	1.64	4.25
TMH1	−0.27	−0.44
TMH2	−6.77	1.80
ECL1	16.11	19.93
TMH3	−155.54	−121.57
TMH4	−1.46	−2.42
ECL2	−74.05	−56.60
TMH5	−13.85	−7.94
TMH6	−23.90	−16.59
TMH7	−66.71	−48.88

12.5.2 Energetic Description of 5-HT$_{1B}$R Segments

Another feature analyzed here is the energetic characteristic of a single region of protein. GPCRs exhibit almost the same structure, with seven transmembrane helices linked by three internal and external loops, accounting for 13 structural patterns. It has been shown that each one of these patterns, or segments, can present a different movement after the ligand binding, been related to their effect in protein activation levels [302]. Therefore, we add the binding energies of the amino acids, separating them by segment, as shown in Table 12.1.

The highest attractive binding energy is shown by TMH3 (−121.57), followed by ECL2 (−56.60), TMH7 (−48.88), TMH6 (−16.59), TMH5 (−7.94), TMH4 (−2.42), and TMH1 (−0.44). Also, the most repulsive binding energies are observed in ECL1 (19.93), N-terminal portion (4.25), and TMH2 (1.80).

A comparison of many crystallographic structures shows that TMH1 and TMH2 did not undergo major structural change after ligand binding, whereas transmembrane helices THM3, TMH6, and TMH7 suffer the main changes [308], in agreement with our theoretical findings.

12.5.3 Effects of the Electrostatic Embedding Scheme

The MFCC method has been widely used to calculate binding energies of biomolecular complexes when a quantum investigation of a large number of amino acid residues in a protein is possible with small computational cost and no loss in accuracy is possible.

Despite the good correlation between the MFCC results and the experimental ones, it is known that the environmental contribution of solvent, ions, and other amino acids is very important in molecular calculations. Accordingly, many approaches have been implemented in the MFCC scheme, aiming to take into account the electrostatic effects of the whole system within each step of fragmentation, with the common use of many caps and continuum solvation models [216]. Besides, the use of point charges is largely applied to include the surrounding energies in fragmentation methods and quantum mechanical calculations in hybrid methods. Furthermore, since the early application of MFCC, point charges obtained through classical or quantum methods have been used to include a polarization function on fragmentation steps for protein systems [309].

In view of that, we included the electrostatic embedding (EE-MFCC) scheme into our MFCC analysis on DHE–5-HT$_{1B}$R complex to obtain the individual contribution of the amino acid residues inside a binding pocket with a selected radius, ranking the most relevant interactions in the complex. The employment of this approach makes the system more realistic through polarization of the electron density [310].

In order to give us the idea about the differences between the MFCC and the EE-MFCC approaches, we compare our results using both methods by considering the residues with a highest energetic variation in Figure 12.9, excepting those shown in Figure 12.4. The highest shift is found for the residue D204 (−9.82; all units in kcal/mol), followed by the residues Y359 (8.51), D123 (8.19), and D129 (8.09) – all of them negatively charged amino acid, except the residue Y359.

It is also important to notice the high changes in the binding energies accounted for the residues I137 (0.74 to 5.88) and A135 (0.16 to 5.10). The amine from the main chain of these last two residues is close to the amine N1 of dihydroergotamine and, as depicted by Figure 12.6, they repel each other. For the residue Y359, it is shown that its side chain also repels (i)(C7)H$_3$.

It is noteworthy that the presence of the electrostatic embedding scheme brought, for the majority of the binding energies, a positive feature when it is compared to conventional MFCC, mainly to charged residues where it adds from 9.82 (D204) to 1.75 kcal/mol (R114), in agreement with Reference [311]. It also intensifies the attractivity of some amino acids, such as F351 (−6.23 to −7.86 kcal/mol) and L126 (−6.39 to −7.96 kcal/mol). Despite that, in most cases, the order of importance, as far as the binding energies are concerned, is kept by the two models, as one can see in Figure 12.4, 12.7, 12.8, and 12.9 and Table 12.1.

In a recent study, it was shown that polarization created by partial charges of common force fields, such as CHARMm and AMBER, are not suitable enough to give a good electrostatic description. However, the energetic features depicted here have shown that the use of point charges obtained through CHARMm force field is a reasonable approach to improve the characterization of ligand-binding complexes. It is in agreement with the results found by Wang and co-workers to protein total energy, calculated using a generalized molecular fractionation with conjugate caps embedded in point charges obtained through Amber94 [296].

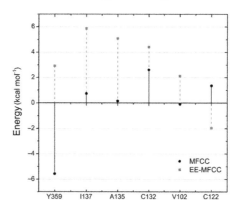

Figure 12.9 The highest energetic shift between EE-MFCC and MFCC results per residue. Dash lines in gray represent the EE-MFCC, and solid black lines represent the MFCC results. After Lima Neto et al. [295]

Overall, our results demonstrate that the electrostatic field using point charges is an efficient way to treat the polarization effect in fragment calculations, with the residues–ligand binding energy becoming more reliable to residues close to the ligand, although most of them do not suffer a high variation, possibly resulting from the small number of polar amino acids and the Coulomb effect.

12.6 Conclusions

Serotonin receptors (5-HTR) are attractive and intensively explored due to their biological and drug target importance. Among them, the $5\text{-HT}_{1B}R$ receptor and its agonist dihydroergotamine (DHE) are heavily used not only for the treatment of migraine attacks but also for many neurological dysfunctions, making it an important subject in the pharmaceutical research and drug market. Therefore, the understanding of the receptor–ligand complex interaction energies, also emphasizing the high affinity of some amino acid residues, may be vital to a successful structure-based drug design (SBDD) against these targets.

To clarify this picture, we have investigated the binding energies profiles of this complex, using fragment-based calculations within a quantum biochemistry framework. We not only performed a common MFCC calculation, but we also considered another MFCC-based scheme to realize our simulations. In this improved scheme, each DHE–residue complex is electrostatically embedded into point charges of the full protein, the so-called electrostatic embedding MFCC (EE-MFCC) method, aiming for the improvement of our description of the protein–ligand binding energy.

In general, our energetic description of the DHE–5-HT$_{1B}$R crystallographic structure was consistent with previous results from crystallographic, computational, and mutation studies. The amino acids used here were selected based on convergence criteria that shows a stabilization after 7.0 Å from dihydroergotamine. We considered 108 DHE–residue complexes, showing that the most important residues affecting the binding mechanism were, in decrescent order D129, D352, D123, E198, D204, F330, L126, F351, I130, V201, V200, T355, and R114. We also depict the energetic description of other experimentally relevant amino acids, as well the regions of ligand and protein segments. We believe that our results are an important step toward in silico quantum biochemical design and the probing of new medications to treat migraine and other diseases related to the 5-HT$_{1B}$ receptor.

13

Antiparkinson Drugs

13.1 Introduction

Parkinson's disease (PD) is a progressive, long-term, and chronic neurodegenerative disorder characterized by the death of nigrostriatal dopaminergic neurons
that worsens over time; the reason for that is still poorly understood [312]. The
symptoms of PD develop gradually, often starting with a slight tremor in one hand
and a feeling of stiffness in the body. The disease is named after the English surgeon
James Parkinson, who published the first detailed description of the disease in 1817.
The disease typically occurs in people older than age 60, of which about 1 percent
are affected. Males are more often affected than females at a ratio of around 3:2.

The cause of Parkinson's disease is generally unknown, but believed to involve
both genetic and environmental factors. It is normally caused by insufficient formation and action of the neurotransmitter dopamine, $C_8H_{11}NO_2$, but secondary causes
may result from toxicity, head trauma, and medical disorders. Clumps of specific
substances within the brain cells, called Lewy bodies, may be an important clue to
the cause of Parkinson's disease.

There is no cure for Parkinson's disease, with treatment directed at improving
symptoms. People who have Parkinson's disease may have the same average life
expectancy as people without it. But when the illness is in its advanced stages,
its symptoms can lead to life-threatening complications, such as problems with
swallowing that can cause Parkinson's patients to aspirate food into the lungs,
leading to pneumonia or other pulmonary conditions, or falling and breaking a
bone or hitting the head.

The five primary symptoms of PD are as follows:

(a) Tremor, or trembling in hands, arms, legs, jaw, and face, the most common
presenting sign, which disappears during voluntary movement of the affected
arm and in the deeper stages of sleep.

(b) Rigidity, or stiffness and resistance to limb movement caused by increased muscle tone, an excessive and continuous contraction of muscles.

(c) Bradykinesia, or slowness of movement, the most handicapping symptom of Parkinson's disease, leading to difficulties with everyday tasks such as dressing and bathing.

(d) Postural instability, or impaired balance and coordination, typical in the later stages of the disease, leading to frequent falls and, secondarily, to bone fractures, loss of confidence, and reduced mobility.

(e) Chewing and eating problems, mainly at the late-stage of Parkinson's disease, affecting the muscles of the mouth, making chewing difficult, leading to choking and poor nutrition.

All these symptoms decrease the quality of life in its sufferers and emerge when a significant proportion of the substantia nigra dopamine neurons have been lost and striatal dopamine has been reduced by 60% to 80% [313].

Since dopamine cannot cross the blood–brain barrier (BBB), the most effective treatment for PD has remained the administration of the oral dopamine precursor levodopa (L-3,4-dihydroxyphenylalanine, $C_9H_{11}NO_4$) since its introduction in the late 1960s. Levodopa is capable of BBB crossing and is converted into dopamine by the enzyme DOPA decarboxylase, mainly to patients at the initial stages of the disease.

Levodopa (LDOPA for short) has a short half-life (about 60 min) due to its rapid and extensive decarboxylation to dopamine and methylation to 3-O-methyldopa by dopa decarboxylase and catechol-O-methyltransferase. Only about 1% of an orally administered levodopa dose enters the brain because of extensive first-pass metabolism and rapid plasma clearance.

Unfortunately, after several years of levodopa treatment, its efficacy is diminished as a consequence of clinical response modifications resulting from a complex interaction between the long-term effects of the drug and the disease itself [314]. Several new levodopa formulations that may provide more continuous dopamine stimulation are being investigated. These include oral long-acting, once-daily pills, transdermal formulations, and continuous infusion.

An innovative delivery strategy for Parkinson's treatment is a skin patch, or a transdermal therapeutic system, which offers considerable advantages over parenteral or oral administration of antiparkinson therapy. The patch could enhance plasma concentration, reduce gastrointestinal variations, and avoid first-pass metabolism, simplifying the daily dosing schedule and ensuring a short plasma elimination half-life of the drug after patch removal.

Furthermore, there are indications that patient compliance may be increased with a transdermal therapeutic system treatment. A large number of nanosystems (nanoparticles, carbon nanotubes, fullerenes, etc.) can be potentially used as carriers for transdermal delivery because of their versatility properties, including good biocompatibility, selective targeted delivery, and controlled release of carried drugs. The ability of fullerenes to penetrate through the skin is widening their application in cellular drug and gene delivery [315].

In the current PD standard treatment, for better results, levodopa is administered in combination with DOPA decarboxylase, like carbidopa, chemically known as (2S)-3-(3,4-di-hydroxyphenyl)-2-hydrazinyl-2-methylpropanoic acid, whose chemical formula is $C_{10}H_{16}N_2O_5$ [316].

Carbidopa inhibitors prevent the breakdown of levodopa in the bloodstream and also delay the conversion of levodopa into dopamine until it reaches the brain, reducing adverse side effects. Furthermore, it can decrease peripheral aromatic L-amino acid decarboxylase (AADC) before crossing the BBB, metabolized by an enzymatic reaction of levodopa. As a result, a greater amount of the levodopa dose is transported to the brain, compensating for the deficiency of dopamine and avoiding the exposure of the brain dopamine receptors to an alternating low and high concentration. Unfortunately, the peripheral levodopa conversion to dopamine by the enzyme DOPA decarboxylase is responsible for the typical gastrointestinal (nausea, emesis) and cardiovascular (arrhythmia, hypotension) side effects.

In this chapter, a quantum chemistry study is carried out in search of the conformational, optoelectronic, and vibrational properties of the levodopa and carbidopa molecules, efficient drugs used in conjunction to the Parkinson's disease treatment [317, 318]. Although they occupy a prominent place in current treatment protocols as antiparkinson drugs, few works were done so far to characterize their physical properties.

13.2 Levodopa Molecule

After more than 50 years, the administration of levodopa (LDOPA) remains the most effective treatment for Parkinson's disease, despite the manifestation of important side effects. The development of carrier systems to increase its rate crossing the blood-brain barrier, to achieve stable therapeutic plasma levels, and to minimize side effects has been a challenge.

Innovative nanosystems to deliver LDOPA are being tested for improved Parkinson's disease therapy. In particular, buckminsterfullerene C_{60} is promising due to its ability to penetrate through the skin and the gastrointestinal tract, as well as its biomedical applications to enhance drug delivery (see Chapter 9) [315].

Pursuing the reduction of LDOPA side effects through C_{60} fullerene-based transdermal and gastrointestinal delivery, we intend here to design an optimized

LDOPA formulation for transdermal and oral administration, not only looking for a minimization of its side effects, but also enhancing the central nervous system bioavailability in order to achieve stable therapeutic plasma levels.

When focusing on transdermal and oral administration, it is necessary to take into account that the $LDOPA@C_{60}$ fullerene adsorbate will be put in an environment with pH varying in the 2.0–8.0 range, which means that the levopoda molecule has its carboxyl and amine groups charged, $COO-$ and NH_3+, respectively.

The first step here is to use molecular dynamics simulation to find a preliminary optimal geometrical configurations for the protonated LDOPA adsorption on C_{60}. Afterward, a second geometry optimization is performed within the scope of the density functional theory (DFT) using the local density approximation (LDA) to obtain the interaction energy of the $LDOPA@C_{60}$ adsorbate as a function of the distance between the centroids of protonated LDOPA and the C_{60} buckminster-fullerene. Charge population analysis is employed to study the electron transfer between the protonated LDOPA and C_{60} molecules, assessing the noncovalent character of their interaction.

Finally, the infrared and Raman spectra of the four smallest energy proto-nated $LDOPA@C_{60}$ configurations were calculated to check for the existence of vibrational signatures of the noncovalent interactions, which can be valuable for comparison with experimental data.

13.2.1 Computational Details

In order to achieve transdermal or oral delivery of LDOPA, it is important to understand the molecular state of levodopa in the skin, gastrointestinal tract, and blood. Measurements of pH changes in the gastrointestinal tract were carried out, showing that gastric pH was highly acidic (range 1.0–2.5). The average pH in the proximal small intestine was 6.6, while the mean pH in the terminal ileum was 7.5, progressively lowering from the right to the left colon and reaching a final average value of 7.0.

On the other hand, normal skin surface pH stays between 4 and 6.5 in healthy people, though it varies at different places. Finally, blood pH is regulated to stay within the narrow range from 7.35 to 7.45, making blood slightly alkaline. Thus, to describe LDOPA delivery through the skin and the gastrointestinal tract, it is neces-sary to consider a pH variation in the 2–8 range. The levodopa protonated states in the 0–14 pH range at 300 K were predicted in this chapter using the MarvinSketch 5.4 software, and are depicted in Figure 13.1. One can observe that near to pH $= 2.0$, the protonated NH_3+ and $COO-$ levodopa populations (see B and C states in Figure 13.1) are about 30% and 70%, respectively. The population

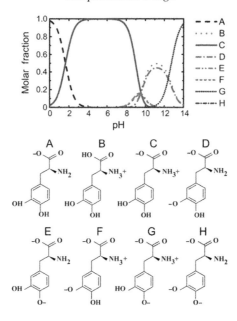

Figure 13.1 (a) Molar fraction curves as a function of pH of 8 (A, B, C, D, E, F, G, H) levodopa protonated states; (b) chemical structure of each molecular states. After Frazao et al. [317]

of the protonated levodopa state C increases remarkably when the pH is changed from 2 to 4, arriving at 100% in the 4–8 pH range and decreasing sharply for pH > 9.

Focusing the NH_3+, COO–C protonated LDOPA (LDOPAc from now on) adsorption on C_{60}, we started by performing a geometry optimization using the Gaussian 09 code, employing the hybrid B3LYP exchange-correlation functional together with a 6-31+G(d,p) basis set to expand the electronic states for isolated molecules of LDOPAc and C_{60}. Convergence thresholds for geometry optimization were: maximum force smaller than 7.7×10^{-4} eV/Å, self-consistent field energy variation smaller than 2.7×10^{-5} Å, and maximum atomic displacement smaller than 3.2×10^{-5} Å.

The LDOPAc optimized geometry is depicted in Figure 13.2a, which also shows the atomic labels adopted from now on. N_5 and O_2, O_4 are the positively and negatively charged atoms, respectively, for the 2–8 pH range considered. Figure 13.2b shows the projection of the calculated electron density of LDOPAc onto an electrostatic potential isosurface. From it, one can see that the most negatively charged atoms are O_2 and O_4, followed by O_1 and O_3; the most positively charged atoms are the nitrogen N_5 and the hydrogen atoms H_{18}, H_{19}, and H_{16}; the ring and remaining carbon atoms are slightly positively charged. By the way, the C_{60} cage has a

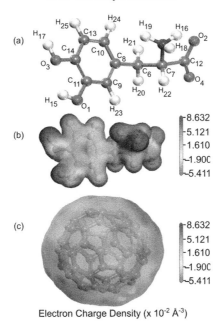

(a)

(b)

-8.632
-5.121
-1.610
-1.900
-5.411

(c)

-8.632
-5.121
-1.610
-1.900
-5.411

Electron Charge Density (x 10^{-2} Å^{-3})

Figure 13.2 The NH_3+, $COO-$ protonated state LDOPAc. (a) its most stable isolated converged structure; (b) the electron density of its most stable isolated converged structure projected onto an electrostatic potential isosurface; (c) the electron density of the C_{60} most stable isolated converged structure projected onto an electrostatic potential isosurface. After Frazao et al. [317]

negative charge being distributed mainly along the C=C bonds, as can be observed in Figure 13.2c. The mean values for the single and double bond lengths between carbons for the converged C_{60} cage are 1.42157 and 1.42152 Å, respectively, which are in good agreement with experimental values, since the weighted average of the two measured bond lengths and the difference between them are 1.439(2) Å and 0.057(6) Å, respectively [319]. The converged C_{60} and LDOPAc structures were used for the calculation of their infrared and Raman vibrational properties, which are discussed later in this chapter.

Bond lengths (in Å), angles, and torsion angles (in degrees) between bonds of the isolated I-LDOPAc and of the four LDOPAc@C_{60} spatial configurations (A, B, C, D) are shown in Tables 13.1, 13.2, and 13.3. While differences in bond length are in general small for most bonds when comparing the adsorbed species of LDOPAc and the LDOPAc in vacuum, statistical analysis shows that the adsorption process tends to increase the bond lengths of A, B, and D LDOPAc@C_{60} by about 0.06%, 0.16%, and 0.08%, respectively, while the C geometry exhibits bond lengths smaller by about 0.07% in comparison with the isolated molecule. As one can see in Table 13.1, the most pronounced decrease in bond length occurs for the

Table 13.1 *Bond lengths of the isolated I-LDOPAc, and the LDOPAc@C_{60} spatial configurations A, B, C, and D. After Frazao et al. [317].*

BL	I	A	B	C	D
O_1–C_{11}	1.343	1.332	1.363	1.340	1.349
O_1–H_{15}	0.982	0.990	0.995	0.981	0.983
O_2–C_{12}	1.273	1.259	1.268	1.233	1.270
O_3–C_{14}	1.361	1.361	1.339	1.354	1.364
O_3–H_{17}	0.974	0.998	0.990	0.975	0.975
O_4–C_{12}	1.226	1.234	1.235	1.260	1.235
N_5–C_7	1.484	1.473	1.483	1.476	1.483
N_5–H_{16}	1.048	1.048	1.049	1.049	1.046
N_5–H_{18}	1.048	1.048	1.048	1.048	1.048
N_5–C_{19}	1.046	1.046	1.047	1.046	1.048
C_6–C_7	1.515	1.516	1.516	1.502	1.513
C_6–C_8	1.488	1.486	1.486	1.483	1.493
C_6–H_{20}	1.101	1.101	1.104	1.105	1.104
C_6–H_{21}	1.105	1.105	1.106	1.117	1.104
C_7–C_{12}	1.558	1.557	1.552	1.557	1.556
C_7–H_{22}	1.099	1.101	1.099	1.105	1.103
C_8–C_9	1.393	1.391	1.394	1.386	1.397
C_8–C_{10}	1.393	1.393	1.398	1.399	1.393
C_9–C_{11}	1.380	1.384	1.378	1.380	1.386
C_9–H_{23}	1.095	1.094	1.096	1.092	1.096
C_{10}–C_{13}	1.394	1.393	1.389	1.381	1.388
C_{10}–H_{24}	1.096	1.096	1.098	1.095	1.096
C_{11}–C_{14}	1.401	1.403	1.403	1.402	1.394
C_{13}–C_{14}	1.376	1.380	1.385	1.381	1.380
C_{13}–H_{25}	1.095	1.093	1.093	1.095	1.096

O_2–C_{12} bond ($\approx 3.1\%$ for the C adsorbate), while the largest bond length increase is observed for the O_3–H_{17} bond length, about 2.5% for the A adsorption geometry.

On the other hand, statistical analysis of Table 13.2, which shows the angles between bonds in isolated I-LDOPAc and the four LDOPAc@C_{60} configurations, reveals that the A and B adsorption geometries do not change significantly in comparison to the isolated I-LDOPAc molecule, with bond–bond angle smaller by $0.03°$ in average for the A configuration and larger by $0.01°$ for the B case (standard deviations of $0.85°$ and $1.9°$, respectively). For the C and D geometries, in contrast, the average bond–bond angle is $0.25°$ and $0.26°$ larger, in comparison to the isolated I-LDOPAc structure, with standard deviations of $1.6°$ and $1.1°$, in respective order. The C adsorbate has the largest variation, an increase of about $4.5°$, for the O_2–C_{12}–C_7 angle, being followed closely by the O_3–C_{14}-C_{11} angle for the B adsorption case. The most pronounced bond–bond angle decrease occurs for the B adsorbate involving the O_1–C_{11}–C_{14} chain $\approx 4.2°$.

Finally, Table 13.3 reveals the torsion angles involving groups of three adjacent bonds. The A and B adsorption geometries display a small average decrease of the torsion angle, of about $2.1°$ for A and $1.0°$ for the B case. The C adsorbate, in comparison, is much more distorted, with large torsion angle variations for practically all torsion angles, except for C_6–C_8–C_{10}–C_{13} and C_8–C_9–C_{11}–O_1, which involve three atoms of the carbon ring. For the D adsorbate, the torsion

Table 13.2 *Bond angles of the isolated l-LDOPAc, and the LDOPAc@C_{60} spatial configurations A, B, C, and D. After Frazao et al. [317].*

Angles	I	A	B	C	D
O_1–C_{11}–C_9	121.179	121.814	124.168	121.194	121.238
O_1–C_{11}–C_{14}	119.074	118.243	114.904	119.241	119.402
O_2–C_{12}–O_4	132.174	131.521	131.511	131.777	131.201
O_2–C_{12}–C_7	111.025	112.746	112.212	115.560	111.992
O_3–C_{14}–C_{11}	113.896	114.828	118.376	114.176	114.204
O_3–C_{14}–C_{13}	125.693	124.876	122.048	125.255	125.375
O_4–C_{12}–C_7	116.746	115.637	116.226	112.648	116.800
N_5–C_7–C_6	110.063	110.861	110.256	112.503	113.196
N_5–C_7–C_{12}	103.423	103.418	103.053	103.599	101.927
C_6–C_7–C_{12}	109.931	108.201	109.710	111.257	112.784
C_6–C_8–C_9	120.736	120.034	120.941	120.564	120.206
C_6–C_8–C_{10}	120.059	120.352	120.477	119.855	120.580
C_7–C_6–C_8	111.831	112.833	112.173	112.918	111.907
C_8–C_9–C_{11}	120.552	120.045	120.116	120.338	120.699
C_8–C_{10}–C_{13}	120.782	120.652	121.711	120.648	120.205
C_9–C_8–C_{10}	118.996	119.576	118.489	119.495	119.158
C_9–C_{11}–C_{14}	119.781	119.940	120.918	119.565	119.353
C_{10}–C_{13}–C_{14}	119.447	119.463	119.157	119.383	120.145
C_{11}–C_{14}–C_{13}	120.436	120.280	119.519	120.569	120.421

Table 13.3 *Torsion angles of the isolated l-LDOPAc, and the LDOPAc@C_{60} spatial configurations A, B, C, and D. After Frazao et al. [317].*

TANG	I	A	B	C	D
H_{15}–O_1–C_{11}–C_9	−54.043	−59.876	−54.518	−172.173	45.608
H_{16}–O_3–C_{14}–C_{11}	−76.888	−85.448	−84.283	−126.090	−101.720
C_6–C_7–N_5–H_{19}	−8.611	−13.444	−9.862	177.950	10.234
N_5–C_7–C_6–C_8	175.750	179.807	178.389	−177.099	178.786
C_7–C_6–C_8–C_9	−175.528	−178.107	−175.760	177.164	−178.717
N_5–C_7–C_{12}–O_2	179.229	178.149	175.304	−179.812	178.516
C_6–C_8–C_9–C_{11}	0.150	−0.965	−1.775	−0.259	0.393
C_6–C_8–C_{10}–C_{13}	0.159	−0.876	0.863	−0.032	0.209
C_8–C_9–C_{11}–O_1	−179.771	−177.833	−177.044	−179.921	−179.163

angle variation is larger for the H_{15}–O_1–C_{11}–C_9 (100°), C_6–C_7–N_5–H_{19} (19°), and H_{17}–O_3–C_{14}–C_{11} (−25°).

To study the LDOPAc adsorption on C_{60}, a set of about 100 initial configurations was generated from the previously converged structures for the isolated molecules. These configurations were obtained by randomly varying the LDOPAc orientation and its distance to the C_{60} and performing, afterward, a classical annealing simulation using the Forcite Plus code. The Universal Force Field (UFF) was selected to perform about a hundred of annealing cycles for each initial geometry. We have adopted the following control parameters for each annealing cycle: mid-cycle temperature of 200 K, 50 heating ramps per cycle, and 100 dynamic steps per ramp. After each cycle, the smallest energy configuration was optimized classically.

The NVE ensemble and the Nose thermostat were chosen to carry out the molecular dynamics with a time step of 1 femtosecond (fs).

The classically optimized geometries obtained from the annealing process were then optimized at the DFT level using the LDA as implemented in the DMOL3 code. It is true that in systems with strong charge-transfer effects, the introduction of contributions from electron density derivatives in the exchange-correlation functional improves the accuracy, but this is not the case of the system we investigate here.

It is a well-known fact that pure DFT methods are unable to provide a good description of systems where noncovalent bonding, such as van der Waals forces, is involved [320]. Besides, hydrogen bonds are not well characterized by the LDA approximation. Nevertheless, the LDA is adequate to obtain intermolecular distances.

After these results, and due to the relatively cheap computational cost of LDA simulations, we have chosen this functional to obtain the interaction energies and geometries of LDOPAc adsorbed on C_{60}, following the LDA parameterization of Perdew and Wang [10]. To study the interaction between LDOPAc and C_{60}, we let the geometry of the isolated l-LDOPAc molecule relax to a total energy global minimum. Only the atoms of the l-LDOPAc molecule were allowed to move during the calculations, the atomic positions of C_{60} being held fixed. Also, in our DFT-LDA simulations, effective DFT semi-core pseudopotentials (DSPP) were used to describe the core electrons.

We define the adsorption energy E_{ads} of LDOPAc on C_{60} as

$$E_{ads} = E_{L@C_{60}} - (E_L + E_{C_{60}}), \qquad (13.1)$$

where $E_{L@C_{60}}$ is the total energy of the LDOPAc@C_{60} complex, E_L is the total energy of the isolated l-LDOPAc molecule, and $E_{C_{60}}$ is the calculated total energy of the isolated buckminsterfullerene C_{60}.

The adsorption energy as a function of the distance d between the C_{60} and l-LDOPAc centroids was evaluated for each optimized LDOPAc@C_{60} configuration by moving rigidly the l-LDOPAc molecule along the axis formed by joining the l-LDOPAc and C_{60} centroids. Both classical UFF (Forcite Plus Universal Force Field) and DFT-LDA (DMOL3) calculated geometries were used to obtain $E_{ads}(d)$ for a variation of d in the 6.0–11 Å range. However, for reasons of computational economy, the LDA adsorption energies of LDOPAc@C_{60} were calculated only when necessary.

A classical molecular dynamics simulation of a single buckminsterfullerene C_{60} molecule interacting with 120 l-LDOPAc molecules inside a cubic box of side $L = 44.386$ Å was also carried out using the GROMACS 4 package [321] at room temperature (300 K) and using the NVT ensemble. The OPLS-AA (all-atom) force field was used to calculate the total energy of this system [322]. The initial geometry for this simulation was obtained after an equilibration time of 1,000 picoseconds (ps) using a time step of 0.0005 ps.

The total energy of the system was minimized using a combination of steepest descent and conjugate gradient techniques. The production time was 1.0 nanosecond (ns) and the nonbonded pair list was updated every five steps with both temperature and pressure controlled using the V-rescale thermostat method [323]. Long-range interactions were dealt with using the particle mesh Ewald (PME) with a cutoff distance of 1.2 nm. The summation technique was implemented to describe electrostatic interactions more accurately. A cutoff distance of 1.2 nm was used to take into account the van der Waals and nonbonded interactions.

13.2.2 Geometry Optimization

The variation of the adsorption energy with the distance between the centroids of I-LDOPAc and the buckminsterfullerene C_{60} after the annealing calculations for about a hundred random initial geometries is depicted in Figure 13.3a. One can identify four clusters of adsorption configurations: the first one for d in the 7.0–7.7 Å range; the second, for d in the 7.7–8.3 Å range; the third, for d in the 8.3–9.0 Å range; and the fourth cluster for d in the 9.0–10.5 Å range. These clusters can be associated with four different sets of similar conformations of the LDOPAc adsorption on the buckminsterfullerene C_{60}, corresponding to four possible adsorption levels of this system. I-LDOPAc molecules can take advantage of the C_{60} spherical symmetry originating different adsorption "orbits" for LDOPAc@C_{60}.

A more detailed description of the adsorption energy as a function of d for some annealed LDOPAc@C_{60} geometries is shown in Figure 13.3b, where the LDOPAc configurations with smallest adsorption energy for each cluster were selected and displaced rigidly along the centroid–centroid line. The adsorption energy profiles obtained in this way look like classical van der Waals potential energy curves and are represented by symbols and lines, the latter being associated with the smallest calculated adsorption energies at each adsorption cluster, whose minima can be interpreted as revealing the distance between the centroids corresponding to the respective "adsorption orbit" radius. The classically calculated energies (classical radii of the orbits) of the four LDOPAc@C_{60} adsorption orbits were −0.44, −0.37, −0.29, and −0.22 eV (7.3, 7.8, 8.7, and 9.8 Å), respectively.

An improved description of the four LDOPAc@C_{60} adsorption configurations is obtained using the quantum DFT-LDA description. Using DFT-LDA, we observe that the adsorption energies increase in comparison with the classical values, and the adsorption energy minimum is shifted to smaller centroid–centroid distances – see Figure 13.3c. As a matter of fact, the DFT-LDA calculated

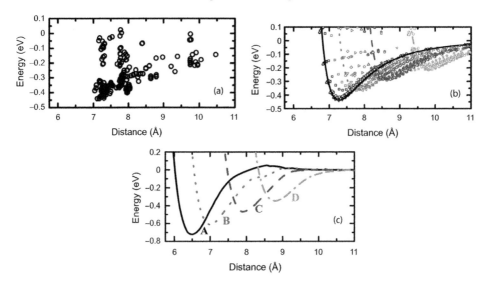

Figure 13.3 (a) Classical adsorption energies as a function of the distance between the I-LDOPAc and the buckminsterfullerene C_{60} centroids obtained after the annealing of a hundred initial LDOPAc@C_{60} spatial configurations; (b) van der Waals–like adsorption potentials obtained classically after the annealing of some of the initial LDOPAc@C_{60} spatial configurations, i.e., those of smaller adsorption energies at each level; (c) DFT-LDA calculated van der Waals–like adsorption potentials for configurations A–D of the LDOPAc adsorbed on C_{60}. They were calculated using as inputs the configurations with the smallest energy at each level obtained classically after the annealing of some of the initial LDOPAc@C_{60} spatial configurations. After Frazao et al. [317]

energies (centroid–centroid distances) of the four distinct optimized LDOPAc@C_{60} adsorbates are -0.72, -0.60, -0.46, and -0.34 eV (6.6, 7.0, 7.9, and 8.8 Å), in this order. To simplify the description of the results, we label these four LDOPAc@C_{60} adsorption geometries as A, B, C, and D, in decreasing order of binding energy (defined here as the negative of the adsorption energy), as it is shown in Figure 13.3c.

We can compare the results for LDOPAc with similar data for the absorption of ascorbic acid (AsA) and ibuprofen on the buckminsterfullerene C_{60}, presented in Chapter 9, the latter also exhibiting distinct adsorption configurations with adsorption energies in the range -0.65 to -0.29 eV and centroid–centroid distance in the 6.3–9.0 Å range [222]. For AsA@C_{60}, the optimal adsorption geometry corresponds to an adsorption energy of -0.54 eV and a centroid–centroid distance of about 6.7 Å [221]. For a temperature of 300 K, $k_B T \approx 26$ meV, suggesting that even at room temperature, the lowest-energy adsorption configurations of AsA@C_{60}, IBU@C_{60} and LDOPAc@C_{60}, would be stable.

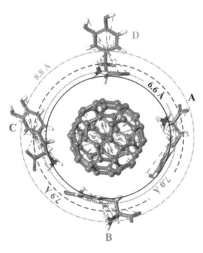

Figure 13.4 A, B, C, and D configurations of the four levels adsorption of LDOPAc on the buckminsterfullerene C_{60}. The DFT-LDA calculated "orbits" are represented by circles of radii 6.6 (solid), 7.0 (dotted), 7.9 (dashed), and 8.8 Å(dotted-dashed). After Frazao et al. [317]

The DFT-LDA calculated configurations A–D of the LDOPAc adsorbed on C_{60} are displayed in Figure 13.4. For the A and B configurations, which are the most stable, one can note that the carbon rings are practically parallel to the buckminsterfullerene C_{60} surface, showing the important role played by the π-stacking interaction in the adsorption mechanism. The most striking difference between the A and B configurations is the positioning of the NH_3+ group, which faces the C_{60} surface in A but not in B. It seems that the charged NH_3+ group interacts electrostatically with the C_{60} surface, improving the binding strength of A LDOPAc@C_{60} in comparison with the B configuration. DFT-LDA calculated adsorption energies were -0.72 and -0.60 eV for A and B, respectively.

In the case of the adsorption geometries C and D, with weaker binding, the carbon rings are not parallel to the C_{60} surface. For the D geometry, one can see that the carbon ring is far away from the fullerene, with the NH_3+, $COO-$ groups closer to the C_{60}, while the C adsorption configuration shows both the NH_3+ and the carbon ring closer to the C_{60} and the $COO-$ group moved away. As a matter of fact, the larger distance of the $COO-$ group from the fullerene surface together with the proximity of the carbon ring explains why the binding of the C configuration of LDOPAc@C_{60} is stronger than for the D configuration, with DFT-LDA calculated adsorption energies of -0.46 and -0.34 eV for C and D, respectively.

Information about the effect of the interaction with the buckminsterfullerene C_{60} on the l-LDOPAc molecular structure is obtained by comparing the bond lengths,

Table 13.4 *Population analysis for I-LDOPAc,
and the LDOPAc@C_{60} spatial configurations
A, B, C, and D. After Frazao et al. [317].*

Group	I	A	B	C	D
HPA COO⁻	−0.521	−0.444	−0.473	−0.429	−0.468
MPA COO⁻	−0.572	−0.493	−0.520	−0.476	−0.521
ESP COO⁻	−0.585	−0.491	−0.577	−0.526	−0.678
HPA NH₃⁺	0.353	0.285	0.372	0.318	0.334
MPA NH₃⁺	0.478	0.486	0.496	0.489	0.492
ESP NH₃⁺	0.419	0.278	0.414	0.288	0.365
HPA PP	−0.051	−0.039	0.004	−0.004	−0.067
MPA PP	−0.068	0.024	0.009	0.004	0.008
ESP PP	−0.086	−0.274	0.141	−0.102	−0.283
HPA HD1	−0.017	−0.027	−0.025	−0.010	−0.022
MPA HD1	−0.082	−0.112	−0.185	−0.082	−0.089
ESP HD1	−0.082	−0.112	−0.185	−0.082	−0.089
HPA HD2	0.026	−0.039	−0.024	0.030	0.024
MPA HD2	−0.187	−0.194	−0.116	−0.181	−0.187
ESP HD2	−0.133	−0.234	−0.114	−0.121	−0.129
HPA PH	0.043	0.070	0.046	0.064	0.048
MPA PH	0.373	0.429	0.387	0.488	0.376
ESP PH	0.302	0.400	−0.036	0.126	0.268

angles between bonds, and torsion angles of the isolated I-LDOPAc molecule and the LDOPAc@C_{60} spatial configurations A, B, C, and D.

13.2.3 Electronic Density of States

Table 13.4 presents the calculated electric charges of selected regions of the I-LDOPAc molecule when isolated and adsorbed on the buckminsterfullerene C_{60} using Hirshfeld population analysis (HPA) [140], Mulliken population analysis (MPA) [139], and an electrostatic fitting of electric charges (ESP) [324].

Mulliken population analysis is limited due to the arbitrary division of the overlap electron population. If the charge transfer is very small, Fukui function indices estimated through MPA are unpredictable sometimes. Differently, HPA tends to be more accurate, producing more realistic Fukui function indices [325], and with good prediction of reactivity trends within a molecule in comparison to MPA, natural bond orbital (NBO) analysis [326], and methods that use the molecular electrostatic potential [327]. Lastly, HPA tries to minimize the loss of information due to the joining of atoms to form a molecule [328]. For this reason, the discussion presented here will pay more attention to the charges obtained from the HPA approach, followed by MPA and ESP.

In order to perform the charge population analysis, we divided the I-LDOPAc molecule in six regions: the COO− group; the NH₃+ group; the propionic region (PP), C_6H_2–C_7H; the first hydroxyl group (HD1), connected to the C_{11} carbon, O_1–H_{15}; the second hydroxyl group (HD2), connected to the C_{14} carbon, O_3–H_{17};

and the phenyl ring (PH). For the isolated I-LDOPAc, the COO– group has an HPA charge of −0.5, while the MPA and ESP charges are slightly more negative, close to −0.6. The NH_3+ group, on the other hand, has a HPA charge of +0.353 and MPA charge of +0.478, with the ESP charge between the HPA and MPA values, of about +0.42. The propionic region has a small negative charge of about −0.051 (HPA), while the HD1 group has a negative charge of −0.017 (HPA) (approximately −0.08 for the MPA and ESP). The second hydroxyl group, on the other hand, has a positive HPA charge of +0.026, but the MPA and ESP values are negative (−0.187 and −0.133, respectively). Finally, the phenyl ring is positively charged with HPA charge of +0.043 and MPA and ESP charges of 0.373 and 0.302, in this order. All in all, the ESP and MPA values have a better correlation in comparison with the HPA-ESP and HPA-MPA combinations in the case of isolated I-LDOPAc.

When the I-LDOPAc assumes the A adsorption geometry interacting with the buckminsterfullerene C_{60}, the COO– and NH_3+ groups become less charged than for the isolated I-LDOPAc. The HPA charge for COO– changes to −0.444 and the NH_3+ charge decreases to +0.285. HPA charges for the PP, HD1, HD2, and PH regions change by a very small amount, with the largest variation observed for the HD2 group, from +0.026 in the isolated I-LDOPAc case, down to −0.039.

For the B adsorbate, the COO– HPA charge is −0.473, and the NH_3+ HPA charge is +0.372. So its charge distribution is closer to the values calculated for isolated I-LDOPAc in comparison to the A adsorption configuration. It is interesting to note, however, that the ESP charge for the PH region is very different between the isolated and the B adsorbate case, with the B geometry having a PH ESP charge of −0.036, while the isolated I-LDOPAc has PH ESP charge of +0.302. This difference does not occur with the A geometry, where the ESP charge of the PH region is +0.4.

Looking now to the C adsorbate, the COO– and NH_3+ HPA charges are −0.429 and +0.318, so these groups are less electrically charged than the isolated I-LDOPAc. The other regions have HPA charges very close to the isolated molecule values, the largest difference occurring for the ESP charge of the phenyl ring, +0.126 for the C adsorbate case and +0.302 for the isolated I-LDOPAc.

Finally, the D adsorbate, which resembles very much the isolated I-LDOPAc molecule, has very similar values for all HPA charges, except for the COO– region (−0.468, in comparison to −0.585 for LDOPAc). For the ESP charges, the most significant change in contrast with LDOPAc occurs at the PP region (−0.283 versus −0.086 in LDOPAc). All in all, there is a good correlation between charge trends observed for all population analysis methods we have employed.

For the HPA population analysis, the most striking differences between the four adsorbate geometries studied here occur for the B adsorbate at regions HD1 and HD2, and for the C adsorbate at the phenyl ring and the COO– group. The charge

Electron Charge Density (x 10^{-2} Å^{-3})

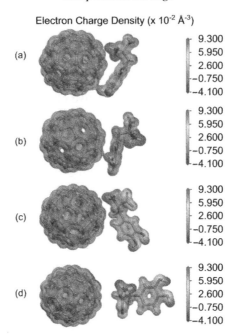

Figure 13.5 Electronic potential isosurface with projected electron charge density surfaces for the (a) A, (b) B, (c) C, and (d) D configurations of the four levels adsorption of LDOPAc on C_{60}. After Frazao et al. [317]

of the NH_3+ region is the least affected by the adsorption on C_{60}. The largest positive net charge change due to the interaction with C_{60} occurs for the C adsorbate (charge increase of $+0.136$, probably related to the positioning of its oxygen atoms, all moved away from the C_{60} surface), being followed by the B ($+0.067$) and D ($+0.016$) adsorbate geometries.

LDOPAc@C_{60} in the A configuration, on the other hand, becomes more negative (by -0.027), in comparison to l-LDOPAc, and has its oxygen atoms nearer to the buckminsterfullerene C_{60}, especially those belonging to the OH groups, when compared with the B, C, and D configurations. Figure 13.5 shows the equivalent electrostatic potential isosurfaces for each adsorbate, with the electron density projected onto them. For the A and B adsorbates, the LDOPAc potential isosurface merges with the C_{60} corresponding isosurface through the HD2 region, indicating that HD2 is the main binding site of LDOPAc to C_{60} for both configurations.

13.2.4 Classical Molecular Dynamics Simulation

Figure 13.6 shows the pair correlation function $g(r)$ of LDOPAc in the case of a single C_{60} fullerene embedded in a unit cell with 100 l-LDOPAc molecules, obtained from classical molecular dynamics simulations at 300 K, as described

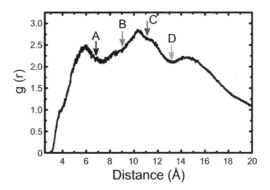

Figure 13.6 Pair correlation function $g(r)$ for the buckminsterfullerene C_{60} immersed in a box with 100 I-LDOPAc molecules obtained through molecular dynamics simulation at room temperature. The arrows indicate the distances corresponding to the total energy minima of the A, B, C, and D adsorption levels. After Frazao et al. [317]

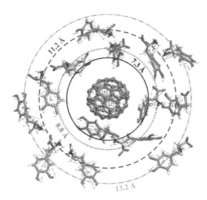

Figure 13.7 Snapshot of the molecular dynamics simulation at room temperature showing a single buckminsterfullerene C_{60} surrounded by several LDOPAc isolated molecules. The "orbit" radius correspond to the structures indicated by the arrows in Figure 13.6. After Frazao et al. [317]

in the computational details section. From it, we can see the formation of at least three LDOPAc solvation shells, the first at about 6 Å, the second at 11 Å, and the last shell near 15 Å.

Figure 13.6 also indicates, through arrows, the distances corresponding to the total energy minima of the A, B, C, and D adsorption levels found in our computations. The A adsorbate is at the first solvation shell, while B is in the inner part and C is in the outer part of the second shell. The D configuration, on the other hand, can be found at the inner part of the third solvation layer.

Looking at Figure 13.7, which reveals a snapshot of the molecular dynamics, one can see some I-LDOPAc molecules. It is easy to note that many molecules assume geometries very similar to the four optimized adsorption levels we have found, reinforcing that the information we have gathered through classical and quantum simulations can be useful to understand the LDOPAc@C_{60} system even under more realistic circumstances, where many molecules and thermal motion are present.

13.2.5 Infrared and Raman Spectra

The results of DFT simulations to obtain the infrared and Raman spectra for the isolated I-LDOPAc molecule and the four LDOPAc@C_{60} spatial configurations (A, B, C, D) are depicted in Figures 13.8 and 13.9, respectively,

Isolated levodopa molecule has 69 normal modes, while the LDOPAc@C_{60} configurations have 249 normal modes. Looking at the top of Figure 13.8, one can see the infrared spectra of these systems in the wavevector range between 0 and 1,000 cm^{-1}. For the A and C adsorbates, the infrared absorption peak at 275 cm^{-1} is shifted to lower frequencies, 249 cm^{-1} (-26 cm^{-1} in comparison to the isolated I-LDOPAc) and 252 cm^{-1} (-23 cm^{-1} relative to I-LDOPAc), while for the B and D configurations, the corresponding frequencies are much closer to the observed for the I-LDOPAc case. The I-LDOPAc peak at 341 cm^{-1} is shifted for the B adsorbate to 367 cm^{-1} ($+26$ cm^{-1}), and to 359 cm^{-1} ($+18$ cm^{-1}) and 353 cm^{-1} ($+12$ cm^{-1}) for the C and D configurations, while the I-LDOPAc peak at 499 cm^{-1} changes by -22 cm^{-1} and -19 cm^{-1} for the C and A adsorbates, respectively.

Finally, the peaks between 780 and 1,000 cm^{-1} for the I-LDOPAc structure are well preserved in the infrared absorption spectra of the A and D adsorbates but change significantly for the B and C adsorption geometries, with the 874 cm^{-1} peak of I-LDOPAc being shifted by 24 (C) and 31 (D) cm^{-1}. The infrared absorption

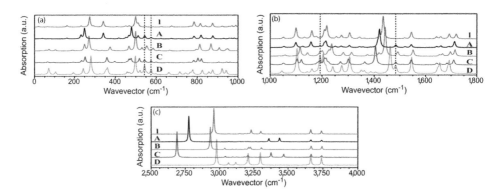

Figure 13.8 Infrared spectra of the isolated I-LDOPAc and and the four LDOPAc@C_{60} spatial configurations (A, B, C, and D) for the wavevector range (in cm^{-1}) of (a) 0–1,000; (b) 1,000–1,800; (c) 2,500–4,000. After Frazao et al. [317]

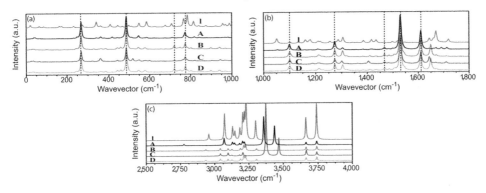

Figure 13.9 The same as in Figure 13.8 but for the Raman spectra case. After Frazao et al. [317]

peaks related to the C_{60} (corresponding frequencies indicated by dashed vertical lines), on the other hand, do not exhibit remarkable changes, with the C_{60} T_{1u} modes at 539 and 589 cm^{-1} (theoretical values, calculated using the same exchange-correlation functional and basis set adopted for the LDOPAc adsorbates) giving rise to absorption bands shifted up and down, respectively, by about 2 to 7 and 5 to 15 cm^{-1}.

Infrared absorption in the 1,000–1,800 cm^{-1} wavevector range is displayed in the middle part of Figure 13.8. The A adsorbate has an infrared absorption spectrum very similar to the I-LDOPAc, except that the I-LDOPAc peak at 1,435 cm^{-1} shifted down to 1,420 cm^{-1}. Small infrared absorption bands in the 1,192–1,193 and 1,483–1,485 cm^{-1} ranges are reminiscent of the isolated C_{60} T_{1u} normal modes at 1,215 and 1,460 cm^{-1}, corresponding to wavevector shifts of about -22 and $+23$ cm^{-1}, in this order.

In comparison, the main differences between the B and the I-LDOPAc configurations occur for the I-LDOPAc normal modes at 1,161 cm^{-1} (B: 1,135), and 1,218 cm^{-1} (B: 1,228). The C_{60} infrared absorption band between 1,192 and 1,193 cm^{-1} is more intense than that observed for the A adsorption geometry. For the C adsorbate, on the other hand, the most striking differences with respect to the I-LDOPAc and the other adsorption geometries are shifts of -40, -30, and -10 cm^{-1} of the I-LDOPAc peaks at 1,161, 1,435, and 1,720 cm^{-1}. Finally, the D-LDOPAc@C_{60} has its infrared vibrational signatures at 1,205, 1,462, and 1,693 cm^{-1} (-13, $+27$, and -27 cm^{-1} relative to the equivalent modes for I-LDOPAc).

Looking now to the infrared spectrum between 2,500 and 4,000 cm^{-1} (bottom part of Figure 13.8), we identify a set of five remarkable absorption peaks for I-LDOPAc at 2,958, 3,231, 3,301, 3,663, and 3,740 cm^{-1}. While the last two infrared absorption peaks do not change significantly for the LDOPAc adsorbed on C_{60}, the first three peaks are significantly shifted for most adsorption geometries (see more on these large shifts in the discussion of the Raman spectra). The 2,958 cm^{-1} I-LDOPAc absorption band decreases its wavevector to 2,776 cm^{-1} (-182 cm^{-1}) for the A adsorbate, while the I-LDOPAc absorption bands at 3,231 and 3,301 cm^{-1} are shifted to 3,360 cm^{-1} ($+129$ cm^{-1}) and 3,437 cm^{-1} ($+136$ cm^{-1}) for the A-LDOPAc@C_{60} configuration, respectively. These large wavevector shifts can be used as molecular signatures in experimental data suggestive of the presence of the A adsorbate geometry.

The B configuration, on the other hand, does not show a very noticeable variation with respect to the isolated form I-LDOPAc, having its largest wavevector shift with respect to the I-LDOPAc peak observed at 2,958 cm^{-1}, which decreases to 2,935 cm^{-1} (-23 cm^{-1}), a smaller difference in comparison to the A configuration, but experimentally observable as well.

The C-LDOPAc@C_{60} case, on the other hand, shows a very large wavevector shift for the I-LDOPAc peaks at 3,231 and 3,301 cm^{-1}, very close to the A shift values, and the largest wavevector variation for the I-LDOPAc peak at 2,958 cm^{-1}, which decreases to 2,690 cm^{-1} (-268 cm^{-1} wavevector shift).

Finally, the D adsorbate shows largest wavevector variations in comparison to the I-LDOPAc molecule for the absorption peaks at 2,978 cm^{-1} (I: 2,958) and 3,209 cm^{-1} (I: 3,231). The other absorption peaks, by the way, occur at wavevectors very close to the corresponding I-LDOPAc values. Taking into account these results, we believe that the calculated wavevector shifts associated with each adsorption geometry in comparison to the isolated molecule provide specific molecular signatures that can be very helpful to explain experimental measurements of the infrared spectra produced by LDOPA@C_{60} systems.

The Raman scattering spectra shown in Figure 13.9 give additional ways to detect the four distinct adsorption geometries obtained from our DFT calculations.

Looking at the top of Figure 13.9, we see the Raman bands in the wavevector range up to 1,000 cm^{-1}. The A-LDOPAc adsorption geometry leads to a Raman spectrum that is dominated by the buckminsterfullerene C_{60} in the 0–1,000 cm^{-1} wavevector range, with very intense scattering bands at 268 to 269 (corresponding to the C_{60}: 266 H_g mode), 486 (C_{60}: 497 A_g mode), and 773–774 (C_{60}: 788 H_g mode). There is also a weak Raman activity band near 720 (C_{60}: 721 H_g mode); all wavevectors in cm^{-1}.

For the isolated I-LDOPAc, the largest difference occurs for the 815 cm^{-1} normal mode, which is shifted up by only $+3$ cm^{-1}. The B adsorbate, on the other hand, has the C_{60}-associated Raman scattering peaks at the same wavevectors we found for the A configuration (indeed, the same is true for the C and D adsorption geometries).

In comparison with the I-LDOPAc, the most noticeable difference, not visible in the infrared spectrum, occurs for (all units in cm^{-1}) the 586 normal mode, which is shifted down to 579 (-7). For the C adsorbate, the peak of I-LDOPAc at 586 is now located at 599, while the peak at 766 is shifted to 761 (as for B). Finally, D-LDOPAc@C_{60} has the most pronounced wavevector shifts in comparison to the I-LDOPAc Raman spectrum peaks at 766 and 586, -24, and -6, in this order.

The middle part of Figure 13.9 shows the Raman scattering spectra for the LDOPAc systems from 1,000 cm^{-1} to 1,800 cm^{-1}. The A adsorbate exhibits strong Raman activities for C_{60} related normal modes, with very strong bands (all units in cm^{-1}) at 1,099–1,100 (isolated C_{60}: 1,127 H_g mode, largest absolute variation, 27), 1,273–1,274 (C_{60}: 1,277 H_g), 1,464–1,468 (C_{60}: 1,455 H_g), 1,529–1,532 (C_{60}: 1,503 A_g, second-largest absolute variation, about 26), and 1,609–1,611 (C_{60}: 1,617 H_g).

The most noticeable differences of A-LDOPAc@C_{60} in comparison to I-LDOPAc (not overlapping with the infrared spectrum) are 1,179 (I) changes to 1,208 cm^{-1} (A), and 1,287 (I) changes to 1,283 cm^{-1} (A). The C_{60}-related high activity peaks for the B, C, and D configurations are also very close to the corresponding A wavevectors. The I-LDOPAc peaks at 1,179, 1,287, and 1,417 cm^{-1} are shifted to 1,209, 1,304, and 1,410 cm^{-1}, respectively, for the B adsorbate, while for the C adsorbate, the most important differences with respect to I-LDOPAc occur for the I peaks at 1,179, which is shifted up to 1,207 cm^{-1}, and 1,417, which is shifted down to 1,389 cm^{-1}. The D configuration has the I-LDOPAc peak at 1,179 shifted up to 1,205 cm^{-1}, while the I-LDOPAc peak at 1,417 is shifted down to 1,406 cm^{-1}.

The highest wavevector domain, between 2,500 and 4,000 cm^{-1}, does not have any C_{60} related Raman activities, the Raman scattering peaks being due entirely to the LDOPAc vibrational normal modes. The 11 isolated I-LDOPAc normal modes in this range are all Raman active, originating from CH, NH, and OH bond length oscillations, having the following wavevectors: 2,958, 3,073, 3,132, 3,149, 3,185, 3,207, 3,218, 3,231, 3,301, 3,663, and 3,740 cm^{-1}.

The adsorbate Raman spectra for wavevectors greater than 2,500 cm^{-1} has the same remarkable wavevector shifts (greater than 100 cm^{-1} in absolute value) in comparison to I-LDOPAc, already noted in the infrared spectra. Wavevector shifts visible only in the Raman data, however, are much smaller (-3 cm^{-1} at worst, for the A adsorbate). Some distinct Raman signatures can be seen for the B, C, and D configurations of LDOPAc@C_{60}. For example, the I-LDOPAc peaks at 3,073, 3,132, and 3,149 cm^{-1} are shifted down to 3,042, 3,098, and 3,127 cm^{-1} for the B adsorbate, while for the C adsorbate, the corresponding wavevectors are 3,043, 3,098, and 3,125 cm^{-1}. The C adsorbate also has a wavevector shift of -11 cm^{-1} for the I-LDOPAc normal mode at 3,218 cm^{-1}. The D adsorbate, on the other hand, has its most pronounced wavevector changes for the I-LDOPAc peaks at 3,132, 3,149, 3,185, and 3,218 cm^{-1} – which are shifted, respectively, to 3,120, 3,130, 3,195, and 3,230 cm^{-1}.

The large wavevector shifts observed in the infrared and Raman spectra for the A adsorbate from 2,958 to 2,776 cm^{-1} (-182 cm^{-1}), from 3,231 to 3,360 cm^{-1} ($+129$ cm^{-1}), and from 3,301 to 3,437 cm^{-1} ($+136$ cm^{-1}), and the corresponding large shifts for the C adsorbate seem to be due to the stretching of N–H bonds at the NH$_3$+ group, which is very close to the fullerene C$_{60}$ surface in these adsorption geometries. Differently, the B adsorption geometry, which has the NH$_3$+ group moved away from the fullerene surface, has smaller wavevector shifts in this region of the vibrational spectrum.

Lastly, the D adsorption geometry has both the NH$_3$+ and COO− near to the C$_{60}$ molecule, as well as the C$_7$–H$_{22}$ atoms, which decreases the strong effect of the NH$_3$+ interaction with the C$_{60}$ fullerene.

13.3 Carbidopa Molecule

Carbidopa is a peripheral inhibitor of DOPA decarboxylase (DDC) that is ingested together with the levodopa. The combination levodopa–carbidopa antiparkinson drug is most used today. It can also reduce some of levodopa's side effects such as nausea and vomiting. If used alone, carbidopa does not affect Parkinson's symptoms.

In this section, annealing calculations will be employed to explore the molecular geometry of carbidopa, to obtain its most stable conformations of smaller energies, by using DFT with the LDA/PWC, GGA/PBE, and GGA/BLYP functionals. Carbidopa's molecular orbital study (HOMO, LUMO, PDOS, and DOS) is then obtained considering its smallest conformation energy. A detailed interpretation of carbidopa's harmonic vibrational frequencies are also presented, through the analysis of its Raman scattering spectroscopy.

13.3.1 Computational Details

To perform the quantum optimization, together with the optical and electronic calculations of the carbidopa, we have used the DMOL3 software within the DFT formalism considering the LDA, and the GGA as the exchange-correlation functionals.

The LDA functional was used in its standard parameterization, as proposed by Perdew and Wang (PWC functional) [10], while the GGA functional was taking into account using the Perdew-Burke-Ernzerhof (PBE functional) method [12], as well as the BLYP functional. We do not adopt any pseudopotential calculation [329–331] to replace the core electrons in each atomic species: all electrons were included in the calculation. Double numerical plus polarization basis set (DNP) was adopted to expand the Kohn-Sham electronic eigenstates with an orbital cutoff radius of 3.5 Å.

The optimized structural parameters were used in the vibrational frequency calculations to characterize all stationary points as minima. We have utilized the gradient corrected DFT with the three-parameter hybrid functional (B3LYP) for the exchange part, and the Lee-Yang-Parr (LYP) correlation functional [13], looking for the vibrational frequencies of the optimized structures.

Vibrational frequencies have been calculated at B3LYP/6-311G(d,p) basis set to expand the electronic states [332], enabling us to make the detailed Raman assignments spectra of the carbidopa molecule. All DFT calculations are obtained by performing a geometrical optimization using the Gaussian 09 code without any constraint on the geometry, using the same convergence values as in the previous levodopa case.

Optical properties are analyzed through their optical absorption spectra by using its 25 lowest singlet states. Gaussian integration was the broadening method used to calculate the optical spectra with a value of 5.0 nm to the smearing width.

13.3.2 Geometry Optimization

The molecular conformers of carbidopa, obtained from Boltzmann jump search method of calculation, are shown in Figure 13.10. The carbidopa molecule contains

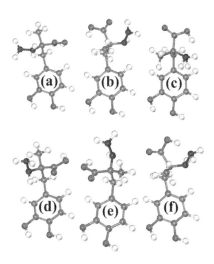

Figure 13.10 The carbidopa molecule conformations of smallest energy (all units in eV): (a) E $= -21,596.63$, (b) E $= -21,596.57$, (c) E $= -21,596.45$, (d) E $= -21,596.31$, (e) E $= -21,596.28$, (f) E $= -21,596.27$.

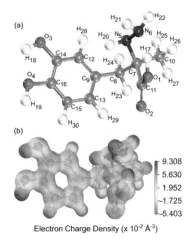

Electron Charge Density (x 10^{-2} Å^{-3})

Figure 13.11 The carbidopa molecule of smallest energy. (a) Its most stable converged structure; (b) the electron density of its most stable converged structure projected onto an electrostatic potential isosurface.

three radicals connected to the benzene ring, one being $C_4O_2N_2H_9$ and the others being the OH hydroxyl group [333].

The first radical is composed by four important groups, the so-called primary amine, secondary amine, carboxyl, and methyl. There are six possible conformers of smaller energy: the one depicted in Figure 13.10a is the conformer with the smallest energy (E = $-21,596.63$ eV), followed by the conformers in Figure 13.10b–13.10f, with energies in the range E = $-21,596.57$ to E = $-21,596.27$ eV.

To represent the carbidopa molecule from now on, we have chosen the geometry with the smallest energy as the most appropriate conformer. The labels adopted to identify each atom of the carbidopa optimized geometry are shown in Figure 13.11a, and they are quite important to analyze their structural measures. Figure 13.11b shows the projection of the calculated electron density of carbidopa onto an electrostatic potential surface. From there, one can see that the most negatively charged atoms are the ones labeled as O_1 and O_2, followed by O_3 and O_4. The most positively charged atoms are the nitrogen N_5 and N_6, as well as the H_{17}, H_{18}, H_{19}, H_{20}, H_{21}, and H_{22} hydrogen atoms; the ring and remaining carbon atoms are slightly positively charged.

The optoelectronic properties depend essentially on the energy gap

$$E_{gap} = |E_{LUMO} - E_{HOMO}|, \tag{13.2}$$

as well as on the electron and hole mobilities.

In particular, E_{gap} is an important parameter because it determines the molecular admittance, which is a measure of the electron density hardness. The calculated

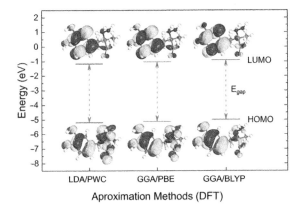

Figure 13.12 The highest occupied molecular orbital (HOMO) and the lowest unoccupied molecular orbital (LUMO) of the carbidopa molecule calculated by using (from left to right) DFT-LDA/PWC; DFT-GGA/PBE; and DFT-GGA/BLYP functionals based on the DFT model, respectively.

highest occupied molecular orbital (HOMO) and lowest unoccupied molecular orbital (LUMO) are depicted in Figure 13.12. Analyzing the frontier region, neighboring orbitals are often closely spaced. In face of that, consideration of only the HOMO and LUMO orbitals may not yield a realistic description of the frontier molecular orbitals.

13.3.3 Electronic Density of States

The total electronic density of states (DOS) was calculated using the DMOL3 program through a linear interpolation of the cuboids formed by the points of the Monkhorst-Pack set, followed by the histogram sampling of the resultant set of the band energies.

Summation of these contributions over all bands produces a weighted DOS calculated, as depicted in Figure 13.13, by considering the three different quantum methods of simulation already mentioned, namely the LDA/PWC, GGA/PBE, and GGA/BLYP functionals. The vertical dashed line represents the Fermi energy level, -5.197, -5.089, and -4.980 eV being the values predicted by the DFT-LDA/PWC, DFT-GGA/PBE, DFT-GGA/BLYP functionals, respectively.

On the other hand, the partial electronic density of states (PDOS) calculations, as shown in the inset of Figure 13.13, are based on the Mulliken population analysis, allowing the contribution from each energy band to a given atomic orbital to be calculated. They show mainly the contribution of the atomic orbitals s, p, d, and f to the molecular one.

As one can see from the insets of Figure 13.13, the most relevant contribution from C, N, and O atoms to the total electronic density of states on the Fermi energy

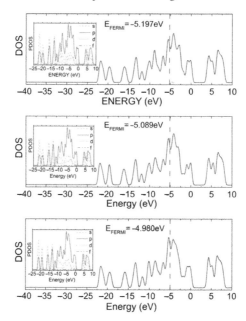

Figure 13.13 The predicted total electronic density of states (DOS, main figure) and the partial electronic density of states (PDOS, inset) of the carbidopa molecule calculated by using the functionals (from top to bottom) DFT-LDA/PWC, DFT-GGA/PBE, and DFT-GGA/BLYP. The vertical dashed lines indicate the Fermi energy. The labels s, p, d, and f represent the atomic orbitals.

is the 2p orbital for the three kinds of calculation, with maximum at -5.633 eV (LDA/PWC), -5.388 eV (GGA/PBE), and -5.333 eV (GGA/BLYP), respectively. The PDOS represents useful semi-qualitative tools for analyzing the electronic structure. It further qualifies these results by resolving the contributions according to the angular momentum of the states.

The energies for the HOMO, LUMO, and E_{gap} using LDA/PWC (GGA/PBE) (GGA/BLYP) calculation yield -5.191, -1.164, and 4.027 (-4.978, -0.891, and 4.087) (-5.098, -1.016, and 4.082) eV, respectively.

The HOMO-LUMO energy gap explains the eventual charge transfer interaction within the molecule, which influences its biological activity. Consequently, the lowering of the HOMO-LUMO energy gap is essentially a consequence of the large stabilization of the LUMO due to the strong electron-acceptor ability of the electron-acceptor group.

13.3.4 Optical Absorption Spectra

We further calculate the optical absorption spectra of carbidopa using the LDA/PWC, GGA/PBE, and GGA/BLYP functionals of the DFT method. The

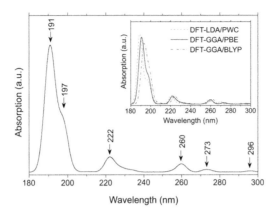

Figure 13.14 Optical absorption spectrum of carbidopa (with Gaussian broadening of 5.00 nm) calculated by using the GGA/PBE functional. The resonance absorption peaks values are labeled. The inset chart shows a comparative optical spectra calculated by using the functionals LDA/PWC (dashed line), GGA/PBE (solid line), and GGA/BLYP (dotted-dashed line), respectively.

single theoretical spectrum displayed in Figure 13.14 refers to the DFT-GGA/PBE functional, from which one can see the labeled peaks of resonance. It depicts six peaks of absorption; the largest resonance one appears at 191 nm, and the smallest one is observed at 296 nm. The remaining ones are shown in the inset of Figure 13.14 (we kept the DFT-GGA/PBE profile for comparison), being the LDA (PBE) (BLYP) spectrum curve depicted as a dashed (solid) (dotted-dashed) line.

Looking at the LDA curve, one can notice that the peaks are shifted backward. The second peak of resonance is less intense and more prominent when compared to the PBE spectrum, which has a second peak located at 197 nm. By contrast, in the BLYP curve the peaks are shifted forward, and the second peak disappears when it is compared to the PBE curve. The major resonance peaks in all curves are assigned between 189 and 194 nm, and the energy of that peak is 6.48 eV.

13.3.5 Raman Spectra

The carbidopa molecule consists of 30 atoms and does not belong to any specific group of symmetry. Hence, the number of normal modes of vibrations for carbidopa is 84. Out of the 84 fundamental vibrations, 76 modes are active by Raman scattering. The harmonic-vibrational frequencies are calculated for carbidopa through the DFT/B3LYP method using the triple-split valence basis set, with the diffuse and polarization functions 6-311G(d,p).

In order to show the Raman peaks, we have plotted in Figure 13.15 the spectra profile for each kind of vibrational analysis. Let us analyze in depth the Raman scattering activity spectrum by dividing the group vibrations separately:

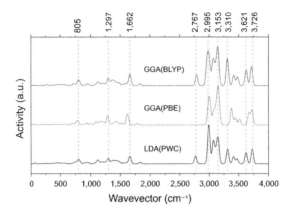

Figure 13.15 Raman scattering activity spectrum of the carbidopa molecule by using the functionals (from top to bottom) GGA/B3LYP, GGA/PBE, and LDA/PWC, respectively.

(a) The C–H vibrations

Carbidopa molecule presents the carbon–hydrogen stretching vibration in the region $3,000$–$3,100$ cm^{-1}, which is a characteristic region for the aromatic compounds. Since carbidopa is a tri-substituted aromatic system, it has three bonds to complete the benzene ring. The expected three C–H vibrations correspond to stretching modes of C_{15}–H_{30}, C_{12}–H_{28}, and C_{13}–H_{29} bonds. These vibrations are calculated at $3,150$, $3,180$, and $3,209$ cm^{-1}, respectively. The C–H out-of-plane bending mode usually appears in the range 670–950 cm^{-1}, and the C–H in-plane bending vibrations in the region 950–$1,300$ cm^{-1}. In face of that, the C–H out-of-plane (C–H in-plane) bending vibrations are observed at 818, 882, and 950 ($1,168$, $1,181$, and $1,214$) cm^{-1}, respectively.

(b) The C=C vibrations

Basically, the C=C stretching vibrations in aromatic compounds are observed in the region $1,430$–$1,650$ cm^{-1}. On the other hand, around the region $1,575$ to $1,625$ cm^{-1}, the presence of conjugate, such as C=C, causes a heavy doublet formation. The six-ring carbon atoms go through coupled vibrations, known as skeletal vibrations, and in the region $1,420$–$1,660$ cm^{-1}, they give the maximum of four bands. Because of that, the peaks at $1,305$, $1,313$, and $1,357$ cm^{-1} are due to the strong C–C skeletal vibrations, while the peaks at $1,500$, $1,547$, $1,643$, and $1,661$ cm^{-1} are attributed to the strong C=C stretching on the carbidopa. The peaks at $1,305$, $1,313$, and $1,357$ cm^{-1} as well as the peaks at $1,500$, $1,547$, $1,643$, and $1,661$ cm^{-1} are due to the semicircle stretching and the quadrant stretching of C–C and C=C bonds, respectively. The peaks that are assigned at 468, 552, and 589 cm^{-1} are due to out-of-plane bending vibrations, and the peaks at 597, 674, and 755 cm^{-1} are due to C–C–C in-plane bending vibrations.

(c) The CO–OH vibrations

Carbonyl groups are sensitive, and both the carbon and oxygen atoms of the carbonyl move during the vibration with nearly equal amplitude. A single band was observed at 1,814 cm^{-1}, due to the C=O stretching vibration. Nevertheless, the OH stretching band is characterized by a broad band appearing near 3,400 cm^{-1}. The band observed at the position 3,230 cm^{-1} is generated by the O–H stretching vibration. The O–H out-of-plane and in-plane bending vibrations are usually observed in the regions 590–720 cm^{-1} and 1,200 to 1,350 cm^{-1}. In carbidopa, the O–H out-of-plane and in-plane bending vibrations are found at 988 and 1,488 cm^{-1}, respectively. The peak in the Raman spectrum at 1,244 cm^{-1} is assigned to the C–O stretching mode, which is a pure mode. The C–O out-of-plane and in-plane bending vibrations for carbidopa are found at 419 and 755 cm^{-1}, respectively.

(d) The methyl group vibrations

The stretching vibrations on the C–H bonds of the methyl group are normally observed in the region 2,840–2,975 cm^{-1}. In the carbidopa molecule, there are strong bands around 3,034 cm^{-1} and around 3,104 cm^{-1}, corresponding to the symmetric and asymmetric stretching modes in the Raman spectrum, respectively. The C–H out-of-plane and in-plane bending vibrations for CH$_3$ methyl group in carbidopa are assigned at 755, 847, and 932 cm^{-1} and 1,401, 1,474, and 1,492 cm^{-1}, respectively. The C–CH$_3$ out-of-plane bending, in-plane bending, and stretching vibrations for carbidopa are found at 246, 310, and 1,114 cm^{-1}, respectively.

(e) The O-H vibrations

In a vibrational spectrum, the strength of the hydrogen bond affects the position of the O–H bands. Normally, the O–H stretching vibrations fall in the region 3,400–3,600 cm^{-1}. Therefore, the theoretically computed stretching values for the O–H group are 3,230, 3,788, and 3,849 cm^{-1}, whereas the O–H out-of-plane and in-plane bending modes are observed at 1,643, 1,661 and 1,814 cm^{-1} and 242, 459 and 988 cm^{-1}, respectively, as pure modes of bending vibrations. The two last peaks in the Raman spectrum are related to the stretching on the O–H group linked to the phenyl ring.

(f) The NH–NH2 vibrations

The H–N stretching modes are more visible in the Raman activities spectrum, being localized around 3,500 cm^{-1}. The NH$_2$ symmetric and asymmetric stretching is observed at 3,476 and 3,553 cm^{-1}, respectively. The H atom of the H–N–N group presents stretching vibration at 3,512 cm^{-1}, and the N–N stretching is seen at 1,017 cm^{-1}. By contrast, the rocking mode is assigned in the range 28–43 cm^{-1}, the wagging mode is found in the range 1,099–1,175 cm^{-1}, and the twisting vibration mode is observed from 1,279–1,356 cm^{-1}. The carbidopa compound presents two joined nitrogen (N) atoms. Therefore, there is an N–N stretching vibration mode on that molecule, being observed at 1,017 cm^{-1}, while the C–N stretching vibration appears at 1,175 cm^{-1}.

13.4 Conclusions

In summary, we have presented the conformational, structural, optoelectronic, and vibrational properties of the levodopa (LDOPA), adsorbed on the C$_{60}$ fullerene, and carbidopa molecules, the most powerful antiparkson drugs of the pharmaceutical industry. A detailed interpretation of their harmonic vibrational frequencies was also presented through the analysis of its Raman scattering spectroscopy. Overall, our results suggest the feasibility of these compounds as a system for oral and transdermal drug delivery for the treatment of Parkinson's disease.

Unfortunately, there is, so far, no experimental data to be compared to our computer simulation results. We strongly hope that, motivated by this and other progress, the theoretical predictions obtained here can be reproduced experimentally. Furthermore, it would be interesting to investigate compound interactions with levodopa and carbidopa, considering different amounts of each type of prodrug, looking for the design of new drugs or carriers.

14

Central Nervous System Disorders

14.1 Introduction

The nervous system is a complex, highly specialized network consisting of the brain, spinal cord, sensory organs, and all of the nerves (neurons) that connect these organs and carry messages to and from the brain and spinal cord to various parts of the body. It includes both the central nervous system (CNS) and the peripheral nervous system (PNS). It organizes, explains, and directs interactions between you and the world around you.

The nervous system allows you to be conscious and have thoughts, memories, and language, controlling

(a) Sight, hearing, taste, smell, and feelings.
(b) Voluntary and involuntary functions, such as movement, balance, and coordination.
(c) The ability to think and reason.

The central nervous system is a part of the nervous system consisting of the brain and the spinal cord, which control most functions of the body and mind (see Figure 14.1). It is better protected than any other system or organ in the body. Its main line of defense is the bones of the skull and spinal column, which are both housed within a protective triple-layered membrane called the meninges, creating a hard physical barrier to injury. A fluid-filled space below the bones, called the syrnix, provides shock absorbance.

The CNS can be roughly divided into white and gray matter. As a very general rule, the brain consists of an outer cortex of gray matter and an inner area housing tracts of white matter. Both types of tissue contain glial cells, which protect and support neurons. White matter mostly consists of axons (nerve projections) and oligodendrocytes, a type of glial cell, whereas gray matter consists predominantly of neurons.

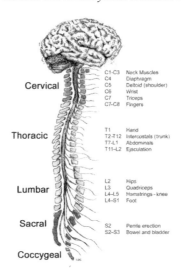

Figure 14.1 The diagram of the central nervous system depicts the brain and the spinal cord, which extends from the back of the brain, down the center of the spine, stopping in the lumbar region of the lower back.

Central nervous system disorders are a group of neurological disorders that affect the structure or function of the brain or spinal cord. They affect approximately one billion of human beings around the world, leading to more hospitalizations than any other group of diseases. The most common general signs and symptoms of a nervous system disorder are a persistent headache, weakness or loss of muscle strength, loss of sight or double vision, impaired mental ability, muscle rigidity, and tremors and seizures.

These neurological disorders may involve the following:

(a) Vascular disorders, such as stroke and transient ischemic attack (TIA).
(b) Infections, such as meningitis, encephalitis and polio.
(c) Structural disorders, such as brain or spinal cord injury, brain or spinal cord tumors, peripheral neuropathy, and Guillain-Barré syndrome.
(d) Functional disorders, such as headache, epilepsy, dizziness, and neuralgia.
(e) Degeneration, such as Parkinson's disease, multiple sclerosis, amyotrophic lateral sclerosis (ALS), Huntington's chorea, and Alzheimer's disease.

Central nervous system disorders have been ascribed to impairments on the ionotropic glutamate receptors (iGluRs) functions, which are ligand-gated ion channels that undergo structural changes after activation, culminating with the channel opening and generating an ion flux through the membrane [334]. They mediate the majority of excitatory synaptic transmission throughout the central nervous system and are key players in synaptic plasticity, which is important for learning and memory.

The three major subclasses of iGluRs can be differentiated according to their amino acid sequence and pharmacology as kainate (GluK1-5), N-methyl-D-aspartic acid (NMDA; GluN1-3), and α-amino-3-hydroxy-5-methyl-4-isoxazolepropionic acid (AMPA; GluR1-4), related to the physiological processes [335].

Regarding the iGluR2-AMPA subclass, a set of willardiines, from *Acacia willardiana* and *Mimosa asperata*, has been used to elucidate its molecular basis of partial agonism [336]. Since 1980, willardiine and its analogs have been tested and proved to be a neurotransmitter composite. The substitution of a single atom at position 5 of the uracil ring of (s)-willardiine by the chemical elements fluorine (F), hydrogen (H), bromide (Br), and iodine (I) has lead to different responses, suggesting the utilization of these four willardiine partial agonist – namely fluorine-willardiine (FW), hydrogen-willardiine (HW), bromine-willardiine (BrW), and iodine-willardiine (IW) – in structure-function studies of the central nervous system disorders [337].

In this chapter, we present an adequate description of the interaction of the four willardiine neurotransmitter composite partial agonists of the central nervous system disorders, described in the preceding paragraph, with the ionotropic glutamate receptor iGluR2-AMPA through quantum biochemistry techniques within the density function theory (DFT) framework [338]. The individual contribution of each amino acid residue was calculated by applying the MFCC scheme, considering residues at the binding site also including other relevant residues. The simulations were performed using the X-ray structure of iGluR2-AMPA co-crystallized with the willardiines FW, HW, BrW, and IW (PDB ID: 1MQI, 1MQJ, 1MQH, and 1MQG, respectively) [336]. A comparison between our theoretical binding energies and the experimental one is also made, and their main features discussed.

14.2 The iGluR2-AMPA Receptors

Ionotropic glutamate-AMPA receptors have a key role in fast synaptic transmission, being well distributed throughout the CNS [339]. This type of receptor has a tetrameric structure organized as a dimer of dimers [340], where each monomer is composed of an extracellular amino-terminal region (ATD; approximately with 400 amino acids), a ligand binding-domain (LBD; with approximately 258 amino acids) and a transmembrane region (TMD; with approximately 169 amino acids) [341], whose molecular structures are depicted in Figure 14.2. In the figure, each color (black, dark gray, gray, and light gray) corresponds to a single chain, and

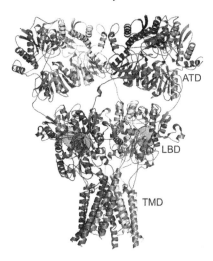

Figure 14.2 The Y-shaped iGluR2-AMPA receptor structure and its monomers composed by the amino-terminal domain (ATD), the ligand-binding domain (LBD) and the transmembrane domain (TMD). Each color (black, dark gray, gray, and light gray) corresponds to a single chain, and the circles in LBD region represent the position of the ligand-binding site in every subunit. The PDB code is 3KG2. After Lima Neto et al. [338]

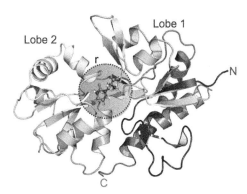

Figure 14.3 Structural representation of the ligand-binding domain (LBD) interface. The binding pocket sphere (BPS) with radius (r) is also shown in this picture as a circle around the willardiine ligand. N and C represents the amino and carboxyl terminal regions. After Lima Neto et al. [338]

the circles in the LBD region represent the position of the ligand-binding site in every subunit.

The iGluR-ligand binding domain (LBD) is formed by polypeptide segments (or lobes) S1 (390–506 amino acids) and S2 (632–775 amino acids), which can be genetically combined and expressed as a soluble protein [336]. Its structural representation is shown in Figure 14.3, which depicts the monomeric structure of the LBD coupled to a willardiine molecule (PDB ID: 1MQH). The letters N and C represent the amino and carboxyl terminal regions, respectively, which are linked to the amino-terminal domain (ATD) and the transmembrane domain (TMD) in the monomer. The circle marks the region of the protein investigated in this work, whose amino acids are selected by increasing the binding region radius (r).

14.3 Crystallographic Data of the Different Types of Willardiines

Crystallographic and electrophysiological studies have significantly advanced the structural and functional characterization of iGluR2-AMPA receptors. Isolated structures of iGluR-LBD in complex with agonists, partial agonists, and antagonists suggest a correlation between the degree of lobe closure and channel activation [342]. When bound to a full agonist, the lobes were closed by approximately $19.1°$ to $21.3°$, leading to channel activation with high efficacy levels [343]. In addition, antagonists have shown $2.5°$ to $9.6°$ of lobe closure, blocking receptor activation. Finally, partial agonists induce an intermediate closure of the LBD ($13.1°–19°$) and the opening of the ion channel.

A comparison between the structures of glutamate coupled to iGluRs-LDB with other full and partial agonists suggested that the α-amino and α-carboxyl groups occupy similar positions in the receptor [344]. This pattern was also observed during the superposition of the crystal structures of glutamate and four partial agonists willardiine, while the substituents attached at γ-position occupy different regions of the receptor [339].

Experimental, computational and crystallographic analysis has been used to describe partial agonism by four-substituted willardiines in iGluR2-AMPA receptors [345]. The crystal structure of iGluR2-LBD with four willardiines provided an opportunity to identify subtle structural differences on the receptor created by a single atom change in the ligand. The analysis of crystallographic structures can be done based on important tools, like the distances and sizes of the connections (*virtual screening*), the de novo design, and molecular docking and molecular dynamics simulations. These methods are limited by the lack of information on the interaction between specific residues of the receptor and the different ligands, which would be quite useful for the design of new drugs. Indeed, the use of quantum mechanics (QM) for in silico drug design has become quite popular in recent years due to its high accuracy in estimating relative binding affinities.

However, the high computational cost of QM methods to calculate the energies of interaction of macromolecules demands a balance between the computer execution time and the accuracy of the results. In view of this, fragmentation methods have been developed to make macromolecules computationally less expensive [24].

Crystallographic data of the willardiines FW, HW, BrW, and IW co-crystalized with iGluR2-AMPA receptors were downloaded from the PDB database (www.rcsb.org) under the following codes (resolutions) 1MQI (1.35 Å), 1MQJ

(1.65 Å), 1MQH (1.8 Å), and 1MQG (2.15 Å) [336]. The state of protonation of all ligands at physiological pH was obtained using the MarvinSketch code version 5.3.2 (Marvin Beans Suite – ChemAxon).

Our quantum chemistry simulation follows the same line already presented in previous chapters: a classical geometry optimization is performed by using the classical force field CHARMm, which is especially parameterized for organic molecules; interaction energies for each residue at the binding site were performed using the DMol3 code, within the DFT-GGA-PBE [12] and DFT-LDA-PWC [10] formalisms; for the noncovalent interactions, we used the DFT + D method [274] following the GGA-PBE-Grimme [19] and LDA-PWC-OBS (OBS is the Ortmann-Bechstedt-Schmidt correction) [274] schemes to calculate the interaction energies of the four willardiine-iGluR2 complexes. The Khon-Sham orbitals for all electrons are calculated by using the 6-311+G(3df, 2pd) basis set [134] with the orbital cutoff radius set to 3.7 Å, and the self-consistent field convergence threshold adjusted to 10^{-6} Ha. Finally, the interaction energy between each willardiine molecule and the amino acid residues were calculated by using the MFCC formalism [22].

A convergence study of the interaction energy as a function of the ligand-binding pocket radius was performed to put a limit to the number of amino acid residues to be analyzed without missing important interactions. We investigated the variation of the total interaction energy considering the contributions of all amino acid residues within a sphere of radius r, with origin in the ligand, capping the dangling bonds of each amino acid residue [346]. Here we label the ligand molecule as L and the ith amino acid residue interacting with the ligand as R^i. The total interaction energy for each willardiine is obtained by adding up the interaction energies with each amino acid residue within a given binding pocket radius. Water molecules were taken into account in the calculation procedures when hydrogen bonds were formed with the residues of interest or with the caps. A hydrogen bond length limit of 2.5 Å was adopted in this case.

14.4 Willardiines Partial Agonism in iGluR2-AMPA Receptors

Complexes with full agonists, partial agonists, and antagonists are useful to relate structural and dynamic properties of the glutamate receptor S1–S2 domain to functional properties that can be only measured for the intact protein. The series of willardiines considered in the present study are particularly suited for this analysis because they differ structurally at just one position, having functional properties that vary according to their electronegativity (potency and binding affinity) and size (efficacy and desensitization) [337].

The partial agonism of four-substituted willardiines can help us understand the mechanisms of activation and desensitization of iGluR2-AMPA receptors. It was proposed that the size of the halogen substituent is directly involved in the modulation of the binding site and, consequently, in the closure of the lobes [336, 339]. Results reported in References [336, 347], which relied on electrophysiological and crystallographic data, have been pivotal to our understanding of the partial agonism mechanism in iGluR2-AMPA, showing that the potency increases with the electronegativity. Besides, the peak current and the extent of receptor desensitization varies according to the size of the substituent, which implies that the extent of lobes closure affects the response of the receptor. Recently, the importance of the willardiine protonation state in the activation of iGluR2-AMPA receptors has been stressed [348].

To give a good explanation about the 4-substituted willardiine interactions with the iGluR2-AMPA receptors, we used the MarvinSketch software to obtain their pK$_a$ curves at 7.3 pH value. In Figure 14.4, we see the molar fraction curves in pH values between 0 and 12 (Figure 14.4a), and the molecular view of the two distinct protonation states of the four willardiine derivatives compounds (Figure 14.4b).

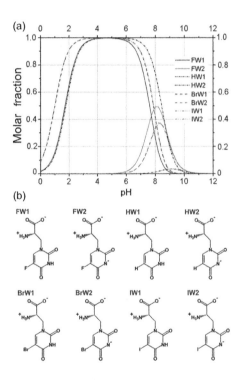

Figure 14.4 Protonation state of four-substituted willardiines as a function of pH. (a) Molar fraction curves for the two distinct molecular states for fluorine-willardiine (FW, solid lines), hydrogen-willardiine (HW, chain-dotted lines), bromine-willardiine (BW, dashed lines) and iodine-willardiine (IW, chain-double dotted lines). (b) Chemical structure of each molecular states. After Lima Neto et al. [338]

Our results show that the uncharged protonation state has a larger contribution in comparison with the charged state, in agreement with [349, 350]. For this reason, all calculations were made here with the N3 atom of the uracil ring in the uncharged state.

14.5 Interaction Energies

We analyzed the binding pocket of the receptor employing the MFCC scheme to obtain the individual contribution of the amino acid residues inside a binding pocket with a selected radius, ranking the most relevant interactions between residues and ligands. The total binding energy was obtained by adding up the individual contributions.

To evaluate the binding interactions through fragment-based quantum mechanics method, it is important to take into account every significant attractive and repulsive amino acid residue that can influence this mechanism. Therefore, instead of taking an arbitrary region of the binding site, we performed a search for an optimal binding pocket radius for which no significant variation in the total binding energy could be observed after a radius increase.

For this task, the binding pocket radius r was varied from 2.0 to 12 Å for all willardiines. In order to describe the most important ligand–residue interactions exhibited by iGluR2-AMPA receptors we have considered the compounds subdivided into two regions, as one can see in Figure 14.5a. Figure 14.5b depicts the electron density isosurface for FW, HW, BrW, and IW willardiines' series to observe how the electrons are distributed around the molecules and how their distribution is affected by the four-substituted halogen atom. Differences in the electron density can be observed in the halogen substituent region.

Figure 14.6a and b show a comparison between the calculated interaction energies $E(r)$ for the four-substituted willardiines considering the binding pocket radius $r = 4.0, 6.0, 8.0, 10.0,$ and 12.0 Å. In Figure 14.6a, using the GGA-PBE-Grimme exchange-correlation functional approach, we found that at the smallest binding pocket radius (4.0 Å), the absolute value of the total four-substituted willardiines interaction energy follows the order HW > BrW > FW > IW, which does not reproduce the corresponding experimental data [351].

For a binding pocket radius greater than 10.0 Å, the absolute value of the binding energies $E(r)$ follows the willardiine sequence FW > HW > BrW > IW. This result is close to the experimental sequence FW > BrW > IW > HW [336, 337], meaning that the binding energy of each halogenated willardiine can be compared to its experimental value, the difference being in the position of the willardiine HW. The same occurs for the LDA-PWC-OBS exchange-correlation functional approach (see Figure 14.6b), considering a smaller binding pocket radius (4.0 Å),

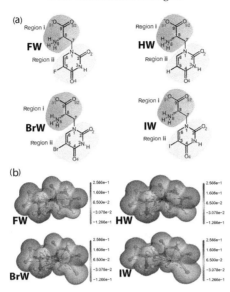

Figure 14.5 Willardiines FW, BrW, IW and HW: chemical representation at pH 7.3 and electron density distribution. (a) Region (**i**) contains the carboxyl and amine functional groups, while region (**ii**) corresponds to the uracil ring and its four-substituents. (b) Electrostatic potencial surface and willardiine centroids. After Lima Neto et al. [338]

although only the radius is greater than 10.0 Å the total binding energy stabilizes as a function of r. One can note that, in this case, the LDA-PWC-OBS scheme is very effective, confirming previously reported data [352].

However, the efficiency of the willardiine HW and the three halogenated willardiine analogs FW, BrW, and IW is correlated to the degree of closure of the low-density lipoprotein (LDL)-iGluR2 lobes [336, 337]. Other mechanisms may influence the recognition and receptor activation, as the interaction between dimers. Besides, the dynamics of the iGluR2-AMPA receptor in complex with agonists, partial agonists, and antagonists are not an easy task, as stressed by many authors. Although not addressing whether or not the closed lobe is required for activation of the channel, it was suggested that the closed lobe form is unstable for these partial agonists in agreement with many experimental works and probably not required for the activation of the ion channel [353]. Ligands with lower potency may be weakly bound to the protein, with the formation of multiple conformations to the ligand and amino acid side chain [337]. Due to the mobility in the protein structure, there are many conformations obtained from the X-ray diffraction of the willardiine-AMPA complex. Thus, it is difficult to obtain the optimal conformation for our analysis – i.e., those that respond to the ligand-binding affinity.

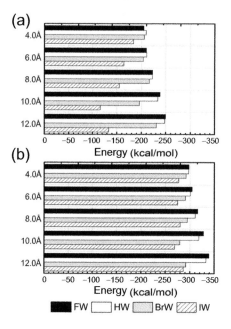

Figure 14.6 Total willardiine interaction energy for FW, HW, BrW, and IW considering pocket radius values of 4.0, 6.0, 8.0, 10.0, and 12.0 Å. (a) GGA-PBE-Grimme exchange-correlation functional results. (b) LDA-PWC-OBS exchange-correlation functional results. The absolute value of $E(r)$ obeys the sequence FW > HW > BrW > IW, reproducing the experimental data [336], excluding the HW case. After Lima Neto et al. [338]

A detailed quantitative analysis to justify this conclusions is the following: the amino acids depicting a higher interaction energy with the HW willardiine are positioned in a similar way to that observed in FW. These amino acids interact with the atoms $(C_9)OO-$, $(N_8)H$, $(C_4)O_4$, and $(N_3)H$ with close proximity in HW, being responsible to a greater interaction energy not observed for BrW and IW since the size/electronegativity of the substituent halogen groups (Br and I) create a repulsion for some closed amino acids, such as M708 and C425. However, M708 and C425 have attractive energy for FW and HW. Furthermore, contrary to what is observed for FW and HW, amino acids that have a higher energy of interaction with BrW and IW are positioned close to the atoms $(C_2)O_2$, $(N_3)H$, and $(C_9)OO-$ (minus signal).

Figure 14.7a to 14.7d show a BIRD graphic panel (binding site, interaction energy, and residues domain) with the interaction energies between the 4-substituted willardiine molecules and the most important amino acid residues at the binding region. The panel depicts

(a) The interaction energy (in kcal/mol) of the residue with the ligands, illustrated by the horizontal bars, from which one can assign quantitatively the role of each

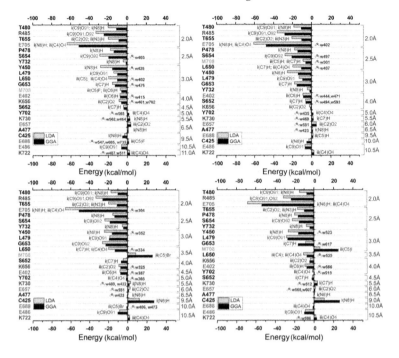

Figure 14.7 Binding site, interaction energy, and residues domain (BIRD) graphic panel showing the most relevant residues of the willardiines (a) FW, (b) HW, (c) BrW, and (d) IW that contribute to the binding of each ligand. After Lima Neto et al. [338]

residue in the binding site – i.e., their effectiveness as well as if they attract or repel the willardiines.

(b) The most important residue contributions to the bonding on the left side.

(c) The region (boldface letters, identified in Figure 14.7) and the atoms of the ligands closer to each residue at the binding site.

The binding energy of the amino acid residues interacting with the willardiine molecules inside the binding pocket will be defined here as the negative of the corresponding interaction energy calculated using the MFCC method.

As one can see from Figure 14.7, 14 amino acid residues are the most important for the stabilization of FW: T480, R485, T655, E705, P478, S654, Y450, L479, L650, G653, M708, E402, K656, and E486 – while five amino acid residues – K730, E657, A477, E688, and K722 – display positive (repulsive) interaction energies. For HW, there are 13 amino acid residues helping the stabilization process – T480, R485, T655, E705, P478, S654, M708, Y450, L479, L650, G653, E402, and E486, while five amino acid residues – K730, E657, A477, E688, and K722 – work against the binding. For the case of BrW, there are 13 amino acid residues that are attractive – T480, R485, T655, E705, P478, S654, Y450, L479, G653, L650, K656, E402, and E486 – while four amino acid residues – M708, K730, E657, and K722 – repel the ligand. Finally, for IW we have 13 attractive amino acid residues – T480, R485, E705, T655, P478, S654, Y450, L479, G653, L650, K656, E402, and E486 – and four repelling amino acid residues – M708, K730, E657, and K722.

Willardiines and glutamate bind to the ligand-binding core in a similar fashion [336]. Regions (**i**) and (**ii**) of willardiine (Figure 14.5a) create a hydrogen bond network with R485, P478, T480, and S654, T655, E705 residues, like other agonists [354]. Ten out of the 20 amino acids listed as the most important residues that contribute to the binding of the willardiines belong to region (**i**), and most of them show attractive interactions, with only two repulsive residues (K730 and A477), in agreement with a previous study [345], which mentions region (**i**) as a recognition site of iGluR2-AMPA receptors. Region (**ii**), which is located at the γ-carboxyl position of glutamate, has seven important interacting residues with three of them (E657, E688, and K722) being non-attractive ones.

As observed in Figure 14.8, E705 is the residue with the most intense binding energy contribution, attracting the ligand through a salt bridge between its side chain and the α-amino group in region (**i**) (N8), and through a hydrogen bond in region (**ii**) (O$_4$). After agonist binding, the residue E705 undergoes a rearrangement that affects the domains' closure. Some works indicate that mutation or neutralization of the residue E705 destabilizes the AMPA-LBD (ligand-binding domain) by decreasing the electrostatic repulsion between S1 and S2 domains [355].

Figure 14.9 displays the electrostatic potential isosurfaces with projected electron densities for the willardiines bound with the most important residues E705, S654, Y450, P478, and R485 at the binding pocket site.

The second-highest energy contribution belongs to R485, whose side chain is facing the α-carboxyl group of all four willardiines in region (**i**) and exhibits ionic and hydrogen bond interactions. The presence of the residue R485 as a docking site to iGluR2-AMPA agonists is confirmed by molecular, theoretical, and crystallographic studies [356]. The residues S654 and T655 form H-bonds in regions (**i**) (O$_{92}$) and (**ii**) (O$_2$ and N$_3$), respectively [336, 337]. The residue T655 forms only two hydrogen bonds, with IW having the lowest energy among other willardiines-T655 complexes (Figure 14.7d). These three residues are positioned in close conformation to other complexes, stabilizing the α-carboxyl (R485 and S654) and γ-carboxyl (T655) regions of the ligands [357].

The residue Y450 forms a cation–π interaction in region (**i**) with the four willardiines. Mutation Y450A results on a dramatic reduction in potency of glutamate to exclude the Y450 steric effect [356]. As we can see, the residue T480 interaction energy varies following the FW > HW > BrW > IW sequence,

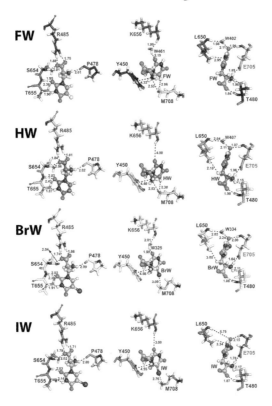

Figure 14.8 Distances between iGluR2-AMPA residues and the willardiines FW, HW, BrW, and IW. Their interaction energies are the most significant to the total binding energy of the four-substituted willardiines. After Lima Neto et al. [338]

like the residue S654, and forms two hydrogen bonds in region (**i**). Molecular dynamics shows that modifications in the residue T480 increases significantly the D1–D2 repulsion, because the backbone amide stabilizes the α-carboxyl group [358]. The residue P478, along with T480, creates hydrogen bond with the α-amino group of the willardiines.

The residue K656 interacts with the willardiines FW and BrW by water-mediated hydrogen bonding (w461 and w325 molecules, respectively), with larger binding energy than K656-HW or IW complexes. Likewise, the residue L650 shows the lowest energy when interacting with IW, being the only L650-willardiine complex without water-mediated hydrogen bonds (see Figure 14.8). It was pointed out that water molecules play a key role in modulating the cleft conformation [359]. This result is confirmed by our methodology if one considers the variation of calculated interaction energy values in residues with small distance shifts and water-mediated hydrogen bonds, such as the residues L650 and K656.

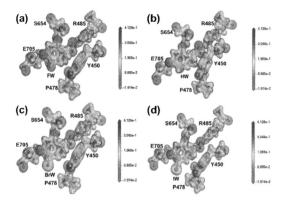

Figure 14.9 Electrostatic potential isosurfaces with the projected electron densities for four-substituted willardiines, (a) FW, (b) HW, (c) BrW, and (d) IW, interacting with the attractive residues E705, Sr654, Y450, P478, and R485. After Lima Neto et al. [338]

Lastly, our analysis shows that the residue M708, which is positioned in the cavity surrounding the halogen substituent, interacts attractively (repulsively) with the willardiines HW and FW (BrW and IW). This suggests that, together with the steric effect promoted by the size of the 4-substituted molecules, the closure of the individual domains might also be affected by the attractive/repulsive effect promoted by the electronegativity of the substituent atom. Thus, the increase in size of the four-substituent and the repulsive effect of the willardiines BrW and IW over the residue Met708 can lead the ligand-binding core to adopt distinct conformations that reduce the domain closure and, consequently, shorten the intra-dimer separation in the ion channel gate.

Among the 20 amino acid residues analyzed in this chapter, E705 and R485 are the residues with largest contribution to the binding of four-substituted willardiines interacting with the atom groups (i) $(N_8)H$, (ii) $(C_4)O_4$, and (i) $(C_9)O_{91},O_{92}$ of the willardiines. The residues Y450, T480, and P478, among others, contribute with attractive interactions, while E657, K730, and K722 repel all four-substituted willardiines. Moreover, it was suggested that the residue Y450 has a key role in the orientation of glutamate through cation–π interaction [360], while the residue M708 seems to interact directly with the four-substituent attracting willardiines HW and FW, repelling the willardiines BrW and IW.

14.6 Conclusions

One of the major goals in medicinal chemistry regarding drug discovery and design is to achieve an accurate description of the binding mechanism between the macromolecules of interest and their ligands. Quantum chemistry methods have proven to be a powerful tool to go in that direction, mainly due to the improvement of DFT methods to describe intermolecular interactions.

The partial agonism of four-substituted willardiines, a neurotransmitter composite very useful in structure-function studies of central nervous system disorders [342], can be used to understand the mechanisms of activation and desensitization of iGluR2-AMPA receptors, key players as mediators of the majority of excitatory synaptic transmission throughout the central nervous system. It was proposed that the size of the halogen substituent is directly involved in the modulation of the binding site and, consequently, in the closure of the lobes and activation of the ionic channel [343]. As a matter of fact, results reported in References [336, 359] have been pivotal to the understanding of the partial agonist mechanism in iGluR2-AMPA receptors. However, it is not clear what protonation state of the four-substituted willardiines is responsible for the activation level observed in those works. Note that recently a reasonable number of papers have showed that partial agonists probably have different lobe orientation distributions whose most stable crystal structures do not represent the entire mechanism of channel activation [361].

To fill this gap, we have performed DFT-dispersion-corrected calculations using the GGA-PBE-Grimme and LDA-PWC-OBS exchange-correlation functionals to find out the interaction energy profile of a set of four-substituted willardiines with iGluR2-AMPA receptors, taking into account more distant amino acid residues in order to obtain a good correlation between experimental and simulation data. The main advantage of the methodology proposed here is the possibility of evaluating what amino acid residues are more relevant to the stabilization of the four-substituted willardiines, which can be useful for drug design. A molecular fractionation scheme (MFCC) has allowed us to infer that modifications in the region (ii), which define the specificity to willardiines (γ-position of Glu), are prone to create new potent agonists or antagonists.

After a convergence study on the size of the binding pocket sphere, we have considered all ligand–residue interactions within a radius of 12 Å from the ligand. Observe that the willardiines FW, BrW, and IW have significant charged states at physiological pH. We tried to take this into account, but, unfortunately, we did not achieve a proper computer convergence of their total binding energies. However, it was already shown that uncharged protonation state is a possible representation of the molecule at physiological pH [348, 350].

Our results suggest that the protonation state with uncharged uracil ring can represent the partial agonist effect on iGluR2-AMPA receptors of the studied willardiines, an important step toward the understanding of the role played by the ionotropic glutamate receptors (iGluRs) in central nervous system disorders.

15

The Biology of Cancer

15.1 Introduction

The World Health Organization (WHO) labels cancer as one of the most pressing health challenges in the world. Statistical data from the National Cancer Institute (NCI) show that one in two men and one in three women will develop this disease.

Cancer is not just one disease but a generic term used to encompass a group of more than 200 diseases sharing common characteristics. It is characterized by an uncontrolled growth of cells in the body, with the potential to invade or spread to other parts of the body via the lymphatic system and bloodstream, creating secondary deposits known as metastases. Cancer is induced by mutations in the genome of a cell population that change the normal function of various classes of protein families – such as cytokines, cell surface receptors, signal transducers, and transcription factors – disturbing cellular activities that are crucial for the development and the maintenance of multicellular organisms, making it one of the most difficult and complex diseases to treat [362]. These changes may be inherited or may develop during the lifetime. Some changes happen for no known reason, while others are due to environmental exposures.

The first stage of a cancer is a noninvasive, non-life-threatening disease, in which frequently the cells grow into a tumor in the location where it started. However, cancer cells do not always form a compact tumor. Leukemia, for example, is a cancer in the blood-forming tissue where cancer cells circulate in the body and behave to some extent like healthy cells. However, in all cases, if not properly treated, these abnormal behavior grow in an uncontrolled way, contrasting with benign tumors, which do not spread. The earlier a cancer is detected, the less likely it is to metastasize, and so the more favorable is the prognosis for the individual. On the other hand, the more widely the cancer spreads, the harder it becomes to eradicate. Possible signs and symptoms include fatigue, a lump, abnormal bleeding, prolonged cough, unexplained weight loss, and a change in bowel or bladder function, to cite just a few.

A healthy person cannot catch cancer from someone who has it because cancer cells from one person are generally unable to live in the body of another healthy person. A healthy person's immune system recognizes foreign cells from another person, including cancer cells, and destroys them. However, there have been rare cases in which organ transplants from people with cancer have been able to cause cancer in the person who got the organ due to his (her) low immune system. If cancer were contagious, we would have cancer epidemics just as we have, for instance, flu epidemics, with a high rate of cancer among the families and friends of cancer patients and among health professionals because of their exposure to the disease. This is not the case.

Cancer is the second-leading cause of death in the world. However, survival rates are increasing for many types of cancer, thanks to improvements in cancer screening and cancer treatment. Cancer can take decades to develop, which is the main reason why most people diagnosed with cancer are 65 years or older. It is not generally possible to prove what caused a particular cancer because the various causes do not have specific fingerprints. The majority, some 90%–95% of cases, are due to genetic mutations from environmental and lifestyle factors; the remaining 5%–10% are due to inherited genetic defect.

There are two kinds of cancer genes: oncogenes, which are cancer-generating genes whose activation causes an uncontrollable distribution of cell tissue, and the tumour-suppressor genes, or anticancer genes, whose cancer-inducing effect is due to the cessation of their activity. Damage to genetic material happens continually in many cells. But the human body contains a defense system developed over a long period, and this is able to repair the damage. If the system breaks down, damaged cells can start to divide uncontrollably, eventually leading to carcinogenesis, a multistage process in which damage to a cell's genetic material changes the cell from normal to malignant. The damage gradually accumulates in the cell's growth regulatory system.

Although these mutations provide a selective advantage to populations of cancer cells, they also increase their divergences from the normal cells, which can allow the recognition by the immune system cells, such as the T lymphocytes (T cells) and B cells. Nevertheless, tumors also evolved to deceive immune cells, including the ability to activate co-inhibitory signaling pathways on T cells by immune checkpoint proteins, such as the cytotoxic T-lymphocite protein 4 (CTLA-4) and the programmed cell death protein 1 (PD-1), leading to a state of immune tolerance [363].

In this chapter, we present a quantum biochemistry method, based on the density functional theory (DFT) model and the molecular fractionation with conjugate caps (MFCC) scheme, to unveil the detailed binding energy features of three different methodologies in the fight against the cancer cells. In the first one, the tissue-selective synthetic agents selective estrogen receptor modulators (SERMs) 4-hydroxytamoxifen (OHT), and raloxifene (RAL), widely used in the breast cancer treatment, co-crystallized with the estrogen receptor α (ERα), is considered [364]. Then, we investigate the importance of integrins in several cell types that affect tumor progression, making them an appealing target for cancer therapy [365]. In particular, the integrin $\alpha_V \beta_3$ plays an important role in angiogenesis and tumor-cell metastasis, besides being evaluated as a target for new therapeutic approaches. Finally, we discuss the efficiency of immunotherapy as the most promising new cancer treatment of recent years. By reawakening and enhancing the immune system to fight cancer, such strategies have achieved impressive clinical responses [366, 367]. Much of the recent excitement has been generated by the recognition that immune checkpoint proteins, like the receptor PD-1, can be blocked by antibody-based drugs with profound effects.

15.2 Estrogen Receptor and Its Agonists/Antagonists

Breast cancer is the second most common malignant neoplasm worldwide and the fifth in cause of deaths. Over the years, some molecular alterations that reflect the biological characteristics of invasive breast carcinomas have been identified and considered as molecular markers, such as the estrogen receptors (ERs) [368]. Expressed in approximately 75% of invasive breast tumors, ERs belong to the nuclear receptor (NR) superfamily of ligand-activated transcription factors, like the glucocorticoid receptor (GR), androgen receptor (AR), retinoic acid receptor (RAR), and orphan receptors [369]. Through these receptors, NR controls the transcription of important genes for developmental, reproductive, neural, skeletal, growth, and cardiovascular processes, transforming them an important drug target.

Similar to other nuclear receptors, estrogen receptors are characterized by presenting a variable N-terminal transactivation domain, a conserved DNA binding domain (DBD), a variable hinge region, a conserved ligand-binding domain (LBD), and a variable C-terminal domain. The ER–ligand-binding domain (ER–LBD) is encoded within a region of about 300 amino acids, and its overall architecture is comprised of helices 1 to 12 (H1–H12) and two β-sheets, S1 and S2 (see Figure 15.1). It possesses the functions of ligand recognition and activation of physiological response. Thus, after ligand docking, there is an LBD rearrangement to the active state with a shift at the position of helix 12. It is followed by a formation of (homo- or hetero-) dimers that recognize specific sequences of the genomic DNA with the binding of many kinds of co-activators or co-repressors, which are related to transcriptional activation/repression, in order to mediate the transcription of various target genes.

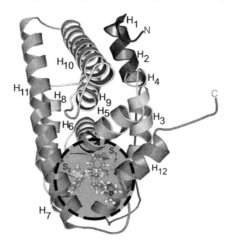

Figure 15.1 Structural representation of the estrogen receptor α ligand-binding domain (ERα-LBD) complexed with 4-hydroxytamoxifen (OHT; PDB ID: 3ERT). The binding pocket sphere with radius r is also shown in this illustration as a circle around the ligand. After Mota et al. [364]

The estrogen receptor (ER) family is composed by estrogen receptor α (ERα) and estrogen receptor β (ERβ), which are expressed in a wide range of tissues and cell types, sharing some level of sequence homology. Estrogen receptor α is encoded by ESR1 gene on chromosome 6 and has been found predominantly in the bone, testes, epididymis, prostate, uterus, ovary, liver, mammary gland, adipose tissue, heart, vascular system, and brain. Estrogen receptor β is encoded by ESR2 gene on chromosome 14 and is present in the bladder, prostate, ovary, colon, adipose tissue, immune system, heart, vascular system, lung, and brain. In breast cancers cases, it is common to see a decrease of ERβ expression and an increase in proliferative response associated to the ERα activation. Thus, many ERα antagonists/modulators have been developed as the selective estrogen receptor modulators (SERMs) to reverse the growth-promoting effect of estrogen in breast cancer.

SERMs, also called antiestrogens, are tissue-selective synthetic agents that modulate estrogen receptor transcriptional activity by acting, depending on the cellular differences, as agonists or antagonists such as cell-signaling pathways, accessory cofactors and DNA response elements. Tamoxifen and raloxifene are SERMs with the broadest utility for health maintenance in women. Tamoxifen, a triphenylethylene derivative, is one of the oldest, most well-known, and most prescribed SERMs with remarkable tissue-specific behavior [370]. It acts as an ER antagonist in the breast and decreases the incidence of invasive and noninvasive breast cancer. However, tamoxifen is associated with an increased incidence of venous thromboembolic events, vasomotor symptoms, stroke, hot flashes, and abnormal endometrial behavior. Raloxifene, a benzothiophene derivative, is used for the prevention of osteoporosis and breast cancer. It acts as an ER agonist in the bone and an antagonist in the breast and endometrium. Resembling tamoxifen, raloxifene use also shows an increase of vasomotor symptoms and the incidence of hot flashes.

Selective estrogen receptor modulators are very effective against breast cancer. However, these drugs have unwanted side effects in nontarget tissue, and the cancer could become resistant to antiestrogen therapy. Thus, a number of experimental and theoretical studies has been carried out to clarify the mechanism of the ER–ligand binding, and a wide repertoire of structurally distinct ligands that bind to ER with different degrees of affinity and potency have been proposed [371].

15.2.1 Structural Representation

Crystallographic X-ray structures obtained by Shiau et al. [372] and Brzozowski et al. [373] represent the complex structure in which agonists and antagonists occupy the same ER–LBD binding pocket. From them, we became aware that the binding of 4-hydroxytamoxifen (OHT) and raloxifene (RAL) is guided by their phenolic ring (see Figure 15.2a, region i) as well as the bulky side chain, which is involved by the hydrophobic core of LBD. However, even sharing the same binding site, agonists and antagonists bring different rearrangements of the receptor that are related to the activation or deactivation of its transcriptional activity.

As shown by crystallographic structures, when estrogen receptor α is bound to antagonists, such as OHT and RAL, there is a reorganization of some helices

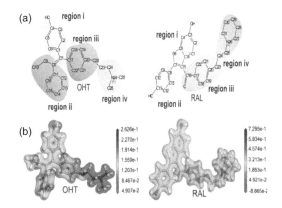

Figure 15.2 Representation of the 4-hydroxytamoxifen (OHT in the left), and raloxifene (RAL in the right) physiological pH and electronic density distribution. (a) Chemical structure subdivided into four parts to help the analysis of its interactions with the estrogen receptor ERα; (b) electron density projected onto an electrostatic potential isosurface showing negatively and positively charged regions in a light and dark gray color, respectively. After Mota et al. [364]

followed by a motion of helix 12 toward the coactivator-binding groove formed by the residues from helices 3, 4, and 5. This, in turn, leads the receptor to a deactivated state by mimicking the coactivator helix binding, creating an impairment of its attachment [372]. Thus, to understand the differences between OHT/RAL@ERα complexes, it is important to notice the binding characteristics of these ligands. Each one of the four regions (i, ii, iii, and iv in Figure 15.2a) of the OHT and RAL interacts with at least one amino acid into our binding pocket radius r (15 Å).

Crystallographic data of the 4-hydroxytamoxifen (OHT) and raloxifene (RAL) co-crystalized with ERα were downloaded from the PDB database (www.rcsb.org) under the codes (resolutions) 3ERT (1.90 Å) and 1ERR (2.60 Å), respectively. The state of protonation of all ligands in physiological pH was obtained using the MarvinSketch code version 5.3.2 (Marvin Beans Suite – ChemAxon). The interaction energy between each SERM molecule and the ERα amino acid residues were calculated by using the MFCC method [22, 23]. The convergence study of the total interaction energy as a function of the ligand-binding pocket radius r was performed in order to put a limit on the number of amino acid residues to be analyzed without missing important interactions. Here, we label the ligand molecule as L and the ith amino acid residue interacting with the ligand as R^i. The C^{i-1} (C^{i+1}) cap is formed from the neighboring residues covalently bonded to the amine (carboxyl) group of the residue R^i along the protein chain. The total interaction energy for RAL and OHT is obtained by adding up the interaction energies with each amino acid residue within a given binding pocket radius. Water molecules were taken into account in the calculation procedures when hydrogen bonds are formed with the residues of interest or with the caps. A hydrogen bond length limit of 2.5 Å was adopted in this case. Energetic calculations for each residue at the binding site were performed using the quantum chemistry approach, previously discussed in this book.

15.2.2 Binding Interaction Energies

Although the first SERMs were identified more than 60 years ago, their molecular basis only began to be unraveled in the last 25 years. Nevertheless, recently, computational and crystallographic studies have reported that the nature of the ligand is a critical determinant of its action in estrogen receptors. Thus, a binding energy scenario, as discussed here, of the complexes between 4-hydroxytamoxifen (OHT)/raloxifen (RAL) and estrogen receptors α (ERα) is a crucial step in this process. As shown in Figure 15.2a, these ligands share some chemical features, but they show distinct agonism/antagonism activity and affinity in different tissues and receptors. Both ligands are subdivided into four regions, as depicted in Figure 15.2a, to facilitate a further analysis. These selected parts are used to describe the most important residues that interact with each SERM region. Figure 15.2b presents the electron density isosurface for the OHT and RAL, in order to clarify how the electrons are distributed around the molecules.

All calculations were performed starting from the crystallographic estrogen receptor α (ERα) structure co-crystallized with 4-hydroxytamoxifen (OHT) and raloxifene (RAL) that have been deposited in the protein data bank (PDB ID: 3ERT and 1ERR, respectively) [372]. The binding energies between of ERα-LDB and two of its antagonists, 4-hydroxytamoxifen (OHT) and raloxifene (RAL), were calculated employing the MFCC scheme to obtain the individual contribution of the amino acid residues inside a binding pocket with a selected radius, ranked from the most relevant interactions, since the determination of ligand characteristics linking their affinity to ER is quite important for the development of new therapeutic agents.

To take into account each significant attractive and repulsive amino acid residue, we perform a search for an optimal binding pocket radius for which no significant variation in the total binding energy could be observed after a radius increase. Figure 15.3 presents a comparison between the calculated total interaction energies $E(r)$ (in kcal/mol) for the OHT and RAL as a function of the binding pocket radius r (in Å), which indicates that the converged binding pocket radius is about 15.0 Å, comprising a total of 138 and 145 amino acids, respectively. Note that the interaction energy follows the order OHT > RAL for each one of pocket radius $r = 4.0$, 6.0, 11.5, 13.0, and 15.0 Å ($-184.470 > -154.410, -164.760 > -121.360, -282.1701 > -250.220, -360.420 > -333.640, -346.230 >$

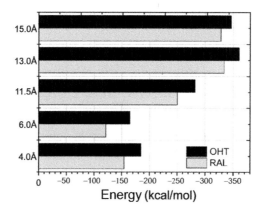

Figure 15.3 Total SERM interaction energy for OHT and RAL, considering the pocket radius $r = 4.0$, 6.0, 11.5, 13.0, and 15.0 Å, calculated by using the GGA-PBE Grimme exchange-correlation functional. The absolute value of $E(r)$ obeys the sequence OHT > RAL, reproducing experimental data. After Mota et al. [364]

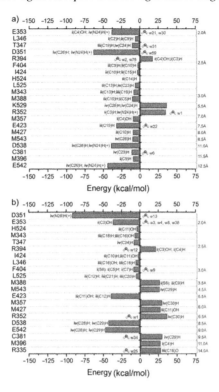

Figure 15.4 Binding interaction energy graphic panel showing the most relevant residues of ERα that contribute to the binding of each ligand: (a) OHT and (b) RAL. After Mota et al. [364]

−327.750 (all units in kcal/mol). These results are in agreement with the experimental data of Clegg et al. [374].

Figure 15.4a and b show a binding interaction energy graphic panel with the interaction energies between the OHT and RAL and the most important amino acid residues at the binding region. The panel depicts the following:

(a) The interaction energy (in kcal/mol) of the residue with the ligands, illustrated by the horizontal bars, from which one can assign quantitatively the role of each residue in the binding site – i.e., their effectiveness as well as whether they attract or repel this SERMs.

(b) The most important residues contributing to the bonding in the left side.

(c) The region (boldface letters, identified in Figure 15.4) and the atoms of the ligands closer to each residue at the binding site.

As one can see from Figure 15.4, the pattern of interaction seems to be simpler for OHT than for RAL. For OHT, the attractive interactions are predominant, but we also observe a few repulsive interactions. The residues that bind more significantly are D351, E542, E353, D538, E423, C381, T347, M388, M543, M396, M427, M357, M343, L525, L346, I424, F404, H524, R394, R352, and K529 – the last four interacting in a repulsive way. For RAL, repulsive interactions appears more frequently, when compared to OHT. Its main residues are D351, E542, D358, E423, E353, F404, T347, M343, L346, L525, H524, I424, R394, R335, M543, M388, M396, M427, M357, C381, and R352, the last nine being repulsive. The difference of attractive and repulsive energies proportion between OHT and RAL may suggest why the first binds tighter than the latter, when considering the overall energy.

Region iv of OHT and RAL is comprised of a positively charged nitrogen that is stabilized by the residue D351 and other hydrophobic contacts. In our calculation, D351 is the residue with the most intense binding energy contribution to both ligands, with interaction energies of -62.92 (OHT) and -91.55 kcal/mol (RAL) (see Figure 15.4). As observed in Figure 15.5, it attracts the ligands through a salt

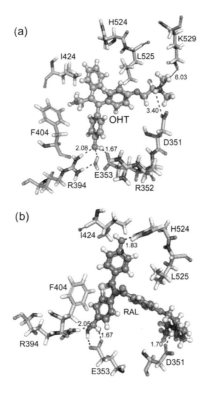

Figure 15.5 Arrangement of the most relevant amino acid residues involved in ERα–SERMs binding. Drug–residue contacts and their distances are depicted by dashed lines. (a) OHT and (b) RAL. After Mota et al. [364]

bridge between its carboxylate side chain and region iv N24(N26)H of OHT (RAL) with a distance of 3.40 (1.70) Å. Crystallographic, mutation, and computational analyses have shown that its position correlates with the entrance of the helix 12 (H12) into the active or unactive form when the ERs (α and β) are attached with agonists or antagonists, respectively [375].

The second-highest binding energy contribution to OHT and RAL (-44.26 and -47.04 kcal/mol, respectively) belongs to the residue E542, which presents different positions in OHT and RAL pocket radius distribution (12.5 and 9 Å, respectively) but are closed to region iv, as depicted in Figure 15.4. It is followed by the residue D538, exhibiting the third-highest binding energy, showing -39.05 (-42.72) kcal/mol to OHT (RAL). Note that E542 and D538 are residues highly conserved that occupy helix 12 (H12) and create a negatively charged area to co-activators, which are, therefore important residues in ER activation or deactivation [372, 375].

A phenolic ring of estrogen is shared by many selective estrogen receptor modulators. It is seen by forming hydrogen bonds with the carboxylate group of E353 (H3) and the guanidinium group of R394 (H6) and a water molecule conserved between E353 and R394 amino acids. As one can see from Figure 15.5, residues E353 and R394 interact in a very similar way with the two SERMs through this hydroxyl phenolic group (region i). It is confirmed by our calculated binding energies between OHT and RAL, respectively, that are closed to each other (see Figure 15.4), with 17.25 (2.08 Å) and 21.54 (2.05 Å) kcal/mol for the residue R394, and -38.06 (1.67 Å) and -35.35 (1.67 Å) kcal/mol for the residue E353 (the fourth most attractive interaction energy). In our calculations, the conserved water molecule was attached to the closer residue R394 (E353) for the complex OHT–ERα (RAL–ERα). Due to the differences in quantity of the crystallographic water molecules, it is named w2 in the first complex and w3 in the second one. In both cases, this water is responsible for forming a water-mediated hydrogen bond, beyond the direct H-bond made by these amino acids. Beyond the H-bond, the energetic results are also an electrostatic effect of the charged region iv that is prevalent in both cases. Thus, even making hydrogen bonds with the same molecular group, we find opposite binding energies. The residue E423 (H8) is the fifth one with the most intense binding energy, with interaction energy of -31.98 and -39.25 kcal/mol, and radius position in 7.5 Å and 5.5 Å to OHT and RAL SERMs, respectively.

To conclude, among the amino acids analyzed here, our results show that the strongest attractive energies for both drugs follow the decreasing sequence D351 > E542 > D538 > E353 > E423, suggesting their importance to the interaction between the ligand and receptor. Except for the residue E423, all these amino acids, by crystallographic and mutation analysis, are relevant to the recognition of ligands and activation or deactivation of ERα transcriptional function. When we consider these specific residues, we observe similar binding energies for the two antagonists. However, we found considerable differences in other residues that may be related to the charge distribution, distance between ligand and residue, and the presence of water molecules. Our results show that the loss of hydroxyl in region ii by OHT decrease its interaction energy with the residue H524, which is a member of the H-bond network in ERα binding pocket (R349, E353, L387, and H524) and is considered to be very important to the ligand's recognition. We also show the relevance of some methionine and cysteine residues (M343, M388, M357, M427, M543, C381, and M396) that are attracted by OHT but repelled by RAL. The existence of more repulsive energies for RAL might suggest a reason why it can remain as an antagonist when bound to ERβ (whereas OHT plays an agonist role), knowing that the LBD of human's ERβ keeps 55% homology when compared to ERα.

15.3 Energetic Description of Cilengitide Bound to Integrin

Integrins are heterodimeric α/β transmembrane receptors that play a central role in the cell–cell adhesion and cell–matrix interaction through binding and association with transmembrane receptors, proteins, and soluble ligands in the extracellular matrix (ECM). In addition to adhesive functions, integrins transduce messages via various signaling pathways and influence proliferation and apoptosis of tumor cells, and they are responsible for the activation of endothelial cells, which provide a repertoire of functions needed for cancer cell migration, invasion, and proliferation [376].

In particular, integrin $\alpha_V\beta_3$ and $\alpha_V\beta_5$ are implicated in tumor neoangiogenesis, cancer development, and the metastatic process, as they are expressed in malignant tumors and intervening in adhesion of tumor cells in a variety of extracellular matrix proteins, allowing these cells to migrate during invasion. The integrin $\alpha_V\beta_3$ (see Figure 15.6) is weakly expressed at low levels in most normal tissues, while high levels of expression are limited to bone, the mid-menstrual cycle endometrium, placenta, inflammatory sites, invasive tumors, and activated endothelial cells during angiogenesis, where it mediates migration during the formation of new vessels, which is essential for adequate nutrient supply for the growing tumor. Also, high levels of this integrin receptor in glioblastoma and melanoma cells have been shown to reinforce tumor proliferation, spread, and adhesion. On the other hand, its inhibition by selective antagonists in preclinical studies has stopped angiogenesis and cell adhesion, leading to tumor regression in experimental breast cancer cells and melanoma and the induction of apoptosis of activated endothelial cells.

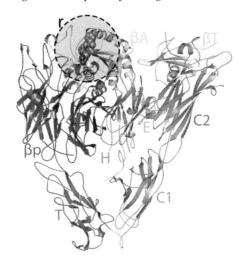

Figure 15.6 Structural representation of the integrin in complex with cilengitide, showing the α_V (dark gray) and β_3 (light gray) subunits assembled into an ovoid head and two nearly parallel tails (PDB ID:1L5G). The integrin α_V consists of a seven-bladed β-propeller (βp) and a tail with a tight (T) and two calf domains (C1 and C2), while the integrin β_3 is composed of a βA domain followed by a hybrid domain (H), four epidermal growth factor domains (E), and a β-tail domain (βT). The binding pocket sphere with radius r is also shown in this illustration as a black dashed circle around the cilengitide ligand. After Lima Neto et al. [365]

Integrin $\alpha_V\beta_3$ has been considered to be an important molecule in biotechnological approaches within cellular, cancer, and anti-angiogenesis therapies [377], and many efforts are applied to understand it and its molecular basis to ligand-binding mechanisms. It is assembled in a short cytosolic tail (<70 residues), a single-pass transmembrane (TM) domain, and a large extracellular domain (>700 residues) containing, in the β_3 subunit, a metal-ion-dependent adhesion site (MIDAS) composed of amino acids D119, S121, S123, E220, and D251 (see Figure 15.7). This receptor recognizes many ECM molecules – such as vitronectin, fibronectin, fibrinogen, laminin, collagen, Von Willebrand's factor, and osteoponin – in a specific RGD (Arg–Gly–Asp) amino acid sequence, which has been used as a scaffold to create numerous RGD-functionalized materials or peptidomimetics for medical applications [378]. Accordingly, RGD-based drugs and peptides containing RGD sequences are used in association with drugs, prodrugs, nanoparticles, and carriers in therapies of cancer and anti-angiogenesis (see Reference [379] and the references therein for a recent review).

Among these compounds, cilengitide (CIL), whose chemical structure is depicted in Figure 15.8, has been shown as a potent and selective antagonist of the integrin $\alpha_V\beta_3$ and $\alpha_V\beta_5$ receptors. It promotes the detachment of glioblastoma and

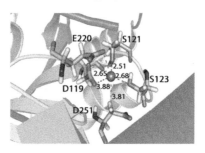

Figure 15.7 Arrangement of the metal-ion-dependent adhesion site (MIDAS) amino acid residues found in integrin $\alpha_V\beta_3$. After Lima Neto et al. [365]

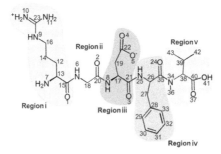

Figure 15.8 Cyclic structure of cilengitide subdivided into five regions to help the analysis of its interactions with integrin $\alpha_V\beta_3$. Regions i–v correspond to arginine, glycine, aspartic acid, D-phenylalanine, and (n-methyl) valine, respectively. After Lima Neto et al. [365]

mesothelioma cells from ECM components and, as a consequence, the activation of apoptosis [376]. In addition, it shows favorable safety and no-dose-limiting toxicities in clinical trials, enhancing radiotherapy efficiency in endothelial cell and non-small cell lung cancer models. In addition, the conformation assumed by the RGD motif in cilengitide is similar to those found in echistatin and fibronectin, suggesting that this cyclic pentapeptide structure can serve as a basis of understanding integrin–ligand interactions. Hence, crystallographic and computational studies have been pivotal to yield good insight into the molecular basis that leads cilengitide (and other small RGD-based molecules) to be recognized by the $\alpha_V\beta_3$ receptor, contributing to the rational design of new RGD-based drugs and peptide-mimetic compounds with RGD motifs specifically recognized.

15.3.1 Drug–Receptor Complex Data: Classical and Quantum Calculations

To perform the calculations in this study, we used the X-ray crystallographic structure of the integrin $\alpha_V\beta_3$ co-crystallized with cilengitide obtained from the PDB database (www.rcsb.org) under the code (resolution) 1L5G (3.2 Å). The state of protonation of the cyclic peptide (ligand) in physiological pH was obtained, as usual, using the MarvinSketch code version 5.3.2.

The interaction energy between the cilengitide molecule and the $\alpha_V \beta_3$ amino acid residues were calculated by using, again, an MFCC-based scheme within a DFT framework.

Energetic calculations for each residue at the binding site were performed using the Gaussian 09 code, within the DFT formalism, by using the GGA functional B97D [19]. Being aware that most biomolecular interactions take place in an electrostatic environment – being, therefore, important to consider the effects of the surrounding molecules in many cases – we use the conductor-like polarizable continuum model (CPCM) with the dielectric constant ε equal to 20 and 40, to represent the environment surrounding each fragment obtained by the MFCC scheme.

To avoid missing important interactions, a convergence study of the total binding energy as a function of the ligand-binding pocket radius was performed, limiting the number of amino acid residues to be analyzed. To this end, we added the individual interaction energy of those amino acid residues within imaginary spheres whose binding pocket radius r achieved convergence when the energy variation in subsequent radii is smaller than 10%.

15.3.2 Interaction Energies of the Ligand and the Amino Acid Residues

To provide a deeper understanding of the cilengitide–integrin $\alpha_V \beta_3$ binding, we energetically inspect the interactions between the ligand and the amino acid residues and select the strongest interactions, considering the dielectric constants $\varepsilon = 20$ and 40, as plotted in Figure 15.9, which depicts a BIRD panel with the highest binding energies found by our calculations. It shows that 15 amino acid residues were selected as the most important for the cilengitide–$\alpha_V \beta_3$ complex.

For $\varepsilon = 20$, 13 residues show attractive (negative) binding energies (in kcal/mol), according to S121 (-14.23) > D218 (-13.57) > R214 (-12.26) > D217 (-4.43) > D150 (-4.18) > A218 (-2.85) > A215 (-2.28) > R216 (-2.03) > S213 (-1.97) > Y122 (-1.95) > D219 (-1.78) > Q180 (-1.33) > K125 (-1.31), while two amino acid residues display repulsion (positive) interaction energies: N215 (2.04) > E220 (6.28).

As we can see from the $\varepsilon = 40$ results, the order of importance of some amino acids change just in the position of Y122, as follows: S121 (-13.66) > D218 (-11.74) > R214 (-11.40) > D217 (-4.57) > D150 (-2.84) > A218 (-3.02) > A215 (-2.30) > R216 (-1.85) > S213 (-1.84) > D219 (-1.29) > Q180 (-1.24) > K125 (-0.86) > Y122 (-0.50). Two amino acid residues display

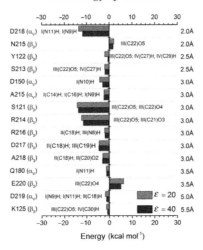

Figure 15.9 Graphic panel showing the most relevant residues that contribute to the integrin $\alpha_V \beta_3$-cilengitide complex. The minimal distances between each residue and the ligand, as well as the region and the atoms of the ligands closer to each residue at the binding site are also shown. After Lima Neto et al. [365]

repulsion (positive) interaction energies: N215 (1.74) > E220 (5.07). Among them, only five residues (D218, D150, A215, Q180, and D219) comprise the β-propeller domain.

Figure 15.10a–c shows some of the main amino acid residues and their respective intermolecular interactions with cilengitide. As observed in Figure 15.10a, charged Region i is surrounded by oxygens from carboxylate moiety of three aspartic acids, D218, D150, and D219 (β-propeller), beyond an oxygen atom from Q180 carbonyl group. In our calculations, D218 is the residue with the second-most attractive binding energy contribution. It attracts the ligand through salt bridges between its negatively charged side chain and two amine groups belonging to the side chain of Region i (arginine), I(N11)H and I(N9)H, at 1.99 and 1.90 Å distances, respectively.

The fifth (sixth) amino acid with the most attractive binding energy for $\varepsilon = 20$ (40) is D150, which is involved in a salt bridge with region i [(N10)H] at a distance of 2.89 Å. The residues D219 and Q180 are the 11th and 12th (10th and 11th) most attractive residues for $\varepsilon = 20$ (40). The side chain of the latter makes contact with I(N11)H at a distance of 3.43 Å, while the former (D219) contacts cilengitide in I(N9)H and II(C18)H at 6.64 Å and 5.17 Å, respectively.

Throughout molecular-docking calculations of nine cyclic and linear compounds, it was possible to determine a map of interactions of each one of these compounds with the integrin $\alpha_V \beta_3$. The key role played by the residues D218 and D150 in the binding mechanism was also observed in other RGD-based compounds. Synthesizing 38 small anticancer agents without the RGD motif, the

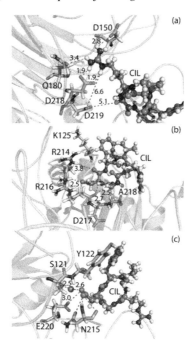

Figure 15.10 The main amino acid residues and their respective intermolecular interactions with the integrin $\alpha_V\beta_3$–cilengitide complex. (a) region i; (b) regions ii and iii; (c) region iii. Regions iv and v are not shown here because they present few amino acid residue contacts (mainly the polar amino acids). Potential hydrogen bonds are indicated by the dashed lines in (a) and (b), while the repulsions are shown by the dashed lines in (c). After Lima Neto et al. [365]

interaction of the residue D150 with them is retained in all compounds used in the docking calculations. Besides, using NMR and molecular docking to characterize RGD-based linear peptides bonded to $\alpha_V\beta_3$, one can show that the change of the phenyl group to phenol makes these molecules more hydrophilic, improving their binding affinity [380]. Also, the superposition of the crystallographic structures of $\alpha_V\beta_3$ and $\alpha_{IIb}\beta_3$, heterodimeric structures with a β_3 subunit, shows that the position of the residue D218 is now occupied by a phenylalanine, creating a hydrophobic impairment that hinders the binding to cilengitide.

Some of the main amino acid residues and their respective intermolecular interactions with the cilengitide–$\alpha_V\beta_3$ complex in Regions ii and iii are shown in Figure 15.10b. Region ii forms 13 contacts with residues from integrin, most of them coming from $C_\alpha H$ [(C18)H]. Among the 15 amino acids chosen as the most energetics, four make contacts with this region, each one showing attractive interactions with cilengitide (R216, D217, A218, and D219). Through molecular docking and molecular dynamic techniques in cyclic peptides into the integrins $\alpha_V\beta_3$ and $\alpha_{IIb}\beta_3$, it was possible to demonstrate that Gly (here represented as region ii) is an important amino acid of the RGD motif since it is liable to make some weak interactions that stabilize the peptide ($C_\alpha H \cdots O{=}C$). The residue R216 forms an H-bond with III(N8)H at a distance of 2.70 Å, while it is involved in non-conventional H-bonds, together with the residue A218, making contacts with II(C18)H and II(C20)O2, respectively.

Meanwhile, the residues R214 and K125 form contacts with region iii(C22) O5 at a distance of 2.5 and 7.0 Å, respectively, as long as R214 also makes an interaction between its NH from the guanidine group and the oxygen carbonyl from the main chain of III(C21)O3 at the distance of 3.8 Å. Furthermore, one way to become a ligand selective to $\alpha_V \beta_1$ is the insertion of properly oriented bulky moieties in the ligand, since the superposition of the $\alpha_V \beta_3$ and $\alpha_5 \beta_1$ integrins shows that the replacement of (β_3)-Arg214 with (β_1)-Gly217 and (β_3)-Arg216 with (β_1)-Leu219 opened up a new space adjacent to the RGD binding site, which is absent in the $\alpha_V \beta_3$ receptor but present in the $\alpha_5 \beta_1$ one.

As one can see from Figure 15.10c, the residue N215 forms a hydrogen bond (1.85 Å) between its amino group of the main chain and the oxygen from carboxylate group of cilengitide [III(C22)O5]. Besides, two oxygens from its carbonyl group of the main chain (3.09 Å) and side chain (3.23 Å) are also in closer contact with the same atom of ligand. Thus, even making a hydrogen bond, a repulsive binding energy prevails. Also, the residue S121 forms two contacts with carboxylate from cilengitide, one weak hydrogen bond with III(C22)O4 (2.8 Å) and a cation-mediated contact with III(C22)O5 (5.1 Å). The latter one (E220) presents an electrostatic repulsion due to the negatively charged carboxylate group [III(C22)O4] is positioned almost face to face to its side chain carboxylate. It is important to notice that though residues S121 and E220 interact with region iii, they show opposite binding energies.

The residue E220 – along with P219, D158, and N215 – constitutes a ligand-associated metal binding site (LIMBS) or, as recently named, a synergistic metal ion binding site (SyMBS). It is a positive regulatory site that is expected to coordinate its orientation toward MIDAS, stabilizing the ligand-binding surface through synergistic effects of low cation concentrations. To understand the effect of LIMBS under integrin adhesiveness, experimental mutations in N215A, D217A, and E220A were made in the $\alpha_4 \beta_7$ integrin [381]. The results showed that these mutations lead to loss of the adhesion function of this integrin, even with 52% of the amino acid sequence identity being conserved between the β_7 and β_3 subunits.

Overall, among the amino acids analyzed in this paper, we found that the binding energies for $\varepsilon = 20$ ($\varepsilon = 40$) follow the decreasing residues sequence: S121 > D218 > R214 > D217 > D150 > A218 > A215 > R216 > S213 > Y122 > D219 > Q180 > K125 > N215 > E220 (S121 > D218 > R214 > D217 >

D150 > A218 > A215 > R216 > S213 > D219 > Q180 > K125 > Y122 >
N215 > E220). Most of these residues are consistent with previous results found
by crystallographic structures, computational, and mutation studies.

15.3.3 Total Binding Energy and Convergence

To evaluate the binding interactions through, fragment-based quantum mechanics
method, it is important to take into account every significant attractive and repulsive
amino acid residue that can influence this mechanism. However, the extracellular
domain is composed of 946 (666) amino acids of the α_V (β_3) subunit, with 438
(243) of them forming the β-propeller (βA) domain. Therefore, instead of taking
an arbitrary region of the binding site, we performed a search for an optimal binding
pocket radius (r) in which a variation less than 10% of the sequential pocket radius
could be observed after a radius increase. For this task, the binding pocket radius r
is varied from 2.0 Å (with binding energy equal to -10.66 for $\varepsilon = 20$ and -9.02 for
$\varepsilon = 40$) to 10 Å (with binding energy equal to -56.69 for $\varepsilon = 20$ and -54.75 for
$\varepsilon = 40$) in order to determine the best value of r, found to be 4.5 Å corresponding
to, a binding energy equal to -54.63 (-50.12) for $\varepsilon = 20$ (40), from which the
convergence was achieved (all binding energies measured in kcal/mol).

According to Figure 15.11, there is a sharp increase in the total binding energy
between $r = 2.5$ Å (-14.58 for $\varepsilon = 20$ and -11.36 for $\varepsilon = 40$) and 3.0 Å (-56.79
for $\varepsilon = 20$ and -51.20 for $\varepsilon = 40$) mainly due to the attractive binding energies
of the residues D150, S121, R214, and D217 to cilengitide. Afterward, a slight
decrease of the total binding energy is seen at the pocket radius $r = 3.5$ Å (-51.84
for $\varepsilon = 20$ and -47.37 for $\varepsilon = 40$), with main contribution of the repulsive energy
of the residue E220. No residue was found in $r = 4.0$ Å, while attractive energy

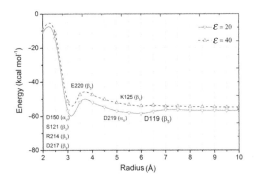

Figure 15.11 The total interaction energy as a function of the binding pocket
radius r calculated using the GGA functional B97D for the dielectric constant
$\varepsilon = 20$ (solid line) and 40 (dotted line). Amino acid residues responsible for the
regions of steepest negative and positive variation are highlighted. After Lima
Neto et al. [365]

of residue D219 increases the total energy in $r = 5.0$ Å (-57.00 for $\varepsilon = 20$ and -52.10 for $\varepsilon = 40$). After that, from $r = 7.5$ Å onward, the total binding energy as a function of r tends to present small oscillations.

Subunits α_V and β_3 are seen in a different number of integrin members, whose binding energetic scenario may distinguish each one of them. The relevance of the β_3 subunit to ligand-binding mechanism is evidenced by the presence of 14 more residues than the α_V one, which is reflected in a binding energy for $\varepsilon = 20$ ($\varepsilon = 40$) approximately 4.00 (11.00) kcal/mol smaller than to the α_V.

In summary, the binding energies of each amino acid at the subunits α_V and β_3 are -26.23 (-21.94) kcal/mol and -30.46 (-32.76) kcal/mol for $\varepsilon = 20$ ($\varepsilon = 40$). Looking to the most energetic residues, we can see that the repulsions are coming from the β_3 subunit, mainly by interactions with negatively charged carboxylate groups of Region iii. Out of the 76 amino acid residues analyzed here, residues S121 and D218 are those with the largest contribution to the binding of cilengitide, interacting with the atom groups III(C22)O4; III(C22)O5 and I(N11)H; I(N9)H of the ligand, respectively. The residues D150, R214, D217, and D219, among others, contribute to attractive interactions, while N215 and E220 repel cilengitide, suggesting that the distribution of negatively charged residues in β_3 surrounding the aspartic acid from RGD may decrease the affinity between the cilengitide and $\alpha_V \beta_3$ integrin.

After adding up the energies of each amino acid interacting with cilengitide, we classified the regions in terms of the binding energies as follows: region iii ($E = -26.88$ for $\varepsilon = 20$ and -28.56 for $\varepsilon = 40$) > region i ($E = -25.48$ for $\varepsilon = 20$ and -21.45 for $\varepsilon = 40$) > region ii ($E = -12.15$ for $\varepsilon = 20$ and -12.18 for $\varepsilon = 40$) > region iv ($E = -3.91$ for $\varepsilon = 20$ and -3.21 for $\varepsilon = 40$) > region v ($E = 0.32$ for $\varepsilon = 20$ and 0.14 for $\varepsilon = 40$) – all binding energies in kcal/mol.

Meanwhile, comparing the number of contacts made with the $\alpha_V \beta_3$ integrin, the rank is as follows: region iii (36) > region i (29) > region ii (13) > region iv (12) > region v (1). It is important to notice that regions iv and v are in the surface of the integrin exposed to solvent and surrounded by the polar residues Y122 and S123 (region iv) and Y178 (region v). Thus, despite being hydrophobic, regions iv and v have larger side chains than region ii, although with fewer contacts (mainly the polar amino acids).

15.4 Cancer Immunotherapy

The molecular identification of cancer antigens helped the creation of new approaches for effective therapies, giving rise to a new era of treatment in which our own immune system evades the block created by malignant cells and fights against them. This so-called cancer immunotherapy treatment is trying to overcome the cancer's ability to resist the immune responses by stimulating the body's own mechanisms to remain effective against the disease [382].

A number of different immunotherapeutic approaches have been investigated, depicting the inhibition of checkpoint proteins as a leading strategy for evading the blockade created by the tumor environment against commonly used therapies. In this sense, immunomodulatory antibodies that directly enhance the function of the lymphocyte T (T cell) by avoiding the activation of its negative regulators (named *molecular checkpoints*) are considered the major breakthrough in cancer immunotherapy [383, 384].

Among these checkpoint proteins, inhibitors that block the programmed cell death receptor 1 (PD-1) pathway showed a powerful clinical potential since the PD-1 receptors expressed in the T cells play a pivotal role in the down regulation of the immune system, and their ligands are overexpressed in malignant cells. Early clinical results using blocking agents against the human cell surface receptor PD-1 and its ligands PDL-1 and PDL-2 point to unprecedented rates of long-lasting antitumor activity in patients with metastatic cancers of different histologies. From the pharmaceutical point of view, however, the clinical development of the PD-1 pathway blockers requires an understanding of the signals that induce the expression of its ligands within the tumor.

Together with its ligands PD-L1/PD-L2, the receptor PD1 forms a family of immune checkpoint proteins that act as co-inhibitory factors, which can halt or limit the development of the T cell response. Through interactions with them, the protein PD-1 negatively affects the function of the T and B cells by inhibiting their immune check, ensuring that the immune system is activated only at the appropriate time in order to minimize the possibility of chronic autoimmune inflammation, leading to a decrease of the cytokine production and antibody formation [385, 386].

On the other hand, tumor cells exploit this immune-checkpoint pathway as a mechanism to evade detection and inhibit the immune response, preventing the immune system from killing cancer cells. Many types of cells can express PD-L1, including tumor cells and immune cells after exposure to cytokines, while PD-L2 is expressed mainly on dendritic cells in normal tissues, making the PD-1/PD-L1 interaction more suitable to cancer therapies. Thus, the search for antibodies that block the interaction between the receptor PD-1 and its ligands will be seen as a major therapeutic advancement in oncology when it shows response often durable without causing serious toxicity effects in most people [387].

The receptor PD-1 (Pdcd1 gene on chromosome 2), a member of the B7-CD28 family, is an immune cell-specific surface inhibitor, mainly expressed in the late effector phase on activated $CD4^+/CD8^+$ T cells, B cells, monocytes, natural killer T cells, and antigen-presenting cells (APC), including dendritic cells. It is a 55-kDa monomeric type I surface transmembrane glycoprotein of the

Ig superfamily, accounting for 288 amino acids, displaying four domains – including a single V-set immunoglobulin superfamily (IgSF) domain, a stalk, a transmembrane domain, and a cytoplasmic domain that contains two tyrosine-based immunoreceptor signaling motifs: the inhibitory motif (ITIM) and the switch motif (ITSM).

However, there are some patients showing resistance to the blockage of the receptor PD-1. Overstimulation of immune responses, on the other hand, it may lead to damage normal, healthy tissue. Therefore, to be more efficient, the design and production of such a pharmaceutical drug require a precise knowledge of not only its biochemical structure but also its binding mechanism with the receptor PD-1, which leads to a better understanding of how this complex influences the cancer tumor and its anti-angiogenesis therapies as well as its microenvironment. Computational tools based on quantum chemistry may be a promising step in this route, further increased by combination with other anticancer therapies.

In spite of that, though clinical results using blocking agents against the human cell surface receptor PD-1 and its ligants PD-L1 and PD-L2 point to unprecedented rates of long-lasting antitumor activity in patients with different types of cancer, some of them present resistance to the blockage of the receptor PD-1 and over-stimulation for the immune responses [388]. Therefore, the quick design of new drugs is essential and requires a precise knowledge of not only of its biochemical structure but also its binding mechanism with the receptor PD-1, leading to a better understanding of how this complex system influences the cancer tumor and its microenvironment.

Due to its relevance for immune system maintenance, as well as the evidence indicating the success of monoclonal antibody therapy against the PD-1/PD-L1 pathway, a number of crystallographic structures of PD-1 and its ligands have been released either in *apo* form, or bound to several ligands (for a review, see Reference [389] and the references therein), which has motivated structure-based drug research through computer simulation approach as a promising step in this route.

However, it was only recently that the first experimentally obtained structure of human PD-1/ligand complex was published, and the first view of it was provided by molecular modeling and molecular dynamic techniques [390]. Besides, molecular docking has been also used to look for a small molecule with the ability to bind at the PD-L1 binding site [391]. This work was pivotal to the development of new inhibitors since it suggests the presence of some PD-L1 amino acid residues that contribute more significantly to their binding to PD-1 (hot spots), which can be used for the design of PD-L1 antagonist drugs.

Currently, a number of cancer immunotherapy agents that target the PD-1 receptor have been developed. Among them, two antibody-based agents targeting PD-1 were approved by the US Food and Drug Administration (USFDA) and other agencies around the world, namely the pembrolizumab (trade name Keytruda) and nivolumab (marketed as Opdivo) drugs. Both drugs are humanized monoclonal antibodies belonging to the immunoglobulin IgG4 subclass with potential immune checkpoint inhibitory and antineoplastic activities. Upon administration, they act as checkpoint inhibitors, blocking the interaction between the programmed cell death protein PD-1 in T cells with their ligands, which results in the activation of T cell–mediated immune responses against tumor cells, thus allowing the immune system to clear the cancer. They were the first and second anti-PD-1 agents to be approved by the FDA, and today they are used to treat some cancer types where cells express PD-L1, including advanced melanoma, advanced non-small cell lung cancer (NSCLC), recurrent or metastatic head and neck squamous cell carcinoma, classical Hodgkin lymphoma (cHL), and more, the list grows with the release of new clinical trials [392].

Although these antibodies are associated with substantial benefits, the immune checkpoint blockade can lead to inflammatory side effects. Besides, there is a necessity for new drugs that target PD-1 [393]. Thus, obtaining a deep understanding of the human PD-1/therapeutic antibody complex is essential for our knowledge about its inhibition mechanism and the design of improved anti-PD-1 therapeutics. In this sense, it has been shown that patients making use of pembrolizumab have experienced better results and fewer side effects than those using the nivolumab drug, making the former a suitable candidate to provide new compounds based on its structure.

Within this context, it is the aim of this section to present the first in silico quantum biochemistry calculation of the noncovalent interaction energies of the amino acid residues of the receptor PD-1 with the residues obtained from its natural ligand PD-L1 and its two drugs inhibitors pembrolizumab and nivolumab, in order to map the common recognition surface, searching for the binding interactions that stabilize these biological complex systems. Our main goal is to supply enough information at the binding interface to substantially contribute to the affinity and specificity between the receptor (PD-1) and the therapeutic antibody (pembrolizumab and nivolumab). The molecular fractionation with conjugated caps (MFCC) scheme is used to compute the interaction energy within the DFT framework, adopting GGA+D to describe the intermolecular forces. It is a standard route to investigate such accurate large biological systems at low computational cost, previously applied to describe molecular interactions at the quantum level related to several biological molecules. There is no other quantum mechanical study relating the binding of these drugs at the PD-1 receptor, despite its highly pharmaceutical relevance.

15.4.1 The Immune Checkpoint Protein PD-1

The programmed cell death protein 1 (PD-1) is an important regulator for the immune tolerance and T cell exhaustion, as it recently emerged as a key target in the treatment of several types of cancer. It is expressed after the T cell activation and binds to the ligands PD-L1 and PD-L2, suppressing immune response against autoantigens and playing an important role in the maintenance of peripheral immune tolerance [394]. However, the ligand PD-L1 is often overexpressed in different tumors – including lymphoma, melanoma, non-small cell lung cancer, and other types – making the PD-1/PD-L1 signaling pathway crucial in dampening the immune surveillance of the tumor. In this context, the target of the PD-1/PD-L1 interaction with the monoclonal antibodies pembrolizumab and nivolumab has demonstrated to be an important strategy for the control and eradication of several types of cancers.

To perform the calculations in this study, we have taken full advantage of the X-ray crystal structure of human-programmed cell death receptor 1 (hPD-1) solved in complex with the extracellular region of human-programmed cell death ligand 1 (hPD-L1; PDB ID: 4ZQK), the Fab fragment of pembrolizumab (PDB ID: 5GGS), and the Fab fragment of nivolumab (PDB ID: 5WT9) at 2.45, 1.99, and 2.41 Å of resolution, respectively [395].

Superposition of the crystallographic structures of PD-1 in complex with the natural ligand PD-L1, as well as with the monoclonal antibodies pembrolizumab and nivolumab, revealed that these three ligands bind PD-1 in different orientations, though a partial complementary binding site was detected [395]. Figure 15.12a depicts the 11 amino acid residues that form the PD-1 common binding site (F63, V64, I126, S127, L128, A129, P130, K131, A132, Q133, and I134), the relevance of some of them having already been proved by experimental and computational studies. Binding modes of pembrolizumab (Figure 15.12b) and nivolumab (Figure 15.12c) to PD-1 are also shown. HC (LC) means the heavy (light) chain of the monoclonal antibodies pembrolizumab and nivolumab. The presence of this similar binding site shows that pembrolizumab and nivolumab could be acting not only as competitive inhibitors of PD-L1, presenting a hot spot that can be used by the pharmaceutical industry to develop new potent drugs with fewer and strong specificity, but also as a powerful strategy to better understand the PD-1 signaling pathway.

The protonation state of the receptor PD-1 and its ligands was adjusted according to the results obtained from the PROPKA 3.1 package. Amino acid main-chain

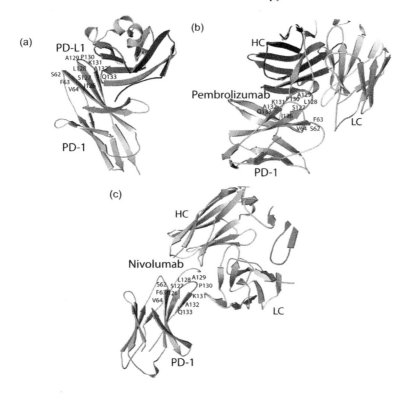

Figure 15.12 Binding mode of (a) PD-L1, (b) pembrolizumab (PEM), and (c) nivolumab (NIV) to PD-1. The residues of the common binding site are pointed out. Here HC (LC) means the heavy (light) chain of the monoclonal antibodies pembrolizumab and nivolumab. After Tavares et al. [367]

atoms were kept fixed while side chains were submitted to a classical geometry optimization – except for the PD-1/ pembrolizumab complex, where only hydrogen and added atoms were optimized. This optimization was performed using the classical CHARMm force field, with the same convergence tolerances used in previous case, namely 10^{-5} kcal/mol (total energy variation), 10^{-3} kcal/(mol Å) (root mean square gradient), and 10^{-5}Å (maximum atomic displacement).

Afterward, it is submitted to a fragmentation scheme based on the MFCC method to calculate the interaction energy between the amino acids from the ligands (at position R^i) and the receptor PD-1 (at position R^j). For each amino acid residue of interest at position R^i, we draw an imaginary sphere with a radius equal to 8.0 Å and evaluate the interaction energy $EI(R^i\text{-}R^j)$ with each residue at position R^j, considering at least one atom inside the sphere.

In order to achieve the structural stability of the complex promoted by interactions with the extended hydration network, all water molecules forming hydrogen bonds with a particular residue were included for completeness in the fragments.

The analysis of the binding scenario was based on the assumption of pair-additive contributions of the amino acids to the total interaction energy within the MFCC method; i.e., the energetic characterization of the full protein–ligand complex is related to the sum of individual contribution of each residue–ligand pair's interaction energy, which can be attractive (negative) or repulsive (positive), excluding many-body interactions [365]. The latter effect was recently considered to describe some protein–ligand complexes [20], in order to correct some of their interaction energies, at a higher computational cost. Furthermore, as it was reported previously, the total binding energies, as those depicted here, were less affected by the many-body corrections, being in a good agreement with the experimental data [266, 396].

15.4.2 The Checkpoint Protein PD-1 in Complex with Pembrolizumab

The immunotherapy treatment with the drug pembrolizumab was approved by the USFDA in 2014 for advanced melanoma, and recently, it has also been approved for an increasing number of cancer types, such as Hodgkin's lymphoma and non-small cell lung cancer. It is believed that the weak frequency in humans and the induction of cell activation, characteristics of the members of the IgG4 subclass, are part of the main basis for the success of this therapeutic compound.

As we can see from Figure 15.13, the pembrolizumab recognition surface on PD-1 is filled by intermolecular direct and water-mediated hydrogen bonds (HBs

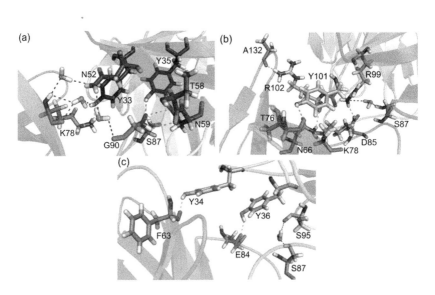

Figure 15.13 Interaction patterns of pembrolizumab/PD-1 recognition surface involving the amino acid residues and the water molecules. (a) and (b) Heavy-chain (HC) residues; (c) light-chain (LC) residues. After Tavares et al. [366]

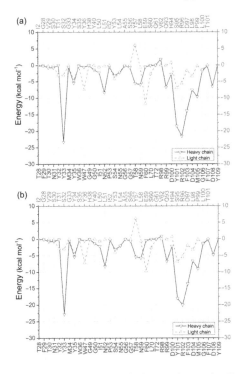

Figure 15.14 Energy profile (in kcal/mol) for each pembrolizumab amino acid residues in recognition surface. (a) The sum of interaction energies of the pembrolizumab residues with each amino acid from PD-1 within a radius of 8.0 Å, using the dielectric constant ε_{20}. (b) The same for ε_{40}. The black solid line represents the heavy-chain energy spectrum, while the gray dashed line is used to depict the light-chain one. After Tavares et al. [366]

and wHBs, respectively), non-conventional hydrogen bonds (nHBs), salt bridges (SB), and hydrophobic contacts. Therefore, mapping the relevant interactions among the pembrolizumab/PD-1 surface complex is highly important to the rationale drug development of new compounds.

In order to characterize the pembrolizumab/PD-1 recognition surface, we have done a structural analysis of the binding site in which 408 pairs of interactions were detected within a range of 8.0 Å. Figure 15.14 shows the sum of all interaction energies from each individual residue of the pembrolizumab drug (R^i), taking into account a dielectric constant $\varepsilon = 20$ (ε_{20}) and 40 (ε_{40}) depicted in Figure 15.14a and b, respectively.

One can note that only two residues (one residue) in the pembrolizumab heavy-chain (HC) energy spectrum are (is) repelled in $\varepsilon = 20$ ($\varepsilon = 40$), while four (three) residues are repelled in its light-chain (LC) one (all binding energies units in kcal/mol), namely $Y32_{HC}$ (ε_{20}: 0.04; ε_{40}: -0.10), $R98_{HC}$ (ε_{20}: 1.80; ε_{40}: 0.85),

$G33_{LC}$ (ε_{20}: 0.008; ε_{40}: -0.04), $Y57_{LC}$ (ε_{20}: 6.06; ε_{40}: 6.17), $G61_{LC}$ (ε_{20}: 0.01; ε_{40}: -0.002), and $H94_{LC}$ (ε_{20}: 0.82; ε_{40}: 0.64).

For instance, the energy value 0.04 kcal/mol calculated for the residue Y32 (HC and LC designated residues, respectively) represents the sum of the binding energies of the four amino acids that compose the receptor PD-1 and are within a radius distance of 8.0 Å from it, namely K78, P89, K131, and A132. Thus, one can see that the repulsive energies are almost insignificant, excluding the residue Y57 that shows the highest one for both dielectric constant values ε_{20} and ε_{40}.

All other residues show attractive interactions (negative energies), with the most intense one being observed for the heavy-chain amino acids $Y33_{HC}$ (ε_{20}: -23.32; ε_{40}: -22.84), $Y101_{HC}$ (ε_{20}: -18.49; ε_{40}: -18.07), $R102_{HC}$ (ε_{20}: -21.58; ε_{40}: -20.03) and $F103_{HC}$ (ε_{20}: -13.91; ε_{40}: -13.62, and light-chain residues $Y36_{LC}$ (ε_{20}: -7.41; ε_{40}: -7.16), $E59_{LC}$ (ε_{20}: -11.69; ε_{40}: -9.69) and $S95_{LC}$ (ε_{20}: -6.99; ε_{40}: -6.85).

Among the 408 residue–residue pairs analyzed here, 260 were pembrolizumab heavy-chain/PD-1 interactions. It is a reflection of the proximity between the HC/LC pembrolizumab with the receptor PD-1 represented in crystallographic structures, where pembrolizumab$_{Fab}$ heavy-chain fragment is closer to the PD-1 receptor than the light-chain one. Besides a greater number of pairs, the sum of the energetic interaction between the pembrolizumab heavy-chain and the PD-1 receptor amino acids also shows lso the higher value, accounting for -142.50 kcal/mol (-138.33 kcal/mol) for the dielectric constant ε_{20} (ε_{40}).

Although the residue $Y33_{HC}$ has been shown to be the most energetic pembrolizumab amino acid residue, it only interacts with 15 residues from the PD-1 receptor. Figure 15.15a depicts the highest interaction energies calculated for $Y33_{HC}$/PD-1 residues. As one can see, the strongest energy (in kcal/mol) of $Y33_{HC}$ is mainly related to its binding with three residues: K78 (ε_{20}: -12.42; ε_{40}: -11.99), Q88 (ε_{20}: -4.45; ε_{40}: -4.51) and P89 (ε_{20}: -3.17; ε_{40}: -3.15).

Meanwhile, the residues $Y101_{HC}$, $R102_{HC}$, and $Y103_{HC}$ are involved with 23, 30, and 21 pairs, respectively. As shown in Figure 15.15b, the residue $Y101_{HC}$ has its most intense interaction with the residue K78 (ε_{20}: -9.43; ε_{40}: -8.92), being followed by the residue T76 (ε_{20}: -4.82; ε_{40}: -4.71), all energies' units in kcal/mol.

The residue $R102_{HC}$ does not show a strong individual interaction with none of the PD-1 residues (Figure 15.15c), its high binding energy being due to a number of minor energies contributions, including those from the residues A132 (ε_{20}: -3.84; ε_{40}: -3.44), I126 (ε_{20}: -2.71; ε_{40}: -2.87), K78 (ε_{20}: -2.43; ε_{40}: -3.12), and N66 (ε_{20}: -2.53; ε_{40}: -2.17). Similar to the residue R102, $Y103_{HC}$ binding energy is composed of the sum of a number of small interactions (Figure 15.15d), with the strongest being associated with the residue V64 (ε_{20}: -3.49; ε_{40}: -3.49).

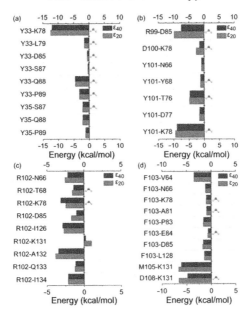

Figure 15.15 Graphical panel presenting the most relevant interactions involving the pembrolizumab heavy-chain residues. (a) Y33 and Y35; (b) R99, D100 and Y101; (c) R102 and (d) F103, M105 and D108, respectively. After Tavares et al. [366]

Figure 15.16a displays how close the residues $Y33_{HC}$, $Y101_{HC}$, $R102_{HC}$, and $Y103_{HC}$ are to the PD-1 receptor. Analyzing it, one can understand the reason why these amino acids present some of the largest number of pair-interaction and binding energies. All of them are involved in a network of hydrogen bonds with the PD-1 residues from CC'FG β-strands and some of its loops, mainly those belonging to the C'D loop, which was described to intrude into the complementary determining region (CDR) of the pembrolizumab drug, a variable portion present in some igG molecules responsible for the recognition of specific antigens. Figure 15.16b depicts some of these interactions, starting with the residue $Y33_{HC}$. Here, one can see that it forms two direct hydrogen bonds with the residues K78 charged amine group and Q88 side-chain oxygen. Besides, the residue $Y33_{HC}$ also makes a water-mediated hydrogen bond with the K78 carbonyl group of the main chain and is involved in nonpolar contacts with P89. The two hydrogen bonds between the residues $Y33_{HC}$-K78 give the major contribution to their high energy.

The interaction network of residues $Y101_{HC}$, $R102_{HC}$ and $Y103_{HC}$ is shown in Figure 15.16c–e. Observe that Figure 15.16c depicts three direct bonds, two water-mediated bonds, and one non-conventional hydrogen bond. The strongest interaction of this residue is related to the pair $Y101_{HC}$–K78, where it creates

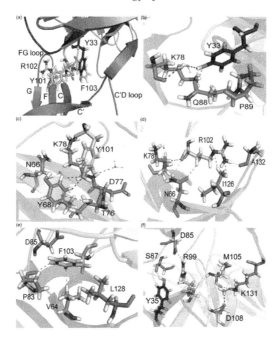

Figure 15.16 Intermolecular interactions of the most energetic pembrolizumab heavy-chain residues. (a) Structural representation of the residues $Y33_{HC}$, $Y101_{HC}$, $R102_{HC}$, and $F103_{HC}$, within the PD-1 binding site; (b)–(f): Interaction of these amino acids with the most relevant residues of the receptor PD-1. Dashed lines in black represent direct, water-mediated, and non-conventional hydrogen bonds, while the gray lines represent $\sigma-\pi$ interactions. After Tavares et al. [366]

a direct hydrogen bond with charged side-chain amine and a non-conventional hydrogen bond with amine of the main chain. Three hydrogen bonds are formed with the residue T76, two with its oxygen atoms (carbonyl and hydroxyl) and a water-mediated bond with the hydroxyl group. It is also involved in one wHB with the residue Y68 side-chain hydroxyl and in some nonpolar contacts with the residues D77 and N66. The residues $R102_{HC}$ and $Y103_{HC}$ are mainly involved in weak interactions. The latter form a $\sigma-\pi$ interaction with the residue V64, a non-conventional hydrogen bond with the residue P83, and some nonpolar contacts with the residues L128 and D85, while the former makes hydrogen bonds with an A132 oxygen atom and the N66 side-chain nitrogen, in addition to non-conventional hydrogen bonds with the side chain of the residues I126 and K78.

The MFCC scheme yields not only the individual pembrolizumab/PD-1 interactions but also important information from the pair interactions that could be otherwise missing – namely $R99_{HC}$–D85 (ε_{20}: -9.75; ε_{40}: -7.43), $Y35_{HC}$–S87 (ε_{20}: -2.13; ε_{40}: -2.08), $M105_{HC}$–K131 (ε_{20}: -6.61; ε_{40}: -6.06), and $D108_{HC}$–K131 (ε_{20}: -6.56; ε_{40}: -4.82), all energies' units in kcal/mol.

The residues $M105_{HC}$ and $D108_{HC}$ make hydrogen bonds with the residue K131. The main-chain oxygen carbonyl of $M105_{HC}$ makes a direct hydrogen bond with the charged amine of this lysine residue. The residue D108 is engaged in a water-mediated hydrogen bond with the same molecular group but presents its negatively charged carboxyl group (D108) in a closer distance to positively charged amine group (see Figure 15.16f). The binding energy of the pair $Y35_{HC}$–S87 is governed by a hydrogen bond formed between the hydroxyl group of this tyrosine residue ($Y35_{HC}$) and the main-chain carbonyl from S87. The pair $R99_{HC}$–D85 is described for some authors as the only salt bridge formed between the pembrolizumab drug and the PD-1 receptor.

The pembrolizumab light chain is a little more distant from the PD-1 receptor than the heavy chain, as it can be inferred from its contribution to the total binding energy (ε_{20}: -44.67; ε_{40}: -47.75), as well as from the number of pairs formed with the receptor (148). This is consistent with previous crystallographic results [393].

According to Figure 15.17, the most intense binding energies (in kcal/mol) for the LC amino acids are associated to the pair $E59_{LC}$–K131 (ε_{20}: -11.90; ε_{40}: -9.70), being followed by the pairs $S95_{LC}$–S87 (ε_{20}: -6.00; ε_{40}: -5.83), $Y53_{LC}$–K131 (ε_{20}: -4.15; ε_{40}: -4.05), $R96_{LC}$–R86 (ε_{20}: -1.47; ε_{40}: -2.62), and $Y36_{LC}$–E84 (ε_{20}: -2.62; ε_{40}: -2.48).

Figure 15.18 depicts some interactions made by the pembrolizumab light-chain residues linked to the PD-1 receptor. As one can see, the residue $Y36_{LC}$ is surrounded by some charged residues (E84, D85, and R86) from the PD-1 receptor

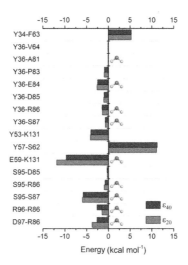

Figure 15.17 Graphical panel depicting the most relevant interactions involving the pembrolizumab light-chain residues Y34, Y36, Y53, Y57, E59, E95, R96, and D97. After Tavares et al. [366]

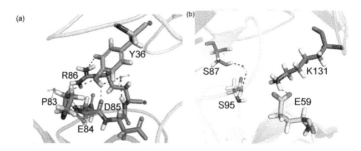

Figure 15.18 Intermolecular interactions of the most energetic pembrolizumab light-chain residues. (a) Interaction of the residue Y36 with the most relevant residues of the receptor PD-1. (b) The same for the residues E59 and S95. Dashed lines represent direct, water-mediated, and non-conventional hydrogen bonds. After Tavares et al. [366]

(Figure 15.18a). However, it only makes a single direct hydrogen bond with the main-chain carbonyl of the residue E84, all the other interactions being non-conventional hydrogen bonds (P83 and R86) and water-mediated hydrogen bonds. It would be inconsistent with the position occupied by this residue among the most energetic one of the pembrolizumab light chain, if the number of pairs formed with $Y36_{LC}$ (10) would not overcome it. The residue making the greatest number of pair interactions is $Y34_{LC}$, with 19 pair interactions. However, the residue $Y34_{LC}$ shows a binding energy less than the $Y36_{LC}$ one, due to an unfavorable interaction with the residue F63 (ε_{20}: 5.27; ε_{40}: 5.30 kcal/mol).

Although only five pair interactions were calculated with the residue $S95_{LC}$ within a radius of 8.0 Å, it shows one of the highest interaction energies of the drug's light chain. Similar to the residue $Y36_{LC}$, the majority of the surrounding residues make weak interactions, excluding S87, which forms a water-mediated hydrogen bond through the hydroxyl group from both side chains (see Figure 15.18b). It is also important to notice the attractive binding energy found in the $R96_{LC}$–R86 pair, even with the positively charged guanidine group from both amino acids, assuming a conformation where they are very close. These two arginine residues are in a T-shaped stacking interaction, which favors an attractive bind. Finally, the $E59_{LC}$–K131 pair forms the second salt bridge of the drugreceptor complex, depicting the highest individual interaction energy in the drug's light chain. This interaction is due to the negatively charged carboxyl group of the residue $E59_{LC}$ and the positively charged amine of the residue K131 side chain. It gives us the idea of the dynamic process that governs the interaction between the Fab fragment of pembrolizumab and the extracellular region of the PD-1 receptor, such as the formation of a new salt bridge between the residues $D108_{HC}$–K131, whose opposite charges from the side chain are very close (3.6 Å).

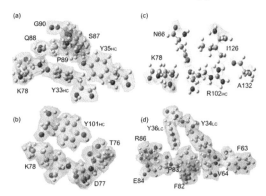

Figure 15.19 Electrostatic potential isosurfaces with the projected electron densities for some pembrolizumab residues interacting with the most attractive residues of the receptor PD-1. After Tavares et al. [366]

For completeness, we display in Figure 15.19 the electrostatic potential isosurface with the projected electron densities for the pembrolizumab amino acids bound to some of the most important residues at the binding pocket site.

To evaluate the total binding interaction energies through fragment-based quantum mechanics method, it is important to take into account every significant attractive and repulsive amino acid residue that can influence this mechanism. Therefore, instead of taking an arbitrary region of the binding site, we performed a search for an optimal binding pocket radius (r) in which a variation less than 10% of the sequential pocket radius could be observed after a radius increase. For this task, the binding pocket radius r is varied from 2.0 Å (ε_{20}: -32.04; ε_{40}: -27.29) to 8.0 Å (ε_{20}: -187.17; ε_{40}: -186.08) in order to determine the best value of r, found to be 6.0 Å corresponding to an energy of -179.94 (-179.60) kcal/mol for $\varepsilon = 20$ (40), from which the convergence was achieved (see Figure 15.20). This result is in agreement with previous works, stating that after 6.0–7.0 Å the molecular interactions are weak [290].

15.4.3 The Checkpoint Protein PD-1 in Complex with Nivolumab

Nivolumab is the generic name of the immunotherapy drug Opdivo, manufactured by Bristol-Myers Squibb. It was approved in 2015 by the FDA as the first immunotherapy drug and second-line treatment for advanced non-small cell lung cancer, but it is also used as an off-label treatment for mesothelioma.

As we can see from Figure 15.21, the nivolumab recognition surface on PD-1 is filled majoritary by nonpolar amino acids in this common binding region (F63, V64, I126, L128, A129, A132, and I134), which decreases the total interaction energy, since the number of strongest polar interactions (e.g., hydrogen bonds

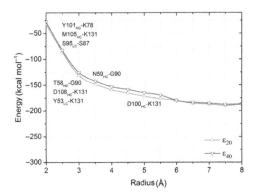

Figure 15.20 The total interaction energy as a function of the binding pocket radius r calculated using the GGA functional B97D for the dielectric constant $\varepsilon = 20$ (light gray line) and 40 (dark gray line). Pairs of amino acids responsible for the regions of steepest negative and positive variation are highlighted. After Tavares et al. [366]

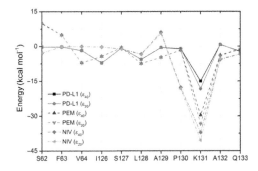

Figure 15.21 Energy profiles for each PD-1 amino acid residue in the common recognition surface. This figure represents the sum of the interaction energies of the PD-1 residues with each amino acid from its ligands PD-L1, pembrolizumab and nivolumab within a radius of 8.0 Å, using the dielectric constant ε_{20} and ε_{40}. After Tavares et al. [367]

and salt bridges) are reduced, in agreement with the energetic results shown in Figure 15.21, where many of them are close to zero for all proteins. Besides, most of the residue–residue pairs are formed between the PD-1 amino acids with those from the drugs' light chain (LC), although they are more distant from the receptor protein than the heavy chain (HC) ones [393].

Furthermore, one can see that for the PD-1/PD-L1 complex, only the residue A132 shows a repulsive energy, with interaction energy equal to 0.75 (0.86) kcal/mol for ε_{40} (ε_{20}). On the other hand, in the complex PD-1/pembrolizumab, two amino acids present a repulsive value (in kcal/mol), namely S62 (ε_{40}: 10.00; ε_{20}: 9.75) and F63 (ε_{40}: 4.92; ε_{20}: 4.93). Concerning the monoclonal antibodies, we found that the interaction with residues from their light chain is mainly responsible for the positive energetic result.

All other amino acids have negative interaction energies (displayed in kcal/mol), with the most intense attraction being observed for the lysine residue K131 in complex with the ligand PD-L1 (ε_{40}: -15.02; ε_{20}: -18.24), the drug pembrolizumab-PEM (ε_{40}: -29.63; ε_{20}: -33.40) and the drug nivolumab-NIV (ε_{40}: -37.08; ε_{20}: -40.49). This lysine residue gives a very important contribution to the formation and stabilization of the complexes because its long side chain with a charged amine group creates a dynamic surface capable to form hydrogen bonds, salt bridges, and cation–π interactions with water molecules and residues from the ligands – such as $N63_{PD-L1}$ (ε_{40}: -1.35; ε_{20}: -1.72), $Q66_{PD-L1}$ (ε_{40}: -5.21; ε_{20}: -5.12), $Y53_{PEM}$ (ε_{40}: -4.05; ε_{20}: -4.15), $E59_{PEM}$ (ε_{40}: -9.70; ε_{20}: -11.90), $M105_{PEM}$ (ε_{40}: -6.06; ε_{20}: -6.62), $D108_{PEM}$ (ε_{40}: -4.82; ε_{20}: -6.56), $D100_{NIV}$ (ε_{40}: -5.77; ε_{20}: -6.58), $D101_{NIV}$ (ε_{40}: -8.24; ε_{20}: -10.27), $Y49_{NIV}$ (ε_{40}: -6.62; ε_{20}: -6.43), $D50_{NIV}$ (ε_{40}: -8.82; ε_{20}: -10.85), as can be seen in Figure 15.22. The residue P130 shows the second-most attractive interaction energy when complexed with the drug NIV (ε_{40}: -17.65; ε_{20}: -18.21), while the complexes with the ligand PD-L1 (ε_{40}: -1.09; ε_{20}: -0.92) and the drug PEM (ε_{40}: -1.63; ε_{20}: -1.57) present a small energy very close to each other.

The single residue that shows superposing energetic values for the three complexes is S127. It accounts for -0.64 (-0.69) kcal/mol for PD-1 bound to PD-L1 considering a dielectric constant ε_{40} (ε_{20}), -0.48 (-0.51) kcal/mol, and

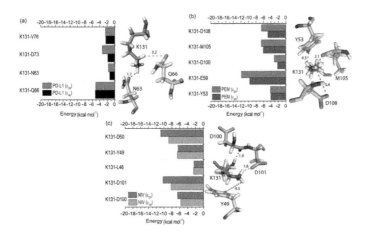

Figure 15.22 Interaction of the Lysine residue K131 with the most relevant amino acids of the: (a) the receptor PD-L1; (b) the pembrolizumab drug; (c) the nivolumab drug. The dashed lines represent hydrogen bonds, cation-π, salt bridges and non-conventional hydrogen bonds. After Tavares et al. [367]

−0.98 (−1.17) kcal/mol when bound to the drugs pembrolizumab and nivolumab, respectively.

Meanwhile, only the residue Q133 presents the interaction energy close among the three ligands, depicting values of −2.14 (−2.34) kcal/mol, −1.17 (−1.27) kcal/mol, and −3.03 (−3.05) kcal/mol for PD-L1, pembrolizumab, and nivolumab complexes, respectively, considering a dielectric constant ε_{40} (ε_{20}). It is easy to see that the hot spot formed by the six residues between S62 and L128 creates a region of low interaction energy, which may allow some degree of flexibility and help in the receptor rearrangement after the pembrolizumab and nivolumab binding to avoid the coupling of natural ligands.

Adding up the interaction energies for each residue–residue pair, we have found the total interaction energies for every ligand, namely −44.34 (−47.27) for PD-1/PD-L1 considering the dielectric constant ε_{40} (ε_{20}). Likewise, we found −47.54 (−51.90) for PD-1/pembrolizumab and −65.23 (−69.88) for PD-1/nivolumab complexes, all interaction energies' units in kcal/mol.

15.5 Conclusions

Summarizing, we have discussed in this chapter some interesting points related to the biology of cancer. We started presenting an investigation related to the interactions among the selective estrogen receptor modulators (SERMs) 4-hydroxytamoxifen (OHT) and raloxifene (RAL), widely used in the breast cancer treatment, co-crystallized with the estrogen receptor α (ERα). After a convergence study on the size of the binding pocket sphere, we have taken into account all ligand–residue interactions within a radius of 15 Å from the ligand. We could notice that SERMs-OHT binds more strongly when compared to SERMs-RAL, confirming experimental data [374]. In addition, we have established the main regions of both ligands (regions i and iv) and the residues that play important role on the binding affinity of ERα with their antagonists. Through combining these computation techniques to ligand-based approaches, more accurate models should be possible in order to study and understand receptor-mediated effects.

Afterward, we depicted the roles of integrins in controlling cellular behavior that affects tumor progression, making integrins an appealing target for cancer therapy. Although therapeutic strategies for targeting these molecules abound, only a few are used in clinical treatment nowadays, with many of them not going through the early stages of clinical trials. The good news is the appearance of the integrin antagonist cilengitide, a cyclic RGD (Arg–Gly–Asp amino acid sequence) and the first integrin inhibitor in clinical phase III development for oncology and in phase II for several other tumors. This drug is the first anti-angiogenic small molecule targeting the integrins $\alpha_V\beta_5$, $\alpha_V\beta_3$, and $\alpha_V\beta_1$. Its design and production require a precise knowledge of not only of its biochemical structure but also its binding mechanism with the integrin, leading to a better understanding of how this complex influences the cancer tumor and its anti-angiogenesis therapies, as well as its microenvironment. Computational tools based on quantum chemistry may be a promising step in this route, further increased by combination with other anticancer therapies.

Finally, we discussed the first in silico quantum biochemistry approach related to cancer immunotherapy, a kind of cancer treatment that boosts the body's natural defenses to fight the abnormal cells, considered today as the most promising treatment of recent years. Within the quantum DFT framework, we estimate the interaction energies between the amino acid residues of the programmed cell death receptor-1 (PD-1) and those of its natural ligand PD-L1, as well as the two USFDA-approved monoclonal antibodies, namely pembrolizumab and nivolumab, respectively. Since both drugs have shown great improvement in the treatment of several cancer types, our results could favor the development of new compounds targeting the PD-1/PD-ligands, especially those based on the hot spots, by avoiding structural features that are in minor relationship with this binding scenario. The method employed here provided a good characterization of important molecular features by including an accurate electronic description and enhancing the differences of interaction intensity among the amino acid pairs.

All in all, the quantum chemistry computational methods used in this chapter emerged as a simple and efficient alternative to unveil the drug's amino acid residues that play the most important role on the binding affinity of the receptor–ligand complex. No doubt, taking into account the small cost/benefit of the operation, the in silico approach, not only in clinical oncology but also in pharmaceutical medicine in general, is an important initial step toward the development of efficient cancer pharmaceutical drugs.

16

Concluding Remarks

16.1 Introduction

Computational quantum chemistry, a rapidly growing field, is one of the most popular and successful approaches to calculate the structures and properties of molecules and solids, and it is also widely used in the design of new drugs and biological materials. It covers a broad range of topics and is normally complemented by information obtained by biological/chemical/physical experiments. Its main advantage is that can be safely performed on any system, even those that do not exist, whereas many experiments are limited to relatively stable molecules. In some cases, it can predict hitherto unobserved important phenomena. Nowadays, it is routinely applied for calculating, e.g., the interaction binding energies of molecules in biology and chemistry, as well as the electronic band structure of solids in physics.

The methods used cover both static and dynamic situations, ranging from very approximate to highly accurate, the latter usually feasible for small systems only. In all cases, the computer time increases rapidly with the size of the system being studied. Generally, the methods are based entirely on quantum mechanics and basic physical constants (ab initio methods) being complemented by empirical or semiempirical approaches by using additional ad hoc parameters. Nevertheless, both ab initio and empirical approaches involve approximations, ranging from simplified forms of the first-principles equations (easier or computationally faster to solve) to fundamental approximations of the underlying equations, limiting the size of the system. Among the ab initio methods, for the last 30 years, density functional theory (DFT) has been the dominant one, very widely used for the simulation of biological molecules. DFT owes its versatility to the generality of its fundamental concepts and the flexibility one has in implementing them. Despite this flexibility and generality, DFT is based on a quite rigid conceptual framework; its theoretical footing is based on the Hohenberg-Kohn (H-K) theorems [5] and the Kohn-Sham equations [6].

Unfortunately, accuracy normally can only be improved with greater computational cost. Because, in some cases, the quantum mechanical details are less important than the long-time phase space behavior of molecules, classical approximations to the potential energy surface may be used. They are computationally less intensive than the quantum electronic calculations, enabling longer simulations of larger molecular systems. Among them, classical molecular dynamics (MD) is widely used, as it is based on Newton's equations of motion for a system of interacting particles, where the forces between the particles and their potential energies are often calculated using interatomic potentials or molecular mechanics force fields [397]. This is the case in conformational studies of proteins and protein–ligand binding thermodynamics, as discussed in previous chapters.

16.2 Past Achievements

In principle, quantum mechanics provides a basis for all of physics and chemistry, and, indeed, at one time, the scientific community believed that all problems had been solved. Today, we can see that this is by no means the case, since in all practical situations, the quantum solution is very time consuming, even with the advent of high-speed computers. However, quantum mechanics is particularly valuable in providing a conceptual basis for our understanding of problems in biology, chemistry, and physics. The nature of the covalent bond, for instance, cannot be properly understood on the basis of classical mechanics by means of the exchange energy, a basic component of the binding energy.

The first hints of the chemical basis of life were noted a long time ago, leading to a series of insights that living organisms comprise a hierarchy of structures: organs, which are composed of individual cells, which are themselves formed of organelles of different chemical compositions, and so on. British mathematician and logician Alan Turing, often called the father of computing, used early computers to implement a model of biological morphogenesis (the development of pattern and form in living organisms) in the early 1950s. At the same time, investigations into the nature of the molecules responsible for the biochemical reactions culminate with the discovery of the molecular structure of the deoxyribonucleic acid (DNA) by Watson and Crick [27].

The discovery of the structure of DNA and its relationship to DNA function had a tremendous impact on all subsequent biochemical investigations, basically defining the paradigm of modern biochemistry and molecular biology. This established the primary importance of molecular structure for an understanding of the function of biological molecules and the need to investigate the relationship between structure and function in order to advance our understanding of the fundamental processes of life.

The α-helix and the β-sheet, the secondary structures in proteins now known to form the backbones of tens of thousands of proteins, were proposed by Pauling et al. in 1951 [163]. They deduced these fundamental building blocks from properties of small molecules, known both from crystal structures and from Pauling's resonance theory of chemical bonding that predicted planar peptide groups. Their prediction was soon confirmed by Perutz [398], who had the first glimpse of the secondary structure at low resolution. This landmark work marked the dawn of theoretical studies of biomolecules. It was followed by the prediction of the allowed conformations of amino acids, the basic building block of proteins, in 1963 [399]. By the 1960s, computers had been applied to deal with much more varied sets of analysis, namely those examining protein structure. These developments marked the rise of computational biology as a field, originating from studies centered around protein crystallography, in which scientists found computers indispensable for carrying out laborious Fourier analyses to determine the 3-D structure of proteins.

Since then, many breakthrough achievements in biology, chemistry, and physics were obtained, leading to several Nobel Prizes in these area. In particular, the 1998 Nobel Prize in Chemistry was given to Walter Kohn (*for his development of DFT*) and John A. Pople (*for development of computational methods in quantum chemistry*), signifying the widespread acceptance of computation as a valid tool for investigating chemical phenomena. Furthermore, in 2013, the Nobel Prize in Chemistry was awarded to M. Karplus, M. Levitt, and A. Warshel, *for the development of multiscale models for complex chemical systems.*

With its extension to biomolecular systems, the range of possible applications of computational chemistry was greatly expanded. Though still a relatively young field, quantum chemistry simulation is now pervasive in all aspects of the biological and pharmaceutical sciences. These methods have aided in the interpretation of experimental data, and will continue to do so, allowing for the more rational design of new experiments, thereby facilitating investigations in the whole health sciences. Computational methods will also allow access to information beyond those obtainable via experimental techniques. Indeed, quantum chemistry computer-based approaches for the study of virtually any chemical or biological phenomena may represent one of the most powerful tool now available to scientists, allowing investigations at an unprecedented level of detail.

16.3 The Road Ahead

Nanobiotechnology is the application of nanotechnologies in biological fields. Biologists, chemists, physicists, and those akin to them define nanotechnology as a branch of their own subjects. Collaborations between fields are common, playing a vital role in developing and implementing many useful tools, like pharmaceutical drug design and delivery principles, disease diagnostic devices, biosensors, and any other medical applications. However, nanobiotechnology is still in its infancy.

Thanks to nanoelectronics, health science is undergoing deep changes by exploiting the traditional strengths of the semiconductor industry, namely miniaturization and integration. While conventional electronics have already found many applications in biomedicine – like medical monitoring of vital signals, biophysical studies of excitable tissues, implantable electrodes for brain stimulation, pacemakers, and limb stimulation – the use of nanomaterials and nanoscale applications surely should bring, in a near future, a further push toward implanted electronics in the human body.

The possibility of developing new sophisticated nanodevices integrating human-made nanostructures with biomolecules – such as the nucleic acid analogs, as well as the integration of 3-D DNA arrays, nanoparticles, and nanoelectronic components with some precision – is seeking for new theoretical analysis of the emerging physical properties of these complex structures, similar to those discussed in this book. By the way, the direct electrical interfacing at the biomolecular level discussed here (see Chapter 6) surely opens the possibility of monitoring and controlling critical biological functions and processes in unprecedented ways, giving rise to a vast array of possibilities such as medical monitoring devices, improved drug-delivery systems, and patient monitoring systems, to cite just a few.

DNA structural disorder and dynamic fluctuations may make the measurement of the conductivity of DNA strands a daunting task for experimentalists. In particular, dynamic fluctuations occurring on the picosecond and nanosecond time scales are expected to have a strong effect on the ability of DNA to transport charge. Therefore, any new model of charge transport in DNA must consider the effects of static and dynamic disorder, as static disorder attenuates long-range charge transport and dynamic disorder allows the formation of transient pathways. Besides, biological polymers have a very complicate compositiond of their subunits and can support long-living nonlinear excitations. Charge transport across solvated DNA wires is dominated by conformational fluctuations, implying that transport approaches based on band-like coherent transport or static geometries cannot give a real description of charge motion in these systems.

In contrast to traditional methods of drug design, which rely on trial-and-error testing of chemical substances on cultured cells or animals and matching the apparent effects to treatments, rational drug design begins with a hypothesis that the modulation of a specific biological target may have therapeutic value. Its most basic sense involves the design of molecules that are complementary in shape and charge to the biomolecular target with which they interact and, therefore, will bind to it. The introduction of powerful new technologies should greatly accelerate the pace of new drug discovery. Although genomics should, in principle, increase the number of potential drug targets and provide a greater understanding of cellular events contributing to the pathology of disease, it has not been observed in practice, primarily because of the underlying complexity of cellular signaling processes. Genomics has, however, provided the tools of contemporary drug discovery and the pharmacogenomic pathways to personalized medicine and has greatly influenced the nature of synthetic organic chemistry. In the future, genomics and the tools of molecular biology will have a corresponding impact on drug-delivery processes and mechanisms through the introduction of drug-delivery machines capable of both synthesis and activation by disease-specific signals.

The use of quantum chemistry computational methods emerged as a simple and efficient alternative to unveil the drug's amino acids residues that play the most important role on the binding affinity of the receptor–ligand complex. No doubt, taking into account the small cost/benefit of the operation, the in silico approach in pharmaceutical medicine is an important initial step toward the development of efficient pharmaceutical drugs. Besides, it is a vast field in which the different sides of basic research and practice are combined and inspire each other. Large numbers of the available tools provide a much-developed basis for the design of ligands and inhibitors with preferred specificity.

During the selection process of novel drug candidates, many essential steps are taken to eliminate compounds that have side effects and also to show interactions with other drugs. No doubt, in silico drug-designing software plays an important role to provide innovative proteins or drugs in biotechnology or in the pharmaceutical field. Drug-designing software and programs are used to examine molecular modeling of gene, gene expression, gene sequence analysis, and 3-D structure of proteins.

Unfortunately, a lot of information related to the drug is unavailable even for the best-studied systems. In silico simulations thus always have to use a level of mathematical abstraction, which is dictated by the extent of our biological knowledge, molecular details of the network, and the specific questions that are addressed. These computational methods entailed the creation of increasingly sophisticated techniques for the comparison of strings of symbols that benefited from the formal study of algorithms and the study of dynamic programming in particular. Indeed, efficient algorithms always have been of primary concern in computational biology, given the scale of data available, and biology has, in turn, provided examples that have driven much advanced research in computer science. Examples include graph algorithms for genome mapping (the process of locating fragments of DNA on chromosomes) and for certain types of DNA and peptide sequencing methods, clustering algorithms for gene expression analysis and phylogenetic reconstruction, and pattern matching for various sequence search problems. However, with the development of powerful software and more efficient hardware, one should expect that its importance will be rewarded in a near future.

16.4 Conclusions

For centuries, man has searched for miracle cures to end suffering caused by disease and injury. Many researchers believe that nanotechnology applications in medicine may be humankind's first "giant step" toward this goal springing from the visionary idea that tiny nanorobots and related machines could be designed, manufactured, and introduced into the human body to perform cellular repairs at the molecular level. To this end, computational and theoretical studies of biological molecules, using biological data to develop algorithms or models to understand better their behavior as well as their relationship with other biological systems, should play an important role, as they have already advanced significantly in recent years and are predicted to progress rapidly in the future. These advances have been partially fueled by the ever-increasing number of available structures of proteins, nucleic acids, and carbohydrates, but at the same time, significant methodological improvements have been made in the area of chemistry and physics relevant to biological molecules.

These advances have allowed computational studies of biochemical processes to be performed with greater accuracy and under conditions leading to direct comparison to experimental studies. Examples include improved force fields, treatment of long-range atom–atom interactions, and a variety of algorithmic advances. The combination of these advances with the exponential increases in computational resources has greatly extended and will continue to expand the applicability of computational approaches to biomolecules. These models may describe what biological tasks are carried out by particular nucleic acid or peptide sequences; which gene (or genes), when expressed, produce a particular phenotype or behavior; what sequence of changes in gene or protein expression or localization lead to a particular disease; and how changes in cell organization influence cell behavior.

The medical advances that may be possible range from diagnostic to therapeutic and everything in between – encompassing treatment and prevention of illness and traumatic injuries, the decrease of pain, as well as the preservation and improvement of the human health by molecular tools and molecular knowledge. The medical use of nanomaterials and nanoelectronic biosensors and any possible future applications of molecular nanotechnology are the main tools toward these achievements. Perhaps the most important task here is to frame the biomedical problems as a computational problem. This often means looking at a biological system in a new way, challenging current assumptions or theories about the relationships between parts of the system, or integrating different sources of information to make a more comprehensive model than had been attempted before.

To summarize, it is our hope that the present book will help to expand the accessibility of computational approaches to the vast community of scientists investigating biological systems. We believe that it is a comprehensive and up-to-date account of the main electronic, thermodynamic, optical, vibrational, and pharmaceutical properties of some biological molecules and their interactions, using a theoretical/computational framework based on quantum chemistry. Notwithstanding its large impact on the economical aspects of the million-dollar drug market, quantum chemistry computer simulation pervades scientific activities in many fields, ranging from cosmology to the investigation of living bodies. The continuous extension of these theoretical/computational models to systems of large complexity calls for improvements in the existing scientific literature. A well-known example is provided by the current expansion of computer simulations that require the use of new software not used for the study of simpler systems. These considerations come out naturally when one considers models of theoretical chemistry/physics from this viewpoint.

At any standard, however, we believe that this book can be a valuable source for readers who are looking for specialized topics in biophysics, biochemistry, and related areas.

References

[1] Herbert Goldstein. *Classical mechanics*. Pearson, Harlow, 3rd edition, 2011.

[2] Z. B. Maksic and W. J. Orville-Thomas, editors. *Pauling's legacy, modern modelling of the chemical bond*. Elsevier, Amsterdam, 1st edition, 1999.

[3] Attila Szabo and Neil S. Ostlund. *Modern quantum chemistry*. Dover Publications Inc., New York, 1st edition, 1996.

[4] R. O. Jones. Density functional theory: Its origins, rise to prominence, and future. *Rev. Mod. Phys.*, 87(3):897–923, 2015.

[5] P. Hohenberg and W Kohn. Inhomogeneous electron gas. *Phys. Rev.*, 136(3B):B864–B871, 1964.

[6] W. Kohn and L. J. Sham. Self-consistent equations including exchange and correlation effects. *Phys. Rev.*, 140(4A):A1133–A1138, 1965.

[7] David M. Ceperley and B. J. Alder. Ground state of the electron gas by a stochastic method. *Phys. Rev. Lett.*, 45(7):566, 1980.

[8] John P. Perdew and Alex Zunger. Self-interaction correction to density-functional approximations for many-electron systems. *Phys. Rev. B*, 23(10):5048, 1981.

[9] A. D. Becke. Density-functional exchange-energy approximation with correct asymptotic behavior. *Phys. Rev. A*, 38(6):3098–3100, 1988.

[10] J. P. Perdew and Y. Wang. Accurate and simple analytic representation of the electron-gas correlation energy. *Phys. Rev. B*, 45:13244, 1992.

[11] John P. Perdew. Density-functional approximation for the correlation energy of the inhomogeneous electron gas. *Phys. Rev. B*, 33(12):8822–8824, 1986.

[12] John P. Perdew, Kieron Burke, and Matthias Ernzerhof. Generalized gradient approximation made simple. *Phys. Rev. Lett.*, 77(3):3865–3868, 1996.

[13] Chengteh Lee, Weitao Yang, and Robert G. Parr. Development of the Colle-Salvetti correlation-energy formula into a functional of the electron density. *Phys. Rev. B*, 37(2):785–789, 1988.

[14] John P. Perdew, Jianmin Tao, Viktor N. Staroverov, and Gustavo E. Scuseria. Meta-generalized gradient approximation: Explanation of a realistic nonempirical density functional. *J. Chem. Phys.*, 120(15):6898–6911, 2004.

[15] John P. Perdew, Viktor N. Staroverov, Jianmin Tao, and Gustavo E. Scuseria. Density functional with full exact exchange, balanced nonlocality of correlation, and constraint satisfaction. *Phys. Rev. A*, 78(5):052513, 2008.

[16] Carlo Adamo and Vincenzo Barone. Toward reliable density functional methods without adjustable parameters: The PBE0 model. *J. Chem. Phys.*, 110(13):6158–6170, 1999.

[17] Jochen Heyd and Gustavo E. Scuseria. Efficient hybrid density functional calculations in solids: Assessment of the HeydScuseria-Ernzerhof screened Coulomb hybrid functional. *J. Chem. Phys.*, 121(3):1187, 2004.

[18] Narbe Mardirossian and Martin Head-Gordon. How accurate are the Minnesota density functionals for noncovalent interactions, isomerization energies, thermochemistry, and barrier heights involving molecules composed of main-group elements? *J. Chem. Theor. Comput.*, 12(9):4303–4325, 2016.

[19] Stefan Grimme. Semiempirical GGA-type density functional constructed with a long-range dispersion correction. *J. Comput. Chem.*, 27(15):1787–1799, 2006.

[20] Jens Antony and Stefan Grimme. Fully ab initio protein-ligand interaction energies with dispersion corrected density functional theory. *J. Comput. Chem.*, 33(21):1730–1739, 2012.

[21] Alexandre Tkatchenko and Matthias Scheffler. Accurate molecular van der Waals interactions from ground-state electron density and free-atom reference data. *Phys. Rev. Lett.*, 102(7):6–9, 2009.

[22] Da W. Zhang and J. Z. H. Zhang. Molecular fractionation with conjugate caps for full quantum mechanical calculation of protein-molecule interaction energy. *J. Chem. Phys.*, 119(7):3599, 2003.

[23] Da W. Zhang, X. H. Chen, and John Z. H. Zhang. Molecular caps for full quantum mechanical computation of peptide-water interaction energy. *J. Comput. Chem.*, 24(15):1846–1852, 2003.

[24] Mark S. Gordon, Dmitri G. Fedorov, Spencer R. Pruitt, and Lyudmila V. Slipchenko. Fragmentation methods: A route to accurate calculations on large systems. *Chem. Rev.*, 112(1):632–672, 2012.

[25] S. Datta. *Quantum transport: Atom to transistor*. Cambridge University Press, Cambridge, 2005.

[26] G. Cuniberti, K. Richter, and G. Fagas. *Introducing molecular electronics*. Springer, Berlin, 2005.

[27] J. D. Watson and F. H. C. Crick. Molecular structure of nucleic acids. *Nature*, 171(4356):737–738, 1953.

[28] E. Chargaff, S. Zamenhof, and C. Green. Human desoxypentose nucleic acid: Composition of human desoxypentose nucleic acid. *Nature*, 165(4202):756, 1950.

[29] E. L. Albuquerque and M. G. Cottam. Superlattice plasmon-polaritons. *Phys. Rep.*, 233(2):67–135, 1993.

[30] E. L. Albuquerque and M. G. Cottam. *Polaritons in periodic and quasiperiodic structures*. Elsevier, Amsterdam, 2004.

[31] P. Carpena, P. Bernaola-Galván, P. Ch. Ivanov, and H. E. Stanley. Metal-insulator transition in chains with correlated disorder. *Nature*, 418(6901):955–959, 2002.

[32] G. Cuniberti, L. Craco, D. Porath, and C. Dekker. Backbone-induced semiconducting behavior in short DNA wires. *Phys. Rev. B*, 65(24):241314, 2002.

[33] N. C. Seeman. Nucleic acid junctions and lattices. *J. Theor. Biol.*, 99(2):237–247, 1982.

[34] N. C. Seeman. DNA in a material world. *Nature*, 421(6921):427, 2003.

[35] H. Gu, J. Chao, S. J. Xiao, and N. C. Seeman. Dynamic patterning programmed by DNA tiles captured on a DNA origami substrate. *Nat. Nanotechnol.*, 4(4):245, 2009.

[36] E. L. Albuquerque, U. L. Fulco, V. N. Freire et al. DNA-based nanobiostructured devices: The role of quasiperiodicity and correlation effects. *Phys. Rep.*, 535(4):139–209, 2014.

[37] D. Shechtman, I. Blech, D. Gratias, and J. W. Cahn. Metallic phase with long-range orientational order and no translational symmetry. *Phys. Rev. Lett.*, 53(20):1951, 1984.

[38] M. Senechal. *Quasicrystals and geometry*. Cambridge University Press, Cambridge, 1995.

[39] D. Levine and P. J. Steinhardt. Quasicrystals: A new class of ordered structures. *Phys. Rev. Lett.*, 53(26):2477, 1984.

[40] C. G. Bezerra and E. L. Albuquerque. Localization and scaling properties of spin waves in quasi-periodic magnetic multilayers. *Physica A*, 255(3–4):285–292, 1998.

[41] C. G. Bezerra, J. M. de Araújo, C. Chesman, and E. L. Albuquerque. Self-similar magnetoresistance of Fibonacci ultrathin magnetic films. *Phys. Rev. B*, 60(13):9264, 1999.

[42] D. H. A. L. Anselmo, M. G. Cottam, and E. L. Albuquerque. Localization and scaling properties of magnetostatic modes in quasiperiodic magnetic superlattices. *J. Phys.: Condens. Matter*, 12(6):1041, 2000.

[43] F. Axel and H. Terauchi. High-resolution X-ray-diffraction spectra of Thue-Morse GaAs-AlAs heterostructures: Towards a novel description of disorder. *Phys. Rev. Lett.*, 66(17):2223, 1991.

[44] E. L. Albuquerque and M. G. Cottam. Theory of plasmon-polaritons in Fibonacci-type superlattices with two-dimensionl electron gas layers. *Solid State Commun.*, 81(5):383–386, 1992.

[45] M. S. Vasconcelos, P. W. Mauriz, F. F. de Medeiros, and E. L. Albuquerque. Photonic band gaps in quasiperiodic photonic crystals with negative refractive index. *Phys. Rev. B*, 76(16):165117, 2007.

[46] S. Ostlund, R. Pandit, D. Rand, H. J. Schellnhuber, and E. D. Siggia. One-dimensional Schrödinger equation with an almost periodic potential. *Phys. Rev. Lett.*, 50(23):1873, 1983.

[47] M. Kohmoto, L. P. Kadanoff, and C. Tang. Localization problem in one dimension: Mapping and escape. *Phys. Rev. Lett.*, 50(23):1870, 1983.

[48] K. Nakamura. *Quantum chaos: A new paradigm of nonlinear dynamics*. Cambridge University Press, Cambridge, 1993.

[49] P. A. Lee and T. V. Ramakrishnan. Disordered electronic systems. *Rev. Mod. Phys.*, 57(2):287, 1985.

[50] S. Ostlund and R. Pandit. Renormalization-group analysis of the discrete quasiperiodic Schrödinger equation. *Phys. Rev. B*, 29(3):1394, 1984.

[51] U. Grimm and M. Baake. Aperiodic ising models. In R. V. Moody, editor, *The mathematics of long-range aperiodic order*. Kluwer, Dordrecht, 1997.

[52] P. W. Mauriz, M. S. Vasconcelos, and E. L. Albuquerque. Specific heat properties of electrons in generalized Fibonacci quasicrystals. *Physica A*, 329(1–2):101–113, 2003.

[53] P. W. Mauriz, E. L. Albuquerque, and M. S. Vasconcelos. Electronic specific heat properties in one-dimensional quasicrystals. *Physica A*, 294(3–4):403–414, 2001.

[54] J. A. McCammon and S. C. Harvey. *Dynamics of proteins and nucleic acids*. Cambridge University Press, Cambridge, 1988.

[55] E. M. Boon, A. L. Livingston, N. H. Chmiel, S. S. David, and J. K. Barton. DNA-mediated charge transport for DNA repair. *Proc. Natl. Acad. Sci. U. S. A.*, 100(22):12543–12547, 2003.

[56] E. L. Albuquerque, M. S. Vasconcelos, M. L. Lyra, and F. A. B. F. de Moura. Nucleotide correlations and electronic transport of DNA sequences. *Phys. Rev. E*, 71(2):021910, 2005.

[57] F. J. Dyson. The S matrix in quantum electrodynamics. *Phys. Rev.*, 75(11):1736, 1949.

[58] B. P. W. de Oliveira, E. L. Albuquerque, and M. S. Vasconcelos. Electronic density of states in sequence dependent DNA molecules. *Surf. Sci.*, 600(18):3770–3774, 2006.

[59] H. Sugiyama and I. Saito. Theoretical studies of GG-specific photocleavage of DNA via electron transfer: Significant lowering of ionization potential and 5-localization of HOMO of stacked GG bases in B-Form DNA. *J. Am. Chem. Soc.*, 118(30):7063–7068, 1996.

[60] E. Maciá, F. Triozon, and S. Roche. Contact-dependent effects and tunneling currents in DNA molecules. *Phys. Rev. B*, 71(11):113106, 2005.

[61] Y. A. Berlin, A. L. Burin, and M. A. Ratner. Elementary steps for charge transport in DNA: Thermal activation vs. tunneling. *Chem. Phys.*, 275(1–3):61–74, 2002.

[62] R. G. Sarmento, E. L. Albuquerque, P. D. Sesion Jr, U. L. Fulco, and B. P. W. de Oliveira. Electronic transport in double-strand poly (dG)-poly (dC) DNA segments. *Phys. Lett. A*, 373(16):1486–1491, 2009.

[63] E. Maciá. Electronic structure and transport properties of double-stranded Fibonacci DNA. *Phys. Rev. B*, 74(24):245105, 2006.

[64] D. Klotsa, R. A. Römer, and M. S. Turner. Electronic transport in DNA. *Biophys. J.*, 89(4):2187–2198, 2005.

[65] R. G. Sarmento, U. L. Fulco, E. L. Albuquerque, E. W. S. Caetano, and V. N. Freire. A renormalization approach to describe charge transport in quasiperiodic dangling backbone ladder (DBL)-DNA molecules. *Phys. Lett. A*, 375(45):3993–3996, 2011.

[66] A. V. Malyshev. DNA double helices for single molecule electronics. *Phys. Rev. Lett.*, 98(9):096801, 2007.

[67] E. Winfree, F. Liu, L. A. Wenzler, and N. C. Seeman. Design and self-assembly of two-dimensional DNA crystals. *Nature*, 394(6693):539, 1998.

[68] D. Porath, A. Bezryadin, S. de Vries, and C. Dekker. Direct measurement of electrical transport through DNA molecules. *Nature*, 403(6770):635, 2000.

[69] A. Y. Kasumov, M. Kociak, S. Gueron et al. Proximity-induced superconductivity in DNA. *Science*, 291(5502):280–282, 2001.

[70] R. Landauer. Spatial variation of currents and fields due to localized scatterers in metallic conduction. *IBM J. Res. Dev.*, 1(3):223–231, 1957.

[71] M. Büttiker. Voltage fluctuations in small conductors. *Phys. Rev. B*, 35(8):4123, 1987.

[72] Y. Asai. Theory of electric conductance of DNA molecule. *J. Phys. Chem. B*, 107(19):4647–4652, 2003.

[73] A. D. Stone, J. D. Joannopoulos, and D. J. Chadi. Scaling studies of the resistance of the one-dimensional Anderson model with general disorder. *Phys. Rev. B*, 24(10):5583, 1981.

[74] L. M. Bezerril, D. A. Moreira, E. L. Albuquerque et al. Current–voltage characteristics of double-strand DNA sequences. *Phys. Lett. A*, 373(37):3381–3385, 2009.

[75] M. S. Xu, S. Tsukamoto, S. Ishida et al. Conductance of single thiolated poly (GC)-poly (GC) DNA molecules. *Appl. Phys. Lett.*, 87(8):083902, 2005.

[76] M. L. de Almeida, G. S. Ourique, U. L. Fulco et al. Charge transport properties of a twisted DNA molecule: A renormalization approach. *Chem. Phys.*, 478:48–54, 2016.

[77] E. Maciá. Electrical conductance in duplex DNA: Helical effects and low-frequency vibrational coupling. *Phys. Rev. B*, 76(24):245123, 2007.

[78] R. G. Endres, D. L. Cox, and R. R. P. Singh. Colloquium: The quest for high-conductance DNA. *Rev. Mod. Phys.*, 76(1):195, 2004.

[79] M. Zoli. Helix untwisting and bubble formation in circular DNA. *J. Chem. Phys.*, 138(20):205103, 2013.

[80] H. Fernando, G. A. Papadantonakis, N. S. Kim, and P. R. LeBreton. Conduction-band-edge ionization thresholds of DNA components in aqueous solution. *Proc. Natl. Acad. Sci. U. S. A.*, 95(10):5550–5555, 1998.

[81] A. A Voityuk. Electronic couplings and on-site energies for hole transfer in DNA: Systematic quantum mechanical/molecular dynamic study. *J. Chem. Phys.*, 128:115101, 2008.

[82] E. L. Albuquerque, M. L. Lyra, and F. A. B. F. de Moura. Electronic transport in DNA sequences: The role of correlations and inter-strand coupling. *Physica A*, 370(2):625–631, 2006.

[83] F. Shao and J. K. Barton. Long-range electron and hole transport through DNA with tethered cyclometalated iridium (III) complexes. *J. Am. Chem. Soc.*, 129(47):14733–14738, 2007.

[84] S. B. Baylin. DNA methylation and gene silencing in cancer. *Nat. Rev. Clin. Oncol.*, 2(S1):S4, 2005.

[85] Z. D. Smith and A. Meissner. DNA methylation: Roles in mammalian development. *Nat. Rev. Genet.*, 14(3):204, 2013.

[86] M. L. de Almeida, J. I. N. Oliveira, J. X. Lima Neto et al. Electronic transport in methylated fragments of DNA. *Appl. Phys. Lett.*, 107(20):203701, 2015.

[87] A. D. Becke. A new mixing of Hartree-Fock and local density-functional theories. *J. Chem. Phys.*, 98(2):1372–1377, 1993.

[88] T. H. Dunning Jr. Gaussian basis sets for use in correlated molecular calculations. I. The atoms boron through neon and hydrogen. *J. Chem. Phys.*, 90(2):1007–1023, 1989.

[89] W. Yu, L. Liang, Z. Lin et al. Comparison of some representative density functional theory and wave function theory methods for the studies of amino acids. *J. Comput. Chem.*, 30(4):589–600, 2009.

[90] H. Yin, Y. Ma, J. Mu, C. Liu, and M. Rohlfing. Charge-transfer excited states in aqueous DNA: Insights from many-body Greens function theory. *Phys. Rev. Lett.*, 112(22):228301, 2014.

[91] R. Maul, M. Preuss, F. Ortmann, K. Hannewald, and F. Bechstedt. Electronic excitations of glycine, alanine, and cysteine conformers from first-principles calculations. *J. Phys. Chem. A*, 111(20):4370–4377, 2007.

[92] K. Dedachi, T. Natsume, T. Nakatsu et al. Charge transfer through single- and double-strand DNAs: Simulations based on molecular dynamics and molecular orbital methods. *Chem. Phys. Lett.*, 436(1–3):244–251, 2007.

[93] F. A. B. F. de Moura, M. L. Lyra, and E. L. Albuquerque. Electronic transport in poly (CG) and poly (CT) DNA segments with diluted base pairing. *J. Phys.: Condens. Matter*, 20(7):075109, 2008.

[94] M. Hilke. Noninteracting electrons and the metal-insulator transition in two dimensions with correlated impurities. *Phys. Rev. Lett.*, 91(22):226403, 2003.

[95] H. Mehrez and M. P. Anantram. Interbase electronic coupling for transport through DNA. *Phys. Rev. B*, 71(11):115405, 2005.

[96] C. Tsallis, L. R. da Silva, R. S. Mendes, R. O. Vallejos, and A. M. Mariz. Specific heat anomalies associated with Cantor-set energy spectra. *Phys. Rev. E*, 56(5):R4922, 1997.

[97] L. R. da Silva, R. O. Vallejos, C. Tsallis, R. S. Mendes, and S. Roux. Specific heat of multifractal energy spectra. *Phys. Rev. E*, 64(1):011104, 2001.

[98] P. W. Mauriz, E. L. Albuquerque, and M. S. Vasconcelos. Specific heat properties of polariton modes in quasicrystals. *Phys. Rev. B*, 63(18):184203, 2001.

[99] I. S. Yang and A. C. Anderson. Specific heat of deoxyribonucleic acid at temperatures below 5K. *Phys. Rev. B*, 35(17):9305, 1987.

[100] D. X. Macedo, I. Guedes, and E. L. Albuquerque. Thermal properties of a DNA denaturation with solvent interaction. *Physica A*, 404:234–241, 2014.

[101] D. A. Moreira, E. L. Albuquerque, P. W. Mauriz, and M. S. Vasconcelos. Specific heat spectra of long-range correlated DNA molecules. *Physica A*, 371(2):441–448, 2006.

[102] D. A. Moreira, E. L. Albuquerque, and C. G. Bezerra. Specific heat spectra for quasiperiodic ladder sequences. *Eur. Phys. J. B*, 54(3):393–398, 2006.

[103] G. M. Mrevlishvili. Low-temperature heat capacity of biomacromolecules and the entropic cost of bound water in proteins and nucleic acids (DNA). *Thermochim. Acta*, 308(1–2):49–54, 1998.

[104] D. A. Moreira, E. L. Albuquerque, and D. H. A. L. Anselmo. Specific heat spectra of non-interacting fermions in a quasiperiodic ladder sequence. *Phys. Lett. A*, 372(31):5233–5238, 2008.

[105] E. Schrödinger. *What is life? The physical aspect of the living cell and mind.* Cambridge University Press, Cambridge, 1944.

[106] D. H. A. L. Anselmo, A. L. Dantas, and E. L. Albuquerque. A multifractal analysis of optical phonon excitations in quasicrystals. *Physica A*, 362(2):289–294, 2006.

[107] M. Gell-Mann and C. Tsallis. *Nonextensive entropy: Interdisciplinary applications.* Oxford University Press, Oxford, 2004.

[108] C. Tsallis. Possible generalization of Boltzmann-Gibbs statistics. *J. Stat. Phys.*, 52(1–2):479–487, 1988.

[109] I. N. de Oliveira, M. L. Lyra, and E. L. Albuquerque. Specific heat anomalies of non-interacting fermions with multifractal energy spectra. *Physica A*, 343:424–432, 2004.

[110] I. N. de Oliveira, M. L. Lyra, E. L. Albuquerque, and L. R. da Silva. Bosons with multifractal energy spectrum: Specific heat log periodicity and Bose-Einstein condensation. *J. Phys.: Condens. Matter*, 17(23):3499, 2005.

[111] D. A. Moreira, E. L. Albuquerque, L. R. da Silva, and D. S. Galvao. Low-temperature specific heat spectra considering nonextensive long-range correlated quasiperiodic DNA molecules. *Physica A*, 387(22):5477–5482, 2008.

[112] M. L. Lyra and C. Tsallis. Nonextensivity and multifractality in low-dimensional dissipative systems. *Phys. Rev. Lett.*, 80(1):53, 1998.

[113] A. V. Coronado and P. Carpena. Study of the log-periodic oscillations of the specific heat of Cantor energy spectra. *Physica A*, 358(2–4):299–312, 2005.

[114] M. Peyrard. Nonlinear dynamics and statistical physics of DNA. *Nonlinearity*, 17(2):R1, 2004.

[115] M. Peyrard and A. R. Bishop. Statistical mechanics of a nonlinear model for DNA denaturation. *Phys. Rev. Lett.*, 62(23):2755, 1989.

[116] T. Dauxois, M. Peyrard, and A. R. Bishop. Thermodynamics of a nonlinear model for DNA denaturation. *Physica D*, 66(1–2):35–42, 1993.

[117] T. Dauxois and M. Peyrard. Entropy-driven transition in a one-dimensional system. *Phys. Rev. E*, 51(5):4027, 1995.

[118] M. Joyeux and S. Buyukdagli. Dynamical model based on finite stacking enthalpies for homogeneous and inhomogeneous DNA thermal denaturation. *Phys. Rev. E*, 72(5):051902, 2005.

[119] G. Weber. Sharp DNA denaturation due to solvent interaction. *Europhys. Lett.*, 73(5):806, 2006.

[120] A. Sulaiman, F. P. Zen, H. Alatas, and L. T. Handoko. The thermal denaturation of the Peyrard-Bishop model with an external potential. *Phys. Scr.*, 86(1):015802, 2012.

[121] T. Dauxois, M. Peyrard, and A. R. Bishop. Entropy-driven DNA denaturation. *Phys. Rev. E*, 47(1):R44, 1993.

[122] T. Dauxois and M. Peyrard. Energy localization in nonlinear lattices. *Phys. Rev. Lett.*, 70(25):3935, 1993.

[123] T. Dauxois, M. Peyrard, and A. R. Bishop. Dynamics and thermodynamics of a nonlinear model for DNA denaturation. *Phys. Rev. E*, 47(1):684, 1993.

[124] A. Wildes, N. Theodorakopoulos, J. Valle-Orero, S. Cuesta-Lopez, J. L. Garden, and M. Peyrard. Thermal denaturation of DNA studied with neutron scattering. *Phys. Rev. Lett.*, 106(4):048101, 2011.

[125] R. Owczarzy, Y. You, B. G. Moreira et al. Effects of sodium ions on DNA duplex oligomers: improved predictions of melting temperatures. *Biochemistry*, 43(12):3537–3554, 2004.

[126] T. S. van Erp and M. Peyrard. The dynamics of the DNA denaturation transition. *Europhys. Lett.*, 98(4):48004, 2012.

[127] I. Jelesarov and H. R. Bosshard. Isothermal titration calorimetry and differential scanning calorimetry as complementary tools to investigate the energetics of biomolecular recognition. *J. Mol. Recognit.*, 12(1):3–18, 1999.

[128] D. D. Eley and D. I. Spivey. Semiconductivity of organic substances. Part 9. Nucleic acid in the dry state. *Trans. Faraday Soc.*, 58:411–415, 1962.

[129] Eliot F. Gomez, Vishak Venkatraman, James G. Grote, and Andrew. J. Steckl. Exploring the potential of nucleic acid bases in organic light emitting diodes. *Adv. Mater.*, 27(46):7552–7562, 2015.

[130] M. K. Y. Chan and Gerbrand Ceder. Efficient band gap prediction for solids. *Phys. Rev. Lett.*, 105(19):196403, 2010.

[131] M. B. da Silva, T. S. Francisco, F. F. Maia, Jr. et al. Improved description of the structural and optoelectronic properties of DNA/RNA nucleobase anhydrous crystals: Experiment and dispersion-corrected density functional theory calculations. *Phys. Rev. B*, 96(8):085206, 2017.

[132] Eliot F. Gomez, Vishak Venkatraman, James G. Grote, and Andrew J. Steckl. DNA bases thymine and adenine in bio-organic light emitting diodes. *Sci. Rep.*, 4:7105, 2014.

[133] Erich Runge and Eberhard K. U. Gross. Density-functional theory for time-dependent systems. *Phys. Rev. Lett.*, 52(12):997, 1984.

[134] Bernard Delley. From molecules to solids with the DMol^3 approach. *J. Chem. Phys.*, 113(18):7756–7764, 2000.

[135] D. L. Barker and R. E. Marsh. The crystal structure of cytosine. *Acta Cryst.*, 17(12):1581–1587, 1964.

[136] Go So Parry. The crystal structure of uracil. *Acta Cryst.*, 7(4):313–320, 1954.

[137] Kristian Berland, Valentino R. Cooper, Kyuho Lee et al. Van der Waals forces in density functional theory: A review of the vdW-DF method. *Rep. Prog. Phys.*, 78(6):066501, 2015.

[138] Bernd G. Pfrommer, Michel Côté, Steven G. Louie, and Marvin L. Cohen. Relaxation of crystals with the quasi-Newton method. *J. Comput. Phys.*, 131(1):233–240, 1997.

[139] Robert S. Mulliken. Electronic population analysis on LCAO-MO molecular wave functions. I. *J. Chem. Phys.*, 23(10):1833–1840, 1955.

[140] Fred L. Hirshfeld. Bonded-atom fragments for describing molecular charge densities. *Theor. Chim. Acta*, 44(2):129–138, 1977.

[141] C. S. Wang and W. E. Pickett. Density-functional theory of excitation spectra of semiconductors: Application to Si. *Phys. Rev. Lett.*, 51(7):597, 1983.

[142] A. Klamt, C. Moya, and J. Palomar. A comprehensive comparison of the IEFPCM and SS(V)PE continuum solvation methods with the COSMO approach. *J. Chem. Theor. Comput.*, 11(9):4220–4225, 2015.

[143] F. F. Maia, Jr., V. N. Freire, E. W. S. Caetano et al. Anhydrous crystals of DNA bases are wide gap semiconductors. *J. Chem. Phys.*, 134(17):05B601, 2011.

[144] Jiří Šponer, Kevin E. Riley, and Pavel Hobza. Nature and magnitude of aromatic stacking of nucleic acid bases. *Phys. Chem. Chem. Phys.*, 10(19):2595–2610, 2008.

[145] Ram Kinkar Roy, Sourav Pal, and Kimihiko Hirao. On non-negativity of Fukui function indices. *J. Chem. Phys.*, 110(17):8236–8245, 1999.

[146] Toon Verstraelen, Sergey V. Sukhomlinov, Veronique Van Speybroeck, Michel Waroquier, and Konstantin S Smirnov. Computation of charge distribution and electrostatic potential in silicates with the use of chemical potential equalization models. *J. Phys. Chem. C*, 116(1):490–504, 2012.

[147] S. D. Silaghi, M. Friedrich, C. Cobet et al. Dielectric functions of DNA base films from near–infrared to ultra–violet. *Phys. Status Solidi B*, 242(15):3047–3052, 2005.

[148] Bharat Bhushan, editor. *Springer handbook of nanotechnology*. Springer, Heidelberg, 1st edition, 2005.

[149] Arieh Aviram and Mark A. Ratner. Molecular rectifiers. *Chem. Phys. Lett.*, 29(2):277–283, 1974.

[150] F. L. Carter, R. E. Siatkowski, and H. Wohltjen, editors. *Molecular electronic devices*. North Holland, Amsterdam, 1st edition, 1988.

[151] M. C. Petty, M. R. Bryce, and D. Bloor, editors. *Introduction to molecular electronics*. Oxford University Press, Oxford, 1st edition, 1995.

[152] Heinz-Bernhard Kraatz, Irene Bediako-Amoa, Samuel H Gyepi-Garbrah, and Todd C Sutherland. Electron transfer through H-bonded peptide assemblies. *J. Phys. Chem. B*, 108(52):20164–20172, 2004.

[153] J. I. N. Oliveira, E. L. Albuquerque, U. L. Fulco, P. W. Mauriz, and R. G. Sarmento. Electronic transport through oligopeptide chains: An artificial prototype of a molecular diode. *Chem. Phys. Lett.*, 612:14–19, 2014.

[154] Shuichi Kojima, Yukino Kuriki, Yoshihiro Sato et al. Synthesis of α-helix-forming peptides by gene engineering methods and their characterization by circular dichroism spectra measurements. *Biochim. Biophys. Acta*, 1294(2):129–137, 1996.

[155] L. M. Bezerril, U. L. Fulco, J. I. N. Oliveira et al. Charge transport in fibrous/not fibrous α 3-helical and (5Q, 7Q)α3 variant peptides. *Appl. Phys. Lett.*, 98(5):053702, 2011.

[156] G. A. Mendes, E. L. Albuquerque, U. L. Fulco et al. Electronic specific heat of an α3-helical polypeptide and its biochemical variants. *Chem. Phys. Lett.*, 542:123–127, 2012.

[157] J. I. N. Oliveira, E. L. Albuquerque, U. L. Fulco et al. Conductance of single microRNAs chains related to the autism spectrum disorder. *Europhys. Lett.*, 107(6):68006, 2014.

[158] Robert M. Metzger. Unimolecular electrical rectifiers. *Chem. Rev.*, 103(9):3803–3834, 2003.

[159] K. B. Bravaya, O. Kostko, S. Dolgikh et al. Electronic structure and spectroscopy of nucleic acid bases: Ionization energies, ionization-induced structural changes, and photoelectron spectra. *J. Phys. Chem. A.*, 114:12305–12317, 2010.

[160] J. Huang and M. Kertesz. Validation of intermolecular transfer integral and bandwidth calculations for organic molecular materials. *J. Chem. Phys.*, 122:234707, 2005.

[161] M. J. G. Peach, P. Benfield, T. Helgaker, and D. J. Tozer. Excitation energies in density functional theory: An evaluation and a diagnostic test. *J. Chem. Phys.*, 128:044118, 2008.

[162] David M. Cardamone and George Kirczenow. Single-molecule device prototypes for protein-based nanoelectronics: Negative differential resistance and current rectification in oligopeptides. *Phys. Rev. B*, 77(16):165403, 2008.

[163] Linus Pauling, Robert B. Corey, and Herman R. Branson. The structure of proteins: Two hydrogen-bonded helical configurations of the polypeptide chain. *Proc. Natl. Acad. Sci. U. S. A.*, 37(4):205–211, 1951.

[164] Paul N. Day and Ruth Pachter. A study of aqueous glutamic acid using the effective fragment potential method. *J. Chem. Phys.*, 107(8):2990–2999, 1997.

[165] K. M. C. Davis, D. D. Eley, and R. S. Snart. Semiconductivity in proteins and hæmoglobin: Enhanced semiconductivity in protein complexes. *Nature*, 188(4752):724, 1960.

[166] Boris Rakvin, Nadica Maltar-Strmečki, Chris M. Ramsey, and Naresh S. Dalal. Heat capacity and electron spin echo evidence for low frequency vibrational modes and lattice disorder in L-alanine at cryogenic temperatures. *J. Chem. Phys*, 120(14):6665–6673, 2004.

[167] Evgeni B. Starikov. Many faces of entropy or bayesian statistical mechanics. *ChemPhysChem*, 11(16):3387–3394, 2010.

[168] George A. Linhart. Correlation of heat capacity, absolute temperature and entropy. *J. Chem. Phys.*, 1(11):795–797, 1933.

[169] Rosalind C. Lee, Rhonda L. Feinbaum, and Victor Ambros. The *c. elegans* heterochronic gene lin-4 encodes small RNAs with antisense complementarity to lin-14. *Cell*, 75(5):843–854, 1993.

[170] Brett S. Abrahams and Daniel H. Geschwind. Advances in autism genetics: On the threshold of a new neurobiology. *Nat. Rev. Genet.*, 9(5):341, 2008.

[171] Kittusamy Senthilkumar, Ferdinand C. Grozema, Célia Fonseca Guerra et al. Absolute rates of hole transfer in DNA. *J. Am. Chem. Soc.*, 127(42):14894–14903, 2005.

[172] R. L. Willett, K. W. Baldwin, K. W. West, and L. N. Pfeiffer. Differential adhesion of amino acids to inorganic surfaces. *Proc. Natl. Acad. Sci. U. S. A.*, 102(22):7817–7822, 2005.

[173] Michael A. Stroscio and Mitra Dutta. Integrated biological-semiconductor devices. *Proc. IEEE*, 93(10):1772–1783, 2005.

[174] J. R. Cândido-Júnior, F. A. M. Sales, S. N. Costa et al. Monoclinic and orthorhombic cysteine crystals are small gap insulators. *Chem. Phys. Lett.*, 512(4–6):208–210, 2011.

[175] A. M. Silva, B. P. Silva, F. A. M. Sales et al. Optical absorption and DFT calculations in L-aspartic acid anhydrous crystals: Charge carrier effective masses point to semiconducting behavior. *Phys. Rev. B*, 86(19):195201, 2012.

[176] E. W. S. Caetano, U. L. Fulco, E. L. Albuquerque et al. Anhydrous proline crystals: Structural optimization, optoelectronic properties, effective masses and frenkel exciton energy. *J. Phys. Chem. Solids*, 121:36–48, 2018.

[177] S. N. Costa, F. A. M. Sales, V. N. Freire et al. L-serine anhydrous crystals: structural, electronic, and optical properties by first-principles calculations, and optical absorption measurement. *Cryst. Growth Des.*, 13(7):2793–2802, 2013.

[178] A. M. Silva, S. N. Costa, F. A. M. Sales et al. Vibrational spectroscopy and phonon-related properties of the L-aspartic acid anhydrous monoclinic crystal. *J. Phys. Chem. A*, 119(49):11791–11803, 2015.

[179] A. M. Silva, S. N. Costa, B. P. Silva et al. Assessing the role of water on the electronic structure and vibrational spectra of monohydrated L-aspartic acid crystals. *Cryst. Growth Des.*, 13(11):4844–4851, 2013.

[180] Hendrik J. Monkhorst and James D. Pack. Special points for Brillouin-zone integrations. *Phys. Rev. B*, 13(12):5188, 1976.

[181] K. A. Kerr and J. P. Ashmore. Structure and conformation of orthorhombic L-cysteine. *Acta Cryst.*, B29(10):2124–2127, 1973.

[182] Heloisa N. Bordallo, Elena V. Boldyreva, Jennifer Fischer et al. Observation of subtle dynamic transitions by a combination of neutron scattering, X-ray diffraction and DSC: A case study of the monoclinic L–cysteine. *Biophys. Chem.*, 148(1–3): 34–41, 2010.

[183] A. M. Fox, editor. *Optical properties of solids*. Oxford University Press, Oxford, 1st edition, 2001.

[184] Paul G. Higgs and Ralph E. Pudritz. A thermodynamic basis for prebiotic amino acid synthesis and the nature of the first genetic code. *Astrobiology*, 9(5):483–490, 2009.

[185] Thomas J. Kistenmacher, George A. Rand, and Richard E. Marsh. Refinements of the crystal structures of DL-serine and anhydrous L-serine. *Acta Cryst.*, B30(11):2573–2578, 1974.

[186] G. N. Ramachandran and Gopinath Kartha. Structure of collagen. *Nature*, 176(4482):593, 1955.

[187] Tu Lee and Yu Kun Lin. The origin of life and the crystallization of aspartic acid in water. *Cryst. Growth Des.*, 10(4):1652–1660, 2010.

[188] R. H. A. Plimmer, editor. *The chemical constitution of the proteins*. Longmans, Green and Co., London, 1st edition, 1908.

[189] J. L. Derissen, H. J. Endeman, and A. F. Peerdeman. The crystal and molecular structure of L-aspartic acid. *Acta Cryst.*, B24(10):1349–1354, 1968.

[190] Adriana Matei, Natalia Drichko, Bruno Gompf, and Martin Dressel. Far-infrared spectra of amino acids. *Chem. Phys.*, 316(1–3):61–71, 2005.

[191] Rui P. Lopes, Rosendo Valero, John Tomkinson, M. Paula M. Marques, and Luís A. E. Batista de Carvalho. Applying vibrational spectroscopy to the study of nucleobases-adenine as a case-study. *New J. Chem.*, 37(9):2691–2699, 2013.

[192] George I. Makhatadze. Heat capacities of amino acids, peptides and proteins. *Biophys. Chem.*, 71(2–3):133–156, 1998.

[193] Deepti Kapoor, Navnit K. Misra, Poonam Tandon, and V. D. Gupta. Phonon dispersion and heat capacity of poly (L-aspartic acid). *Eur. Polym. J.*, 34(12):1781–1791, 1998.

[194] K. Umadevi, K. Anitha, B. Sridhar, N. Srinivasan, and R. K. Rajaram. L-aspartic acid monohydrate. *Acta Cryst.*, E59(7):o1073–o1075, 2003.

[195] Susan Jones and Janet M. Thornton. Principles of protein-protein interactions. *Proc. Natl. Acad. Sci. U. S. A.*, 93(1):13–20, 1996.

[196] Benno Schwikowski, Peter Uetz, and Stanley Fields. A network of protein-protein interactions in yeast. *Nat. Biotechnol.*, 18(12):1257–1261, 2000.

[197] Jean-Christophe Rain, Luc Selig, Hilde De Reuse et al. The protein-protein interaction map of *Helicobacter pylori*. *Nature*, 409(6817):211–215, 2001.

[198] Juwen Shen, Jian Zhang, Xiaomin Luo et al. Predicting protein-protein interactions based only on sequences information. *Proc. Natl. Acad. Sci. U. S. A.*, 104(11):4337–4341, 2007.

[199] Ora Schueler-Furman, Chu Wang, Phil Bradley, Kira Misura, and David Baker. Progress in modeling of protein structures and interactions. *Science*, 310(5748):638–642, 2005.

[200] Helen M. Berman. The protein data bank: A historical perspective. *Acta Cryst.*, A64(1):88–95, 2008.

[201] Helen M. Berman, John Westbrook, Zukang Feng et al. The protein data bank. *Nucleic Acids Res.*, 28(1):235–242, 2000.

[202] Helen Berman, Kim Henrick, and Haruki Nakamura. Announcing the worldwide protein data bank. *Nat. Struct. Mol. Biol.*, 10(12):980, 2003.

[203] Christopher Markosian, Luigi Di Costanzo, Monica Sekharan et al. Analysis of impact metrics for the Protein Data Bank. *Sci. Data*, 5:180212, 2018.

[204] Herman J. C. Berendsen, David van der Spoel, and Rudi van Drunen. GROMACS: A message-passing parallel molecular dynamics implementation. *Comput. Phys. Commun.*, 91(1–3):43–56, 1995.

[205] David A. Pearlman, David A. Case, James W. Caldwell et al. AMBER, a package of computer programs for applying molecular mechanics, normal mode analysis, molecular dynamics and free energy calculations to simulate the structural and energetic properties of molecules. *Comput. Phys. Commun.*, 91(1–3):1–41, 1995.

[206] James C. Phillips, Rosemary Braun, Wei Wang et al. Scalable molecular dynamics with NAMD. *J. Comput. Chem.*, 26(16):1781–1802, 2005.

[207] Leandro Martínez, Ricardo Andrade, Ernesto G Birgin, and José Mario Martínez. PACKMOL: A package for building initial configurations for molecular dynamics simulations. *J. Comput. Chem.*, 30(13):2157–2164, 2009.

[208] Ignasi Buch, Toni Giorgino, and Gianni De Fabritiis. Complete reconstruction of an enzyme-inhibitor binding process by molecular dynamics simulations. *Proc. Natl. Acad. Sci. U. S. A.*, 108(25):10184–10189, 2011.

[209] Aravindhan Ganesan, Michelle L. Coote, and Khaled Barakat. Molecular dynamics-driven drug discovery: Leaping forward with confidence. *Drug Discov. Today*, 22(2):249–269, 2017.

[210] J. Warwicker. Improved continuum electrostatic modelling in proteins, with comparison to experiment. *J. Mol. Biol.*, 236(3):887–903, 1994.

[211] Xueyu Song. An inhomogeneous model of protein dielectric properties: Intrinsic polarizabilities of amino acids. *J. Chem. Phys.*, 116(21):9359–9363, 2002.

[212] Biman Bagchi. Water dynamics in the hydration layer around proteins and micelles. *Chem. Rev.*, 105(9):3197–3219, 2005.

[213] Claudia N. Schutz and Arieh Warshel. What are the dielectric constants of proteins and how to validate electrostatic models? *Prot.: Struct., Funct., Bioinf.*, 44(4):400–417, 2001.

[214] Michael J. E. Sternberg, Fiona R. F. Hayes, Alan J. Russell, Paul G. Thomas, and Alan R. Fersht. Prediction of electrostatic effects of engineering of protein charges. *Nature*, 330(6143):86–88, 1987.

[215] Hugh Nymeyer and Huan-Xiang Zhou. A method to determine dielectric constants in nonhomogeneous systems: Application to biological membranes. *Biophys. J.*, 94(4):1185–1193, 2008.

[216] A. C. V. Martins, P. de Lima Neto, E. W. S. Caetano, and V. N. Freire. An improved quantum biochemistry description of the glutamate-GluA2 receptor binding within an inhomogeneous dielectric function framework. *New J. Chem.*, 41(14):6167–6179, 2017.

[217] Ozlem Keskin, Nurcan Tuncbag, and Attila Gursoy. Predicting protein-protein interactions from the molecular to the proteome level. *Chem. Rev.*, 116(8):4884–4909, 2016.

[218] Duncan E. Scott, Andrew R. Bayly, Chris Abell, and John Skidmore. Small molecules, big targets: Drug discovery faces the protein–protein interaction challenge. *Nat. Rev. Drug Discov.*, 15(8):533, 2016.

[219] Harold W. Kroto, James R. Heath, Sean C. O'Brien, Robert F. Curl, and Richard E. Smalley. C60: Buckminsterfullerene. *Nature*, 318(6042):162–163, 1985.

[220] Harold W. Kroto. C60: Buckminsterfullerene, the celestial sphere that fell to earth. *Angew. Chem. Int. Ed.*, 31(2):111–129, 1992.

[221] S. G. Santos, J. V. Santana, F. F. Maia, Jr et al. Adsorption of ascorbic acid on the C60 fullerene. *J. Phys. Chem. B*, 112(45):14267–14272, 2008.

[222] André Hadad, David L. Azevedo, Ewerton W. S. Caetano et al. Two-level adsorption of ibuprofen on C60 fullerene for transdermal delivery: Classical molecular dynamics and density functional theory computations. *J. Phys. Chem. C*, 115(50):24501–24511, 2011.

[223] Diego S. Dantas, Jonas I. N. Oliveira, José X. Lima Neto et al. Quantum molecular modelling of ibuprofen bound to human serum albumin. *RSC Adv.*, 5(61):49439–49450, 2015.

[224] M. Milanesio, R. Bianchi, Piero Ugliengo, Carla Roetti, and D. Viterbo. Vitamin C at 120 K: Experimental and theoretical study of the charge density. *J. Mol. Struct.: Theochem*, 419(1–3):139–154, 1997.

[225] M. A. Mora and F. J. Melendez. Conformational ab initio study of ascorbic acid. *J. Mol. Struct.: Theochem*, 454(2–3):175–185, 1998.

[226] Jason R. Juhasz, Luca F. Pisterzi, Donna M. Gasparro, David R. P. Almeida, and Imre G. Csizmadia. The effects of conformation on the acidity of ascorbic acid: A density functional study. *J. Mol. Struct.: Theochem*, 666:401–407, 2003.

[227] Alicia Jubert, María Leticia Legarto, Néstor E. Massa, Leonor López Tévez, and Nora Beatriz Okulik. Vibrational and theoretical studies of non-steroidal anti-inflammatory drugs ibuprofen [2-(4-isobutylphenyl) propionic acid]; naproxen [6-methoxy-α-methyl-2-naphthalene acetic acid] and tolmetin acids [1-methyl-5-(4-methylbenzoyl)-1h-pyrrole-2-acetic acid]. *J. Mol. Struct.*, 783(1–3):34–51, 2006.

[228] Xiao Min He and Daniel C. Carter. Atomic structure and chemistry of human serum albumin. *Nature*, 358(6383):209–215, 1992.

[229] Jamie Ghuman, Patricia A. Zunszain, Isabelle Petitpas et al. Structural basis of the drug–binding specificity of human serum albumin. *J. Mol. Biol.*, 353(1):38–52, 2005.

[230] Katrin Stierand and Matthias Rarey. Drawing the PDB: Protein-ligand complexes in two dimensions. *ACS Med. Chem. Lett.*, 1(9):540–545, 2010.

[231] Hiroshi Watanabe, Sumio Tanase, Keisuke Nakajou et al. Role of arg-410 and tyr-411 in human serum albumin for ligand binding and esterase-like activity. *Biochem. J.*, 349(3):813–819, 2000.

[232] Justin P. Gallivan and Dennis A. Dougherty. A computational study of cation-π interactions vs salt bridges in aqueous media: Implications for protein engineering. *J. Am. Chem. Soc.*, 122(5):870–874, 2000.

[233] Meng-Xia Xie, Mei Long, Yuan Liu, Chuan Qin, and Ying-Dian Wang. Characterization of the interaction between human serum albumin and morin. *Biochim. Biophys. Acta, Gen. Subj.*, 1760(8):1184–1191, 2006.

[234] Hiroki Sato, Victor Tuan Giam Chuang, Keishi Yamasaki et al. Differential effects of methoxy group on the interaction of curcuminoids with two major ligand binding sites of human serum albumin. *PloS One*, 9(2):e87919, 2014.

[235] Gonzalo Colmenarejo. In silico prediction of drug-binding strengths to human serum albumin. *Med. Res. Rev.*, 23(3):275–301, 2003.

[236] Klefah A. K. Musa and Leif A. Eriksson. Theoretical study of ibuprofen phototoxicity. *J. Phys. Chem. B*, 111(46):13345–13352, 2007.

[237] John W. Gofman, Frank Lindgren, Harold Elliott et al. The role of lipids and lipoproteins in atherosclerosis. *Science*, 111(2877):166–186, 1950.

[238] Konrad Bloch. The biological synthesis of cholesterol. *Science*, 150(3692):19–28, 1965.

[239] Roger S. Blumenthal. Statins: Effective antiatherosclerotic therapy. *Am. Heart J.*, 139(4):577–583, 2000.

[240] Peter Libby. Cholesterol and atherosclerosis. *Biochim. Biophys. Acta*, 1529:299–309, 2000.

[241] Conrad B. Blum. Comparison of properties of four inhibitors of 3-hydroxy-3-methylglutaryl-coenzyme a reductase. *Am. J. Cardiol.*, 73(14):D3–D11, 1994.

[242] Roner F. da Costa, Valder N. Freire, Eveline M. Bezerra, Benildo S. Cavada, Ewerton W. S. Caetano, José L. de Lima Filho, and Eudenilson L. Albuquerque. Explaining statin inhibition effectiveness of HMG-CoA reductase by quantum biochemistry computations. *Phys. Chem. Chem. Phys.*, 14(4):1389–1398, 2012.

[243] Eva S. Istvan and Johann Deisenhofer. Structural mechanism for statin inhibition of HMG-CoA reductase. *Science*, 292(5519):1160–1164, 2001.

[244] Teresa Carbonell and Ernesto Freire. Binding thermodynamics of statins to HMG-CoA reductase. *Biochemistry*, 44(35):11741–11748, 2005.

[245] Qing Y. Zhang, Jian Wan, Xin Xu et al. Structure-based rational quest for potential novel inhibitors of human HMG-CoA reductase by combining CoMFA 3D QSAR modeling and virtual screening. *J. Comb. Chem.*, 9(1):131–138, 2007.

[246] Vinicius B. da Silva, Carlton A. Taft, and Carlos H. T. P Silva. Use of virtual screening, flexible docking, and molecular interaction fields to design novel HMG-CoA reductase inhibitors for the treatment of hypercholesterolemia. *J. Phys. Chem. A*, 112(10):2007–2011, 2008.

[247] Emily A. Kee, Maura C. Livengood, Erin E. Carter, Megan McKenna, and Mauricio Cafiero. Aromatic interactions in the binding of ligands to HMG-CoA reductase. *J. Phys. Chem. B*, 113(44):14810–14815, 2009.

[248] Pier Luigi Silvestrelli. van der Waals interactions in density functional theory using Wannier functions. *J. Phys. Chem. A*, 113(17):5224–5234, 2009.

[249] F. Ortmann, Wolf Gero Schmidt, and Friedhelm Bechstedt. Attracted by long-range electron correlation: Adenine on graphite. *Phys. Rev. Lett.*, 95(18):186101, 2005.

[250] Charlotte Harrison. The patent cliff steepens. *Nat. Rev. Drug Discov.*, 10(1):12–14, 2011.

[251] Peter H. Jones, Michael H. Davidson, Evan A. Stein et al. Comparison of the efficacy and safety of rosuvastatin versus atorvastatin, simvastatin, and pravastatin across doses (STELLAR Trial). *Am. J. Cardiol.*, 92(2):152–160, 2003.

[252] M. C. Smith, N. Burns, J. R. Sayers et al. Bacteriophage collagen. *Science*, 279(5358):1834, 1998.

[253] Matthew D. Shoulders and Ronald T. Raines. Collagen structure and stability. *Annu. Rev. Biochem.*, 78:929–958, 2009.

[254] Darwin J. Prockop and Kari I. Kivirikko. Collagens: Molecular biology, diseases, and potentials for therapy. *Annu. Rev. Biochem.*, 64(1):403–434, 1995.

[255] Rachel Z. Kramer, Jordi Bella, Barbara Brodsky, and Helen M. Berman. The crystal and molecular structure of a collagen-like peptide with a biologically relevant sequence. *J. Mol. Biol.*, 311(1):131–147, 2001.

[256] Maria Schumacher, Kazunori Mizuno, and Hans Peter Bächinger. The crystal structure of the collagen-like polypeptide (glycyl-4 (r)-hydroxyprolyl-4 (r)-hydroxyprolyl) 9 at 1.55 å resolution shows up-puckering of the proline ring in the Xaa position. *J. Biol. Chem.*, 280(21):20397–20403, 2005.

[257] Rachel Z. Kramer, Jordi Bella, Patricia Mayville, Barbara Brodsky, and Helen M. Berman. Sequence dependent conformational variations of collagen triple-helical structure. *Nat. Struct. Mol. Biol.*, 6(5):454–457, 1999.

[258] Alfonso De Simone, Luigi Vitagliano, and Rita Berisio. Role of hydration in collagen triple helix stabilization. *Biochem. Biophys. Res. Commun.*, 372(1):121–125, 2008.

[259] Ian Streeter and Nora H. de Leeuw. Atomistic modeling of collagen proteins in their fibrillar environment. *J. Phys. Chem. B*, 114(41):13263–13270, 2010.

[260] Navaneethakrishnan Krishnamoorthy, Magdi H. Yacoub, and Sophia N. Yaliraki. A computational modeling approach for enhancing self-assembly and biofunctionalisation of collagen biomimetic peptides. *Biomaterials*, 32(30):7275–7285, 2011.

[261] R. Parthasarathi, B. Madhan, V. Subramanian, and T. Ramasami. Ab initio and density functional theory based studies on collagen triplets. *Theor. Chem. Acc.*, 110(1):19–27, 2003.

[262] Midas I. Hsien Tsai, Yujia Xu, and J. J. Dannenberg. Completely geometrically optimized DFT/ONIOM triple-helical collagen-like structures containing the ProProGly, ProProAla, ProProDAla, and ProProDSer triads. *J. Am. Chem. Soc.*, 127(41):14130–14131, 2005.

[263] C. R. F. Rodrigues, J. I. N. Oliveira, U. L. Fulco et al. Quantum biochemistry study of the T3-785 tropocollagen triple-helical structure. *Chem. Phys. Lett.*, 559:88–93, 2013.

[264] Katyanna S. Bezerra, Jonas I. N. Oliveira, José X. Lima Neto et al. Quantum binding energy features of the T3-785 collagen-like triple-helical peptide. *RSC Adv.*, 7(5):2817–2828, 2017.

[265] G. S. Ourique, J. F. Vianna, J. X. Lima Neto et al. A quantum chemistry investigation of a potential inhibitory drug against the dengue virus. *RSC Adv.*, 6(61):56562–56570, 2016.

[266] Katyanna Sales Bezerra, J. X. Lima Neto, Jonas Ivan Nobre Oliveira et al. Computational investigation of the $\alpha 2 \beta 1$ integrin-collagen triple helix complex interaction. *New J. Chem.*, 42(20):17115–17125, 2018.

[267] Todd J. Dolinsky, Jens E. Nielsen, J. Andrew McCammon, and Nathan A. Baker. PDB2PQR: An automated pipeline for the setup of Poisson-Boltzmann electrostatics calculations. *Nucleic Acids Res.*, 32(suppl_2):W665–W667, 2004.

[268] Alex D. MacKerell, Jr., Donald Bashford, M. L. D. R. Bellott et al. All-atom empirical potential for molecular modeling and dynamics studies of proteins. *J. Phys. Chem. B*, 102(18):3586–3616, 1998.

[269] Zygmunt S. Derewenda, Linda Lee, and Urszula Derewenda. The occurence of C-H···O hydrogen bonds in proteins. *J. Mol. Biol.*, 252(2):248–262, 1995.

[270] Justin P. Gallivan and Dennis A. Dougherty. Cation-π interactions in structural biology. *Proc. Natl. Acad. Sci. U. S. A.*, 96(17):9459–9464, 1999.

[271] Jorge A. Fallas, Lesley E. R. O'Leary, and Jeffrey D. Hartgerink. Synthetic collagen mimics: Self-assembly of homotrimers, heterotrimers and higher order structures. *Chem. Soc. Rev.*, 39(9):3510–3527, 2010.

[272] Simona Bronco, Chiara Cappelli, and Susanna Monti. Understanding the structural and binding properties of collagen: A theoretical perspective. *J. Phys. Chem. B*, 108(28):10101–10112, 2004.

[273] Susanna Monti, Simona Bronco, and Chiara Cappelli. Toward the supramolecular structure of collagen: A molecular dynamics approach. *J. Phys. Chem. B*, 109(22):11389–11398, 2005.

[274] F. Ortmann, F. Bechstedt, and W. G. Schmidt. Semiempirical van der Waals correction to the density functional description of solids and molecular structures. *Phys. Rev. B*, 73(20):205101, 2006.

[275] Anton V. Persikov, John A. M. Ramshaw, Alan Kirkpatrick, and Barbara Brodsky. Amino acid propensities for the collagen triple-helix. *Biochemistry*, 39(48):14960–14967, 2000.

[276] Alexander Rich and F. H. C. Crick. The molecular structure of collagen. *J. Mol. Biol.*, 3(5):483–506, 1961.

[277] G. X. Ramachandran and R. Chandrasekharan. Interchain hydrogen bonds via bound water molecules in the collagen triple helix. *Biopolymers*, 6(11):1649–1658, 1968.

[278] Jordi Bella and Helen M Berman. Crystallographic evidence for Cα-H$\cdots$O=C hydrogen bonds in a collagen triple helix. *J. Mol. Biol.*, 264(4):734–742, 1996.

[279] Rajendra S. Bhatnagar, N. Pattabiraman, Keith R. et al. Inter-chain proline: Proline contacts contribute to the stability of the triple helical conformation. *J. Biomol. Struct. Dyn.*, 6(2):223–233, 1988.

[280] Chizuru Hongo, Keiichi Noguchi, Kenji Okuyama et al. Repetitive interactions observed in the crystal structure of a collagen-model peptide, [(Pro-Pro-Gly)$_9$]$_3$. *J. Biochem.*, 138(2):135–144, 2005.

[281] Kenji Okuyama, Hirotaka Narita, Tatsuya Kawaguchi et al. Unique side chain conformation of a leu residue in a triple-helical structure. *Biopolymers*, 86(3):212–221, 2007.

[282] Rudolph L. Juliano and S. Haskill. Signal transduction from the extracellular matrix. *J. Cell Biol.*, 120(3):577–585, 1993.

[283] David A. Calderwood, Iain D. Campbell, and David R. Critchley. Talins and kindlins: Partners in integrin-mediated adhesion. *Nat. Rev. Mol. Cell Biol.*, 14(8):503–517, 2013.

[284] Martin E. Hemler. VLA proteins in the integrin family: Structures, functions, and their role on leukocytes. *Annu. Rev. Immunol.*, 8(1):365–400, 1990.

[285] Iya Znoyko, Naondo Sohara, Samuel S. Spicer, Maria Trojanowska, and Adrian Reuben. Expression of oncostatin M and its receptors in normal and cirrhotic human liver. *J. Hepatol.*, 43(5):893–900, 2005.

[286] Jonas Emsley, C. Graham Knight, Richard W. Farndale, Michael J. Barnes, and Robert C. Liddington. Structural basis of collagen recognition by integrin $\alpha 2\beta 1$. *Cell*, 101(1):47–56, 2000.

[287] Gerlinde R. Van de Walle, Karen Vanhoorelbeke et al. Two functional active conformations of the integrin $\alpha 2\beta 1$, depending on activation condition and cell type. *J. Biol. Chem.*, 280(44):36873–36882, 2005.

[288] Yan Zhao and Donald G. Truhlar. The M06 suite of density functionals for main group thermochemistry, thermochemical kinetics, noncovalent interactions, excited

states, and transition elements: two new functionals and systematic testing of four M06-class functionals and 12 other functionals. *Theor. Chem. Acc.*, 120(1–3):215–241, 2008.

[289] David A. Case, Thomas E. Cheatham III et al. The Amber biomolecular simulation programs. *J. Comp. Chem.*, 26(16):1668–1688, 2005.

[290] Ulf Ryde and Par Soderhjelm. Ligand-binding affinity estimates supported by quantum-mechanical methods. *Chem. Rev.*, 116(9):5520–5566, 2016.

[291] K. M. A. Welch. Drug therapy of migraine. *N. Engl. J. Med.*, 329(20):1476–1483, 1993.

[292] Paul Durham and Spyros Papapetropoulos. Biomarkers associated with migraine and their potential role in migraine management. *Headache*, 53(8):1262–1277, 2013.

[293] Daniela Pietrobon and Jörg Striessnig. Neurological diseases: Neurobiology of migraine. *Nat. Rev. Neurosci.*, 4(5):386–398, 2003.

[294] Fabio Del Bello, Antonio Cilia, Antonio Carrieri et al. The versatile 2-substituted imidazoline nucleus as a structural motif of ligands directed to the serotonin 5-HT$_{1A}$ receptor. *ChemMedChem*, 11(20):2287–2298, 2016.

[295] José X. Lima Neto, Vanessa P. Soares-Rachetti, Eudenilson L. Albuquerque, Vinicius Manzoni, and Umberto L. Fulco. Outlining migrainous through dihydroergotamine-serotonin receptor interactions using quantum biochemistry. *N. J. Chem.*, 42(4):2401–2412, 2018.

[296] Chong Wang, Yi Jiang, Jinming Ma, Huixian Wu et al. Structural basis for molecular recognition at serotonin receptors. *Science*, 340(6132):610–614, 2013.

[297] Tania Córdova-Sintjago, Nancy Villa, Clinton Canal, and Raymond Booth. Human serotonin 5-HT$_{2C}$ G protein-coupled receptor homology model from the β2 adreno-ceptor structure: Ligand docking and mutagenesis studies. *Int. J. Quantum Chem.*, 112(1):140–149, 2012.

[298] Jean Schoenen. Migraine and serotonin: The quest for the holy grail goes on. *Cephalalgia*, 34(3):163–164, 2014.

[299] Stephen D. Silberstein and Douglas C. McCrory. Ergotamine and dihydroergo-tamine: History, pharmacology, and efficacy. *Headache*, 43(2):144–166, 2003.

[300] Bryan L. Roth. Drugs and valvular heart disease. *N. Engl. J. Med.*, 356(1):6–9, 2007.

[301] Bijan K. Rao, Devleena Samanta, Shawn Joshi et al. Receptor-ligand interaction at 5-HT$_3$ serotonin receptors: A cluster approach. *J. Phys. Chem. A*, 118(37):8471–8476, 2014.

[302] John D. McCorvy and Bryan L. Roth. Structure and function of serotonin G protein-coupled receptors. *Pharmacol. Ther.*, 150:129–142, 2015.

[303] G. Zanatta, I. L. Barroso-Neto, V. Bambini-Junior et al. Quantum biochemistry description of the human dopamine D3 receptor in complex with the selective antagonist eticlopride. *Proteomics Bioinf.*, 5:155–162, 2012.

[304] Hans Hebert. The crystal structure and absolute configuration of (-)-dihydroergotamine methanesulfonate monohydrate. *Acta Crystallogr. B*, 35(12):2978–2984, 1979.

[305] Geancarlo Zanatta, Gustavo Nunes, Eveline M. Bezerra et al. Antipsychotic haloperidol binding to the human dopamine D3 receptor: Beyond docking through QM/MM refinement toward the design of improved schizophrenia medicines. *ACS Chem. Neurosci.*, 5(10):1041–1054, 2014.

[306] Alexander Heifetz, Ewa I. Chudyk, Laura Gleave et al. The fragment molecular orbital method reveals new insight into the chemical nature of GPCR-ligand interactions. *J. Chem. Inf. Model.*, 56(1):159–172, 2015.

[307] Geancarlo Zanatta, Gustavo Della Flora Nunes, Eveline M. Bezerra et al. Two binding geometries for risperidone in dopamine D3 receptors: Insights on the fast-off mechanism through docking, quantum biochemistry, and molecular dynamics simulations. *ACS Chem. Neurosci.*, 7(10):1331–1347, 2016.

[308] A. J. Venkatakrishnan, Xavier Deupi, Guillaume Lebon et al. Molecular signatures of G-protein-coupled receptors. *Nature*, 494(7436):185–194, 2013.

[309] Levin Brinkmann, Eugene Heifets, and Lev Kantorovich. Density functional calculations of extended, periodic systems using coulomb corrected molecular fractionation with conjugated caps method (CC-MFCC). *Phys. Chem. Chem. Phys.*, 16(39):21252–21270, 2014.

[310] Jogvan Magnus Haugaard Olsen, Nanna Holmgaard List, Kasper Kristensen, and Jacob Kongsted. Accuracy of protein embedding potentials: An analysis in terms of electrostatic potentials. *J. Chem. Theor. Comput.*, 11(4):1832–1842, 2015.

[311] Jinfeng Liu, Xianwei Wang, John Z. H. Zhang, and Xiao He. Calculation of protein-ligand binding affinities based on a fragment quantum mechanical method. *RSC Adv.*, 5(129):107020–107030, 2015.

[312] K. Lloyd and O. Hornykiewicz. Parkinson's disease: Activity of l-dopa decarboxylase in discrete brain regions. *Science*, 170(3963):1212–1213, 1970.

[313] Stephen J. Kish, Kathleen Shannak, and Oleh Hornykiewicz. Uneven pattern of dopamine loss in the striatum of patients with idiopathic Parkinson's disease. *N. Engl. J. Med.*, 318(14):876–880, 1988.

[314] Urszula Adamiak, Maria Kaldonska, Gabriela Klodowska-Duda et al. Pharmacokinetic-pharmacodynamic modeling of levodopa in patients with advanced Parkinson disease. *Clin. Neuropharmacol.*, 33(3):135–141, 2010.

[315] Nicole F. Steinmetz, Vu Hong, Erik D. Spoerke et al. Buckyballs meet viral nanoparticles: Candidates for biomedicine. *J. Am. Chem. Soc.*, 131(47):17093–17095, 2009.

[316] Donald B. Calne. Treatment of Parkinson's disease. *N. Engl. J. Med.*, 329(14):1021–1027, 1993.

[317] Nilton F. Frazao, Eudenilson L. Albuquerque, Umberto L. Fulco et al. Four-level levodopa adsorption on C60 fullerene for transdermal and oral administration: A computational study. *RSC Adv.*, 2(22):8306–8322, 2012.

[318] N. F. Frazão, E. L. Albuquerque, U. L. Fulco, P. W. Mauriz, and D. L. Azevedo. Conformational, optoelectronic and vibrational properties of the entacapone molecule: A quantum chemistry study. *J. Nanosci. Nanotechnol.*, 16(5):4825–4834, 2016.

[319] Kenneth Hedberg, Lise Hedberg, Donald S. Bethune et al. Bond lengths in free molecules of buckminsterfullerene, C60, from gas-phase electron diffraction. *Science*, 254(5030):410–412, 1991.

[320] Valentino R. Cooper, T. Thonhauser, and David C. Langreth. An application of the van der Waals density functional: Hydrogen bonding and stacking interactions between nucleobases. *J. Chem. Phys.*, 128(20):204102, 2008.

[321] Berk Hess, Carsten Kutzner, David Van Der Spoel, and Erik Lindahl. GROMACS4: Algorithms for highly efficient, load-balanced, and scalable molecular simulation. *J. Chem. Theor. Comput.*, 4(3):435–447, 2008.

[322] William L. Jorgensen, David S. Maxwell, and Julian Tirado-Rives. Development and testing of the OPLS all-atom force field on conformational energetics and properties of organic liquids. *J. Am. Chem. Soc.*, 118(45):11225–11236, 1996.

[323] Giovanni Bussi, Davide Donadio, and Michele Parrinello. Canonical sampling through velocity rescaling. *J. Chem. Phys.*, 126(1):014101, 2007.

[324] U. Chandra Singh and Peter A. Kollman. An approach to computing electrostatic charges for molecules. *J. Comput. Chem.*, 5(2):129–145, 1984.

[325] Robert G. Parr and Weitao Yang. Density functional approach to the frontier-electron theory of chemical reactivity. *J. Am. Chem. Soc.*, 106(14):4049–4050, 1984.

[326] J. P. Foster and F. Weinhold. Natural hybrid orbitals. *J. Am. Chem. Soc.*, 102(24):7211–7218, 1980.

[327] Rosanna Bonaccorsi, Eolo Scrocco, and Jacopo Tomasi. Molecular SCF calculations for the ground state of some three-membered ring molecules:$(CH_2)_3$,$(CH_2)_2NH$, $(CH_2)_2NH_2^+$,$(CH_2)_{20}$,$(CH_2)_2s$,$(CH)_2CH_2$, and N_2CH_2. *J. Chem. Phys.*, 52(10): 5270–5284, 1970.

[328] Roman F. Nalewajski and Robert G. Parr. Information theory, atoms in molecules, and molecular similarity. *Proc. Natl. Acad. Sci. U. S. A.*, 97(16):8879–8882, 2000.

[329] J. M. Henriques, E. W. S. Caetano, V. N. Freire, J. A. P. da Costa, and E. L. Albuquerque. Structural, electronic, and optical absorption properties of orthorhombic CaSnO3 through ab initio calculations. *J. Phys. Condens. Matter*, 19(10): 106214, 2007.

[330] E. Moreira, J. M. Henriques, D. L. Azevedo et al. Structural, optoelectronic, infrared and raman spectra of orthorhombic SrSnO3 from DFT calculations. *J. Solid State Chem.*, 184(4):921–928, 2011.

[331] S. K. Medeiros, E. L. Albuquerque, F. F. Maia, Jr., E. W. S. Caetano, and V. N. Freire. Structural, electronic, and optical properties of CaCO3 aragonite. *Chem. Phys. Lett.*, 430(4–6):293–296, 2006.

[332] Patton L. Fast, Jose Corchado, Maria Luz Sanchez, and Donald G. Truhlar. Optimized parameters for scaling correlation energy. *J. Phys. Chem. A*, 103(17):3139–3143, 1999.

[333] Philip George and Charles W. Bock. A test of the AM1 model for calculating energies and structural properties of benzene, toluene, naphthalene, 1-methyl and 2-methylnaphthalene. *Tetrahedron*, 45(3):605–616, 1989.

[334] Kinning Poon, Linda M. Nowak, and Robert E. Oswald. Characterizing single-channel behavior of GluA3 receptors. *Biophys. J.*, 99(5):1437–1446, 2010.

[335] David Lodge. The history of the pharmacology and cloning of ionotropic glutamate receptors and the development of idiosyncratic nomenclature. *Neuropharmacology*, 56(1):6–21, 2009.

[336] Rongsheng Jin, Tue G. Banke, Mark L. Mayer, Stephen F. Traynelis, and Eric Gouaux. Structural basis for partial agonist action at ionotropic glutamate receptors. *Nat. Neurosci.*, 6(8):803–810, 2003.

[337] Michael K. Fenwick and Robert E. Oswald. NMR spectroscopy of the ligand-binding core of ionotropic glutamate receptor 2 bound to 5-substituted willardiine partial agonists. *J. Mol. Biol.*, 378(3):673–685, 2008.

[338] José X. Lima Neto, Umberto L. Fulco, Eudenilson L. Albuquerque et al. A quantum biochemistry investigation of willardiine partial agonism in AMPA receptors. *Phys. Chem. Chem. Phys.*, 17(19):13092–13103, 2015.

[339] Mark L. Mayer. Glutamate receptors at atomic resolution. *Nature*, 440(7083):456–462, 2006.

[340] Dean R. Madden. Ion channel structure: The structure and function of glutamate receptor ion channels. *Nat. Rev. Neurosci.*, 3(2):91–101, 2002.

[341] Alexander I. Sobolevsky, Michael P. Rosconi, and Eric Gouaux. X-ray structure, symmetry and mechanism of an AMPA-subtype glutamate receptor. *Nature*, 462(7274):745–756, 2009.

[342] Hao Dong and Huan-Xiang Zhou. Atomistic mechanism for the activation and desensitization of an AMPA-subtype glutamate receptor. *Nat. Commun.*, 2:354, 2011.

[343] Albert Y. Lau and Benoît Roux. The hidden energetics of ligand binding and activation in a glutamate receptor. *Nat. Struct. Mol. Biol.*, 18(3):283–287, 2011.

[344] Mark L. Mayer and Neali Armstrong. Structure and function of glutamate receptor ion channels. *Annu. Rev. Physiol.*, 66:161–181, 2004.

[345] Alexander S. Maltsev, Ahmed H. Ahmed, Michael K. Fenwick, David E. Jane, and Robert E. Oswald. Mechanism of partial agonism at the GluR2 AMPA receptor: Measurements of lobe orientation in solution. *Biochemistry*, 47(40):10600–10610, 2008.

[346] Emilia L. Wu, Ye Mei, Ke Li Han, and John Z. H. Zhang. Quantum and molecular dynamics study for binding of macrocyclic inhibitors to human α-thrombin. *Biophys. J.*, 92(12):4244–4253, 2007.

[347] Anne Frandsen, Darryl S. Pickering, Bente Vestergaard et al. Tyr702 is an important determinant of agonist binding and domain closure of the ligand-binding core of GluR2. *Mol. Pharmacol.*, 67(3):703–713, 2005.

[348] Madeline Martinez, Ahmed H. Ahmed, Adrienne P. Loh, and Robert E. Oswald. Thermodynamics and mechanism of the interaction of willardiine partial agonists with a glutamate receptor: Implications for drug development. *Biochemistry*, 53(23):3790–3795, 2014.

[349] Ronald A. Hill, Lane J. Wallace, Duane D. Miller et al. Structure-activity studies for α-amino-3-hydroxy-5-methyl-4-isoxazolepropanoic acid receptors: Acidic hydroxyphenylalanines. *J. Med. Chem.*, 40(20):3182–3191, 1997.

[350] Ahmed H. Ahmed, Christopher P. Ptak, Michael K. Fenwick et al. Dynamics of cleft closure of the GluA2 ligand-binding domain in the presence of full and partial agonists revealed by hydrogen-deuterium exchange. *J. Biol. Chem.*, 288(38):27658–27666, 2013.

[351] Jacob Pøhlsgaard, Karla Frydenvang, Ulf Madsen, and Jette Sandholm Kastrup. Lessons from more than 80 structures of the GluA2 ligand-binding domain in complex with agonists, antagonists and allosteric modulators. *Neuropharmacology*, 60(1):135–150, 2011.

[352] Tamires C. da Silva Ribeiro, Roner F. da Costa, Eveline M. Bezerra et al. The quantum biophysics of the isoniazid adduct NADH binding to its InhA reductase target. *N. J. Chem.*, 38(7):2946–2957, 2014.

[353] Pekka A. Postila, Mikko Ylilauri, and Olli T. Pentikäinen. Full and partial agonism of ionotropic glutamate receptors indicated by molecular dynamics simulations. *J. Chem. Inf. Model.*, 51(5):1037–1047, 2011.

[354] Neali Armstrong and Eric Gouaux. Mechanisms for activation and antagonism of an AMPA-sensitive glutamate receptor: Crystal structures of the GluR2 ligand binding core. *Neuron*, 28(1):165–181, 2000.

[355] Jonas Boström, Anders Hogner, and Stefan Schmitt. Do structurally similar ligands bind in a similar fashion? *J. Med. Chem.*, 49(23):6716–6725, 2006.

[356] Mai Marie Holm, Peter Naur, Bente Vestergaard et al. A binding site tyrosine shapes desensitization kinetics and agonist potency at GluR2. a mutagenic, kinetic, and crystallographic study. *J. Biol. Chem.*, 280(42):35469–35476, 2005.

[357] Robert L. McFeeters and Robert E. Oswald. Structural mobility of the extracellular ligand-binding core of an ionotropic glutamate receptor: Analysis of NMR relaxation dynamics. *Biochemistry*, 41(33):10472–10481, 2002.

[358] Ahmed H. Ahmed, Melissa D. Thompson, Michael K. Fenwick et al. Mechanisms of antagonism of the GluR2 AMPA receptor: Structure and dynamics of the complex of two willardiine antagonists with the glutamate binding domain. *Biochemistry*, 48(18):3894–3903, 2009.

[359] Rongsheng Jin and Eric Gouaux. Probing the function, conformational plasticity, and dimer-dimer contacts of the GluR2 ligand-binding core: Studies of 5-substituted willardiines and GluR2 S1S2 in the crystal. *Biochemistry*, 42(18):5201–5213, 2003.

[360] Okimasa Okada, Kei Odai, Tohru Sugimoto, and Etsuro Ito. Molecular dynamics simulations for glutamate-binding and cleft-closing processes of the ligand-binding domain of GluR2. *Biophys. Chem.*, 162:35–44, 2012.

[361] Ahmed H. Ahmed, Shu Wang, Huai-Hu Chuang, and Robert E. Oswald. Mechanism of AMPA receptor activation by partial agonists dissulfice trapping of closed lobe conformations. *J. Biol. Chem.*, 286(40):35257–35266, 2011.

[362] C. Glenn Begley and Lee M. Ellis. Drug development: Raise standards for preclinical cancer research. *Nature*, 483(7391):531, 2012.

[363] David J. Byun, Jedd D. Wolchok, Lynne M. Rosenberg, and Monica Girotra. Cancer immunotherapy-immune checkpoint blockade and associated endocrinopathies. *Nat. Rev. Endocrinol.*, 13(4):195–207, 2017.

[364] K. B. Mota, J. X. Lima Neto, A. H. Lima Costa et al. A quantum biochemistry model of the interaction between the estrogen receptor and the two antagonists used in breast cancer treatment. *Comput. Theor. Chem.*, 1089:21–27, 2016.

[365] José X. Lima Neto, Katyanna S. Bezerra, Dalila N. Manso et al. Energetic description of cilengitide bound to integrin. *New J. Chem*, 41(19):11405–11412, 2017.

[366] Ana Beatriz M. L. A. Tavares, José X. Lima Neto, Umberto L. Fulco, and Eudenilson L. Albuquerque. Inhibition of the checkpoint protein PD-1 by the therapeutic antibody pembrolizumab outlined by quantum chemistry. *Sci. Rep.*, 8(1):1840, 2018.

[367] Ana Beatriz M. L. A. Tavares, José X. Lima Neto, Umberto L. Fulco, and Eudenilson L. Albuquerque. A quantum biochemistry approach to investigate checkpoint inhibitor drugs for cancer. *New J. Chem.*, 43:7185–7189, 2019.

[368] Ashley G. Rivenbark, Siobhan M. OConnor, and William B. Coleman. Molecular and cellular heterogeneity in breast cancer: Challenges for personalized medicine. *Am. J. Pathol.*, 183(4):1113–1124, 2013.

[369] Kaori Fukuzawa, Kazuo Kitaura, Masami Uebayasi et al. Ab initio quantum mechanical study of the binding energies of human estrogen receptor α with its ligands: An application of fragment molecular orbital method. *J. Comput. Chem.*, 26(1):1–10, 2005.

[370] V. Craig Jordan and Monica Morrow. Tamoxifen, raloxifene, and the prevention of breast cancer. *Endocr. Rev.*, 20(3):253–278, 1999.

[371] G. Gomathi, T. Srinivasan, D. Velmurugan, and R. Gopalakrishnan. A bluish-green emitting organic compound methyl 3-[(E)-(2-hydroxy-1-naphthyl) methylidene] carbazate: Spectroscopic, thermal, fluorescence, antimicrobial and molecular docking studies. *RSC Adv.*, 5(56):44742–44748, 2015.

[372] Andrew K. Shiau, Danielle Barstad, Paula M. Loria et al. The structural basis of estrogen receptor/coactivator recognition and the antagonism of this interaction by tamoxifen. *Cell*, 95(7):927–937, 1998.

[373] Andrzej M. Brzozowski, Ashley C. W. Pike, Zbigniew Dauter et al. Molecular basis of agonism and antagonism in the oestrogen receptor. *Nature*, 389(6652):753–758, 1997.

[374] Nicola J. Clegg, Sreenivasan Paruthiyil, Dale C. Leitman, and Thomas S. Scanlan. Differential response of estrogen receptor subtypes to 1, 3-diarylindene and 2, 3-diarylindene ligands. *J. Med. Chem.*, 48(19):5989–6003, 2005.

[375] Chiduru Watanabe, Kaori Fukuzawa, Shigenori Tanaka, and Sachiko Aida-Hyugaji. Charge clamps of lysines and hydrogen bonds play key roles in the mechanism to fix helix 12 in the agonist and antagonist positions of estrogen receptor α: Intramolecular interactions studied by the ab initio fragment molecular orbital method. *J. Phys. Chem. B*, 118(19):4993–5008, 2014.

[376] Jay S. Desgrosellier and David A. Cheresh. Integrins in cancer: Biological implications and therapeutic opportunities. *Nat. Rev. Cancer*, 10(1):9–22, 2010.

[377] Andrew B. J. Prowse, Fenny Chong, Peter P. Gray, and Trent P. Munro. Stem cell integrins: Implications for ex-vivo culture and cellular therapies. *Stem Cell Res.*, 6(1):1–12, 2011.

[378] Yang Su, Wei Xia, Jing Li, Thomas Walz et al. Relating conformation to function in integrin $\alpha5\beta1$. *Proc. Natl. Acad. Sci. U. S. A.*, 113(27):E3872–E3881, 2016.

[379] Alberto Dal Corso, Luca Pignataro, Laura Belvisi, and Cesare Gennari. $\alpha v\beta3$ integrin-targeted peptide/peptidomimetic-drug conjugates: In-depth analysis of the linker technology. *Curr. Top. Med. Chem.*, 16(3):314–329, 2016.

[380] Francesca Vasile, Gloria Menchi, Elena Lenci et al. Insight to the binding mode of triazole RGD-peptidomimetics to integrin-rich cancer cells by NMR and molecular modeling. *Bioorg. Med. Chem.*, 24(5):989–994, 2016.

[381] JianFeng Chen, Azucena Salas, and Timothy A. Springer. Bistable regulation of integrin adhesiveness by a bipolar metal ion cluster. *Nat. Struct. Mol. Biol.*, 10(12):995–1001, 2003.

[382] Tasuku Honjo. Cancer immunotherapy by PD-1 blockade. *Cancer Sci.*, 109:197–197, 2018.

[383] Spencer C. Wei, Colm R. Duffy, and James P. Allison. Fundamental mechanisms of immune checkpoint blockade therapy. *Cancer Discov.*, 8(9):1069–1086, 2018.

[384] P. S. Chowdhury, K. Chamoto, and T. Honjo. Combination therapy strategies for improving PD-1 blockade efficacy: A new era in cancer immunotherapy. *J. Intern. Med.*, 283(2):110–120, 2018.

[385] Padmanee Sharma and James P. Allison. The future of immune checkpoint therapy. *Science*, 348(6230):56–61, 2015.

[386] Yoshiko Iwai, Junzo Hamanishi, Kenji Chamoto, and Tasuku Honjo. Cancer immunotherapies targeting the PD-1 signaling pathway. *J. Biomed. Sci*, 24(1):26, 2017.

[387] Leila Khoja, Marcus O. Butler, S. Peter Kang, Scot Ebbinghaus, and Anthony M. Joshua. Pembrolizumab. *J. Immunother. Cancer*, 3(1):36, 2015.

[388] Daniel Sanghoon Shin, Jesse M. Zaretsky, Helena Escuin-Ordinas et al. Primary resistance to PD-1 blockade mediated by JAK1/2 mutations. *Cancer Discov.*, 7(2):188–201, 2017.

[389] Krzysztof M. Zak, Przemyslaw Grudnik, Katarzyna Magiera, Alexander Dömling, Grzegorz Dubin, and Tad A. Holak. Structural biology of the immune checkpoint receptor PD-1 and its ligands PD-L1/PD-L2. *Structure*, 25(8):1163–1174, 2017.

[390] Clement Viricel, Marawan Ahmed, and Khaled Barakat. Human PD-1 binds differently to its human ligands: A comprehensive modeling study. *J. Mol. Graph. Model.*, 57:131–142, 2015.

[391] Krzysztof M. Zak, Przemyslaw Grudnik, Katarzyna Guzik et al. Structural basis for small molecule targeting of the programmed death ligand 1 (PD-L1). *Oncotarget*, 7(21):30323–30335, 2016.

[392] Igor N. Ivashko and Jill M. Kolesar. Pembrolizumab and nivolumab: PD-1 inhibitors for advanced melanoma. *Am. J. Health Syst. Pharm.*, 73(4):193–201, 2016.

[393] Shoichiro Horita, Yayoi Nomura, Yumi Sato et al. High-resolution crystal structure of the therapeutic antibody pembrolizumab bound to the human PD-1. *Sci. Rep.*, 6:35297, 2016.

[394] Hiroyuki Nishimura, Masato Nose, Hiroshi Hiai, Nagahiro Minato, and Tasuku Honjo. Development of lupus-like autoimmune diseases by disruption of the PD-1 gene encoding an ITIM motif-carrying immunoreceptor. *Immunity*, 11(2):141–151, 1999.

[395] Shuguang Tan, Hao Zhang, Yan Chai et al. An unexpected n-terminal loop in PD-1 dominates binding by nivolumab. *Nat. Commun.*, 8:14369, 2017.

[396] Aranthya H. Lima Costa, Washington S. Clemente et al. Computational biochemical investigation of the binding energy interactions between an estrogen receptor and its agonists. *New J. Chem.*, 42(24):19801–19810, 2018.

[397] Berni Julian Alder and Thomas Everett Wainwright. Phase transition for a hard sphere system. *J. Chem. Phys.*, 27(5):1208–1209, 1957.

[398] M. F. F. Perutz. New x-ray evidence on the configuration of polypeptide chains: Polypeptide chains in poly-γ-benzyl-l-glutamate, keratin and hæmoglobin. *Nature*, 167(4261):1053, 1951.

[399] G. N. Ramachandran, C. Ramakrishnan, and V. Sasisekharan. Stereochemistry of polypeptide chain configurations. *J. Mol. Biol.*, 7:95–99, 1963.

Index